Organic
and
Bio-organic
Chemistry
of
Carbon Dioxide

Organic and Bio-organic Chemistry of Carbon Dioxide

Edited by
Shohei INOUE
Faculty of Engineering, University of Tokyo, Tokyo, Japan
Noboru YAMAZAKI
Department of Polymer Science, Tokyo Institute of Technology, Tokyo, Japan

A HALSTED PRESS BOOK

KODANSHA LTD.
Tokyo

JOHN WILEY & SONS
New York-Chichester-Brisbane-Toronto

KODANSHA SCIENTIFIC BOOKS

ISBN:0 470 27309-7

Published in Japan by

KODANSHA LTD.
12–21 Otowa 2-chome, Bunkyo-ku, Tokyo 112, Japan

Published by

HALSTED PRESS
a Division of John Wiley & Sons, Inc.
605 Third Avenue, New York, N.Y. 10158, U.S.A.

PRINTED IN JAPAN

Contributors

Numbers in parenthesis indicate the Chapters to which the authors contributed. Editors are indicated by an asterisk.

Kozi ASADA (5), *Assistant Professor of The Research Institute for Food Science, Kyoto University, Uji-shi, Kyoto 611, Japan*

Eiichi HARUKI (2), *Lecturer of Dept. of Applied Chemistry, Osaka Prefectural University, Sakai-shi, Osaka 591, Japan*

Fukuji HIGASHI (4.1), *Assistant Professor of Faculty of Engineering, Tokyo University of Agriculture and Technology, Koganei-shi, Tokyo 184, Japan*

Shohei INOUE* (1,4.2,6), *Professor of Faculty of Engineering, the University of Tokyo, Bunkyo-ku, Tokyo 113, Japan*

Takashi ITO (3), *Assistant Professor of Faculty of Engineering, Yokohama National University, Yokohama-shi, Kanagawa 240, Japan*

Akio YAMAMOTO (3), *Professor of Research Laboratory of Resource Utlization, Tokyo Institute of Technology, Yokohama-shi, Kanagawa 227, Japan*

Noboru YAMAZAKI*[1] (4.1), *Professor of Department of Polymer Science, Tokyo Institute of Technology, Ohokayama, Meguro-ku, Tokyo 152, Japan*

v

Preface

While it may seem strange to discuss carbon dioxide, an inorganic compound, in the context of "organic and bio-organic chemistry," carbon dioxide is one of the essential building blocks of the organic substances produced by photosynthetic processes in green plants and some micro-organisms. The consumption of these organic materials with eventual conversion to carbon dioxide is the basis of human life on this planet.

The accelerating consumption of fossil fuels, products of carbon dioxide fixation in prehistoric times, has reportedly increased the atmospheric carbon dioxide concentration in recent years, probably affecting the world's weather. Diminishing fossil fuel supplies and possible environmental hazards point to the need for greater chemical utilization of carbon dioxide. While industrial applications have been limited so far, simulation of certain efficient biochemical carbon dioxide fixation reactions may lead to new syntheses with potential practical use.

Emphasizing recent developments in the field, this book provides a full overview of the organic and bio-organic chemistry of carbon dioxide. Organometallic and macromolecular synthetic reactions are given special treatment in light of significant new findings in these areas. Other chapters focus on biochemical and model reactions involving carbon dioxide. All are subjects of increasing interest at a time when improved utilization of our planet's carbon resources has become critically important.

The editors wish to express their profound appreciation to the authors for their fine articles, which will certainly contribute much to the enhanced understanding and use of this fundamental small molecule. We are also indebted to Mrs. G. D. Allinson and other members of the staff at Kodansha for their linguistic and editorial assistance in the preparation of this book.

Shohei INOUE
Noboru YAMAZAKI

Contents

ix

1

Introduction

Shohei INOUE
Faculty of Engineering, the University of Tokyo
Bunkyo-ku, Tokyo 113, Japan

Human beings and all other living things on Earth depend for their existence on the process of photosynthesis, the fixation of carbon dioxide by green plants and some micro-organisms, using solar energy. This reaction reduces carbon dioxide, the most oxidized form of carbon, to carbohydrate under moderate conditions. Chemists have long dreamed of imitating this reaction in a test tube.

The chemical utilization of carbon dioxide as a resource for the production of useful substances has been rather limited, however.

At present, the largest industrial organic use of carbon dioxide as a material is in the synthesis of urea from carbon dioxide and ammonia, a process which was first used on a small scale in Germany in 1920. The synthesis involves two reaction steps: very rapid formation of ammonium carbamate (Eq. 1.1), and dehydration to urea (Eq. 1.2) under rigorous conditions such as high temperature and high pressure.

$$CO_2 + 2\ NH_3 \rightleftharpoons [H_2NCOO^-]\,[NH_4^+] \qquad (1.1)$$

$$[H_2NCOO^-]\,[NH_4^+] \longrightarrow H_2NCONH_2 + H_2O \qquad (1.2)$$

These two reaction setps illustrate well the characteristics of the reactivity of carbon dioxide: high reactivity toward the base (nucleophile), and low reactivity of the product thus formed, a carboxylate.

Another industrial process is the synthesis of hydroxybenzoic acid from carbon dioxide and an alkali metal salt of phenol, the Kolbe-Schmitt

1

reaction, discovered in 1860. This synthesis also has two steps: formation of a complex between carbon dioxide and phenolate (Eq. 1.3), and the transfer of a carboxyl group to the aromatic ring (Eq. 1.4).

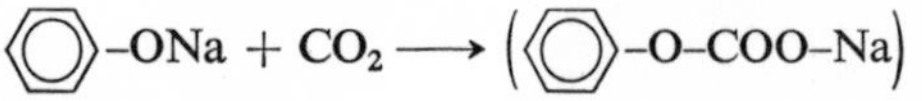

$$\text{〈O〉-ONa} + CO_2 \longrightarrow \left(\text{〈O〉-O-COO-Na} \right) \qquad (1.3)$$

$$\text{〈O〉-O-COO-Na} \longrightarrow \overset{\text{OH}}{\text{〈O〉-COONa}} \qquad (1.4)$$

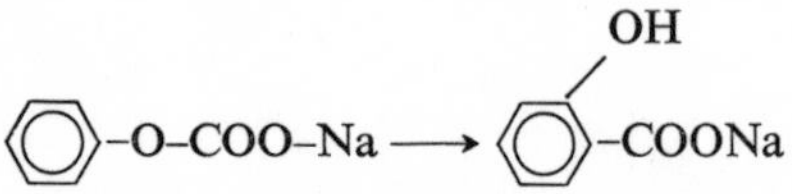

The carbon dioxide-phenolate complex can transfer a carboxyl group to another organic molecule and is regarded as a carrier of activated carbon dioxide.

One of the typical reactions of carbon dioxide is that with organometallic compounds. Even the very early literature concerning the formation of Grignard reagents described the reaction of such compounds with carbon dioxide to form carboxylic acid (Eq. 1.5).

$$R-MgX + CO_2 \longrightarrow R-COO-MgX \qquad (1.5)$$

This reaction is of practical importance in the synthesis of carboxylic acid on a laboratory scale. Related is the carboxylation of an active hydrogen compound with a base.

In the development of organotransition metal chemistry during the 1970s, there were numerous reports of reactions between a transition metal alkyl or hydride and carbon dioxide that gave the corresponding carboxylic and formic acids, as in the Grignard reaction. In addition to these "normal" insertion reactions of carbon dioxide, there has been reported a reaction involving an "abnormal" insertion that formed a metal-carbon bond but not a metal-oxygen bond (Eq. 1.6).

$$Et-Co\ (PPh_3)_2 + CO_2 \longrightarrow Et-O-\underset{\overset{\|}{O}}{C}-Co\ (CO_3)_x(PPh_3) \qquad (1.6)$$

In connection with these, reductive reactions and other catalytic fixations of carbon dioxide with transition metal systems are currently being developed in expectation of novel processes that may lead to significant new syntheses.

Progress was made in the same decade in the syntheses of macromolecules using carbon dioxide as a starting material. The first example was the copolymerization of carbon dioxide and epoxide (Eq. 1.7).

$$\begin{array}{c} RHC\text{–}CHR \\ \diagdown \diagup \\ O \end{array} + CO_2 \longrightarrow \left(\begin{array}{c} -CHR\text{–}CHR\text{–}O\text{–}C\text{–}O- \\ \parallel \\ O \end{array} \right)_x \tag{1.7}$$

This is a catalytic fixation of carbon dioxide with an organozinc system and is thought to involve the insertion of carbon dioxide into a zinc alkoxide as one of the two elementary steps (Eqs. 1.8 and 1.9).

$$\sim\!\!\sim\!\!CHR\text{–}CHR\text{–}O\text{–}ZnX + CO_2$$
$$\longrightarrow \sim\!\!\sim\!\!CHR\text{–}CHR\text{–}O\text{–}C\text{–}O\text{–}ZnX \tag{1.8}$$
$$\parallel$$
$$O$$

$$\sim\!\!\sim\!\!CHR\text{–}CHR\text{–}O\text{–}\underset{\underset{O}{\parallel}}{C}\text{–}O\text{–}ZnX + \underset{\underset{O}{\diagdown\diagup}}{RHC\text{–}CHR}$$
$$\longrightarrow \sim\!\!\sim\!\!CHR\text{–}CHR\text{–}O\text{–}\underset{\underset{O}{\parallel}}{C}\text{–}O\text{–}CHR\text{–}CHR\text{–}O\text{–}ZnX \tag{1.9}$$

Since the discovery of this technique, a number of reactions that produce high molecular weight polymers or oligomers with carbon dioxide as a starting material have been reported, including addition polymerization and condensation polymerization reactions.

Another topic of current interest in the chemistry of carbon dioxide is the development of models for biological carbon dioxide fixation. Such studies will not only increase our understanding of biochemical reactions but will also lead to novel synthetic processes of industrial importance by mimicking the characteristics of certain effective biochemical processes.

The growing interest in "biomimetic" chemistry of carbon dioxide is, of course, the result of studies of biological carboxylations in recent years. Recent research concerned with photosynthesis, a representative of biological carbon dioxide fixation reactions, has uncovered various reactions that must be added to the classical Calvin cycle. Reactions encountered in some micro-organisms are of particular interest for their apparent simplicity and for their significance in biological and chemical evolution. There has also been progress recently in studies of other important biological reactions of carbon dioxide, such as the reversible hydration of carbon dioxide with carbonic anhydrase and the carbon dioxide fixation accompanying carboxyl transfer with biotin enzymes.

Advances in the chemistry and biological chemistry of carbon dioxide will continue to provide much information for the understanding and possible utilization of this fundamental substance.

2

Organic Syntheses with Carbon Dioxide

Eiichi HARUKI
Dept. of Applied Chemistry, Osaka Prefectural University
Sakai-shi, Osaka 591, Japan

When the oil crisis struck in 1973, it came as a great surprise to many people that a large amount of petroleum in the world changes into carbon dioxide every day. It is clear that if this carbon dioxide could be used as a carbon source for various organic syntheses, mankind would benefit tremendously.

Although a number of organic syntheses using carbon dioxide as a starting material have been found, carbon dioxide has been employed for industrial use for urea synthesis, and, on a small scale, for the Kolbe-Schmitt reaction. If we consider recycling of resources, we see that organic synthetic reactions using carbon dioxide are many fewer in number than those using carbon monoxide. With the recent progress in research, however, organic synthetic reactions using carbon dioxide should soon assume a position of equal importance.

In this chapter, we will discuss the following questions: What approaches have recent studies on organic synthesis using carbon dioxide taken? What directions should future research pursue?

As there are a number of reviews[1-6] on carbon dioxide, we will focus the discussion on research since 1970 that has dealt with organic synthetic reactions using carbon dioxide; we have excluded organometallic reactions because they are described in detail in Chapter 3.

2.1.   CARBOXYLATION OF ACTIVE HYDROGEN COMPOUNDS

It is well known that carbon dioxide is readily inserted into various organometallic compounds to produce metal carboxylates, which then produce the corresponding carboxylic acids upon acidification with a strong acid such as hydrochloric or sulfuric acid. The insertion reaction of carbon dioxide in the synthesis of carboxylic acids has been employed in the laboratory to characterize the formation of carbanions. This reaction was first achieved with camphor in 1868 (Eq. 2.1).[7] Another example of this reaction is the carboxylation of cyclopentadecanone, whose carbanion is formed by treatment with triphenylmethylsodium (Eq. 2.2).[8]

$$(CH_2)_{13} \overset{C=O}{\underset{CH_2}{\big\langle}} \xrightarrow[\text{(C}_2\text{H}_5)_2\text{O , -10}^\circ]{(C_6H_5)_3CNa} \xrightarrow{CO_2} (CH_2)_{13} \overset{C=O}{\underset{CHCO_2H}{\big\langle}} \qquad (2.1)$$

$$\xrightarrow{B} \xrightarrow{CO_2} \qquad (2.2)$$

2.1.1. Stiles' Reagent (MMC)

A useful synthetic reaction for carboxylic acids was discovered by Stiles and Finkbeiner.[9] With this, the formation of enolate carbanion and its carboxylation can take place simultaneously by reaction with methyl methoxymagnesium carbonate (MMC) (Eqs. 2.3 and 2.4), which is obtained by treating magnesium methoxide with carbon dioxide in dimethylforma-

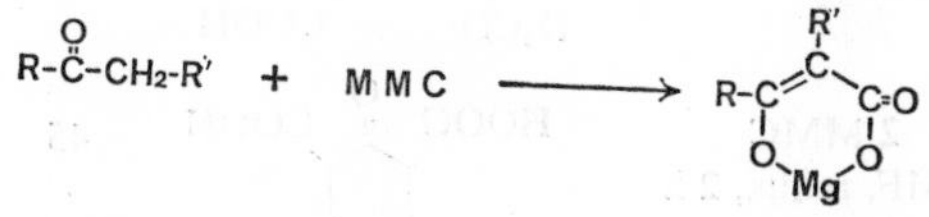

$$
\text{R-}\overset{\overset{\displaystyle O}{\|}}{\text{C}}\text{-CH}_2\text{-R'} \quad + \quad \text{MMC} \quad \longrightarrow \quad \text{(1)} \tag{2.3}
$$

$$
\text{RCH}_2\text{NO}_2 \quad + \quad \text{MMC} \quad \longrightarrow \tag{2.4}
$$

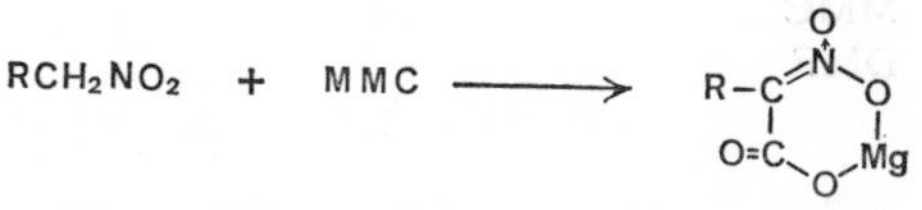

mide (DMF). The reverse reaction is prevented by the formation of a magnesium chelate between the carboxyl group and the enolate. This reaction may be applied to many other substrates with activated carbon atoms, as shown in Table 2.1.

In this reaction, carboxylic acids are obtained by careful acidification of the reaction mixture containing magnesium chelate (1, Eq. 2.3) with a mineral acid such as hydrochloric acid or sulfuric acid, but the resulting carboxylic acids (β-oxocarboxylic acids, α-nitrocarboxylic acids, etc.) are too unstable to be stored even for several days. There are therefore some disadvantages in using these carboxylic acids as starting materials in synthetic reactions except when employing them as reaction intermediates. Direct esterification of the magnesium chelates (1) to eliminate the problem has been reported (Eqs. 2.5 and 2.6).[17,18]

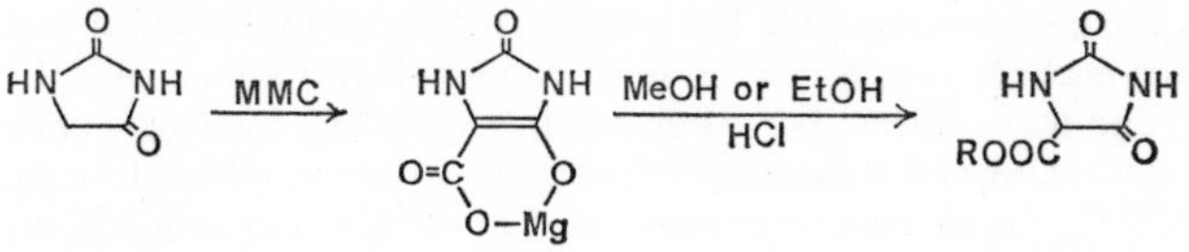

$$
\xrightarrow{\text{MMC}} \xrightarrow[\text{HCl}]{\text{MeOH or EtOH}} \tag{2.5}
$$

$$
\text{H}_3\text{C-}\overset{\overset{\displaystyle O}{\|}}{\text{C}}\text{-CH}_2\text{CH}_2\text{CO}_2\text{H} \xrightarrow{\text{MMC}} \xrightarrow{\text{MeOH·HCl}} \quad \text{92\%} \tag{2.6}
$$

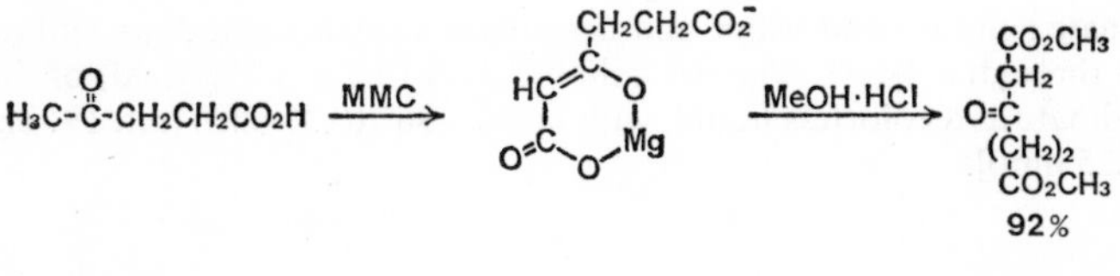

Table 2.1 Carboxylation of active hydrogen compounds with MMC.

Substrate	Reaction conditions	Reaction product	Yield (%)	Reference
$C_6H_5\text{-CO-}CH_3$	MMC DMF	$C_6H_5\text{-CO-}CH_2COOH$	68	10
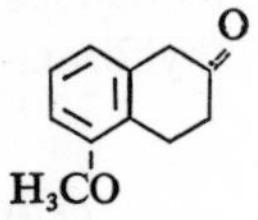 6-methoxy-1-tetralone	MMC DMF, 130°C, 4 h	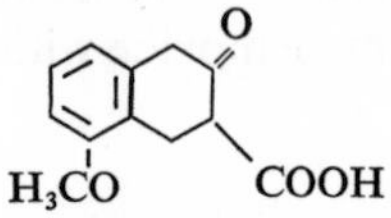 6-methoxy-1-tetralone-2-COOH	44	11
 cyclohexanone	2 MMC DMF, reflux, 2 h	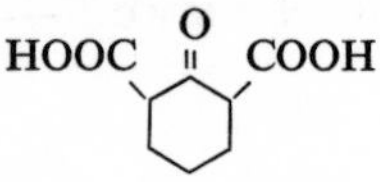 HOOC–(cyclohexanone)–COOH	45	12
	MMC DMF			13
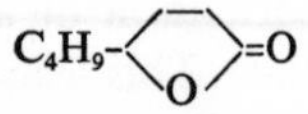 C_4H_9–(dihydrofuranone)=O	MMC DMF	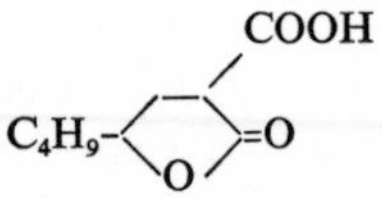 C_4H_9–(dihydrofuranone, COOH)=O	98	14
O_2NCH_3	MMC DMF	$O_2N\text{-}CH_2COOH$	63	15
$O_2N\text{-}CH_2CH_2CH_3$	MMC CH_2N_2 DMF	$O_2N\text{-CH-COOCH}_3$, $\overset{|}{C_2CH_3}$	44	16

Direct esterification plays an important role in the general procedure for carboxylation using MMC.

General procedure *Dimethyl β-ketoadipate.* A solution of 10.6 g (91.4 mmol) of levulinic acid in 360 ml (920 mmol) of 2.56 M MMC in DMF is heated at 135°C for 24 h. The DMF is removed by distillation under vacuum at 60°C. Trituration of the residue with ether after filtration and air drying gives 136 g of yellow solid. The solid is suspended in 820 ml of methanol in a 3–1, three-necked, round-bottom flask equipped with a mechanical stirrer, a condenser, and a gas inlet tube. After cooling to −10°C, hydrogen chloride is passed over the mixture until saturation is reached. After standing overnight and warming to 25°C, the mixture is concentrated at 40°C under reduced pressure. The syrupy residue is poured over ice, and the aqueous solution is extracted four times with chloroform. The organic extracts are then washed with saturated bicarbonate solution and water and dried over anhydrous sodium sulfate. Distillation through a 1-inch Vigreux column gives 15.8 g (92% yield) of dimethyl β-ketoadipate as a colorless liquid, with bp 94°–96° (0.35 mm) [ref. 19: bp 110°–111° (0.25 mm)].

Like MMC, ferric alkoxide may also be used as a reagent in the presence of carbon dioxide to carboxylate a number of active methylene compounds,[20] but it is less effective (Eq. 2.7).

$$\text{Fe(OC}_2\text{H}_5)_3 \;+\; \text{CO}_2 \xrightarrow[\text{DMF}]{} \xrightarrow{80^\circ\text{C.2h}} \text{(2.7)}$$

There are many metal alkoxides other than magnesium and iron, and these compounds are consumed as reagents during carboxylation. However, it has been found that some transition metal complexes serve as catalysts in the fixation of carbon dioxide. The catalytic carboxylation of unsaturated hydrocarbons is described in detail in the following chapter.

2.1.2. Carboxylation of Active Hydrogen Compounds using a PhO⁻–CO₂ Complex

Although the carboxylation of active hydrogen compounds using Stiles' reagent (MMC) is an elegant reaction, especially as a model for biological reactions, the preparation of MMC is incovenient.*

Bottaccio and Chiusoli[22] have described the carboxylation of active hydrogen compounds with an alkali metal phenoxide-carbon dioxide complex in dimethylformamide at room temperature. This carboxylation can be visualized as an intermolecular Kolbe-Schmitt reaction between phenolates and enolates (Eq. 2.8), and it has raised hopes for the industrial

$$\underset{\text{O}}{\overset{\text{O}}{\underset{\|}{\text{C}_6\text{H}_5\text{-C-CH}_3}}} + \text{CO}_2 \xrightarrow[\text{DMF}]{\text{C}_6\text{H}_5\text{OK}} \underset{\|}{\overset{\text{O}}{\text{C}_6\text{H}_5\text{-C-CH}_2\text{-COOK}}} \qquad \text{(2.8)}$$

utilization of carbon dioxide. Various active hydrogen compounds are readily carboxylated with alkali metal phenolates in the presence of carbon dioxide under mild reaction conditions in various organic solvents and with satisfactory yields. Theoretical and fundamental studies of this carboxylation have been reported by Bottaccio *et al.*[23] and Kwan *et al.*[24]

---

*A convenient preparation of MMC was recently reported in a U.S. patent by Sifniades *et al.*[21] A solution of anhydrous magnesium chloride in methanol is saturated with dry carbon dioxide and passed under 40 psig carbon dioxide pressure through a column of Dowex MWA-1 (free base form). The carboxylation of acetone by this reagent gives a mixture of α,ω-acetonedicarboxylic acid and acetoacetic acid in yields of 61.4% and 24%, respectively.

The results of the application of this carboxylation to various substrates are summarized in Table 2.2.

Table 2.2.   Carboxylation of active hydrogen compounds with alkali meal phenoxidet and related compounds.

Substrate	Base	Reaction product	Reference
R_1R_2CHCHO	C_6H_5ONa, C_6H_5OK	$R_1R_2C(CHO)COOH$	25
$H_3-\overset{O}{\overset{\|}{C}}-CH_3$	C_6H_5ONa, $(H_3C)_2C(C_6H_4ONa)$	$HOOCCH_2\overset{O}{\overset{\|}{C}}CH_2COOH$	26, 28, 41, 42
$C_6H_5-\overset{O}{\overset{\|}{C}}-CH_3$	C_6H_5ONa, C_6H_5OK 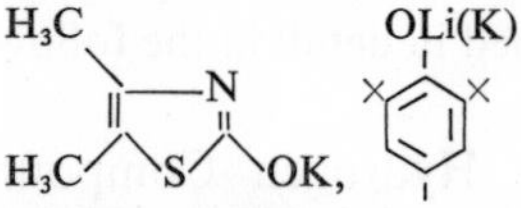 $\underset{H_3C}{\overset{H_3C}{>}}\text{—N}=\langle\rangle\text{, S, OK,}$	$C_6H_5-\overset{O}{\overset{\|}{C}}-CH_2COOH$	22, 41, 43, 44
$H_3C-COOR$	C_6H_5ONa, C_6H_5OK, $(H_3C)_2C(C_6H_4ONa)$ $NaOC_6H_4ONa$	$HOOCCH_2COOR$	26, 27, 29, 30, 31
$C_6H_5-CH_2CN$	C_6H_5ONa, C_6H_5OK	$C_6H_5CH(CN)COOH$	22, 32, 33, 34, 35, 36, 38, 39, 40, 41
$C_6H_5-C\equiv CH$	C_6H_5ONa, C_6H_5OK, 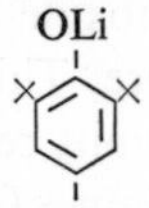 OLi	$C_6H_5-C\equiv C-COOH$	22, 32, 33, 34, 35, 36, 39, 40, 44
Cyclopentadiene	C_6H_5ONa, C_6H_5OK	Tricyclo [5.2.1.0²,⁶] deca-3,8-diene-4, 9-dicarboxylic acid	22, 33, 34, 35, 36, 39
Indene	C_6H_5ONa, C_6H_5OK, $RO-C_6H_4-OK$	Indene-3-carboxylic acid	22, 33, 34, 35, 36, 37, 39
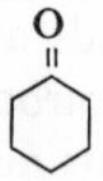 (cyclohexanone)	C_6H_5ONa, C_6H_5OK	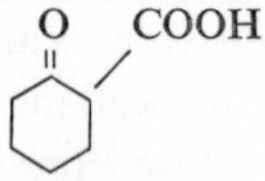 (cyclohexanone-COOH)	41, 43
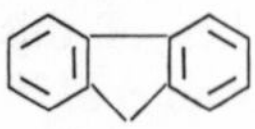 (fluorene)	C_6H_5OK	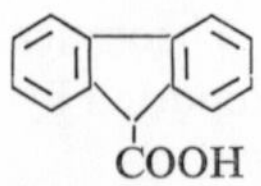 (fluorene-COOH)	40

Another interesting application of this carboxylation is the synthesis of citric acid from acetone (Eq. 2.9).[45]

$$H_3C-\overset{O}{\overset{\|}{C}}-CH_3 + CO_2 \xrightarrow[\text{DMF}]{C_6H_5ONa} NaOOC-CH_2-\overset{O}{\overset{\|}{C}}-CH_2-COONa$$
5.3g

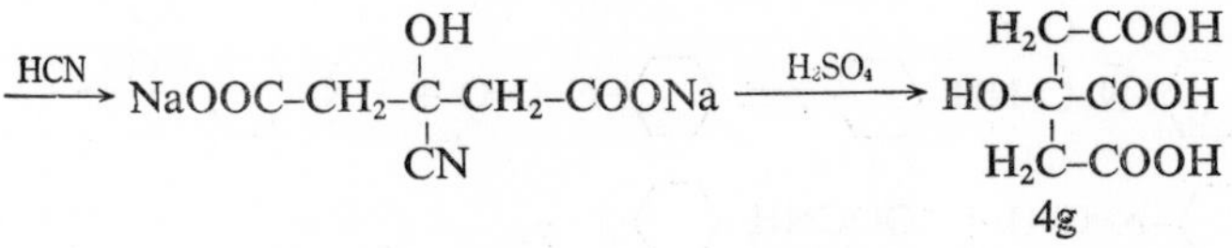

$$\xrightarrow{\text{HCN}} \text{NaOOC-CH}_2\text{-}\overset{\displaystyle \text{OH}}{\underset{\displaystyle \text{CN}}{\text{C}}}\text{-CH}_2\text{-COONa} \xrightarrow{\text{H}_2\text{SO}_4} \text{HO-}\overset{\displaystyle \text{H}_2\text{C-COOH}}{\underset{\displaystyle \text{H}_2\text{C-COOH}}{\text{C}}}\text{-COOH} \qquad (2.9)$$

4g

2.1.3. Carboxylation using Strong Organic Bases

It has been well established that biotin is required as a cofactor in a number of enzymatic carboxylation reactions. In these reactions, carbon dioxide is first transferred to the imidazolone moiety of enzyme-bound biotin to form a CO_2-enzyme-biotin complex and then to a substrate to effect the carboxylation (Eq. 2.10).[46] The detailed mechanism of these reactions remains, unknown, however.[47]

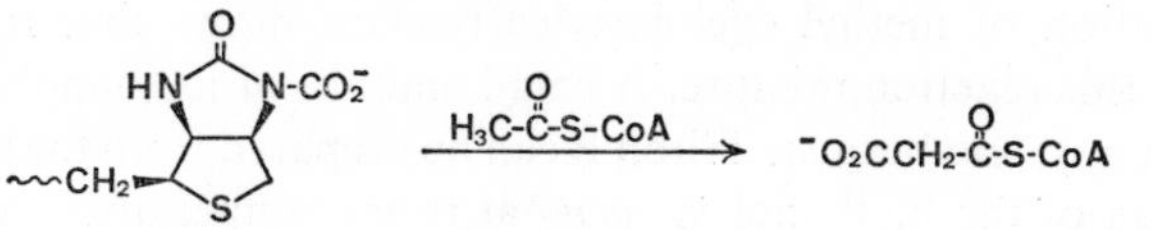

$$(2.10)$$

A. Carboxylation using Biotin Models

As a biotin model, the dicyclohexylcarbodiimide (DCC)-tetra-alkylammonium hydroxide-CO_2 reagent system was developed for the synthesis of *N*-carboxylated urea derivatives. The carboxylation of active hydrogen compounds can be achieved with this reagent system (Eq. 2.11).[48]

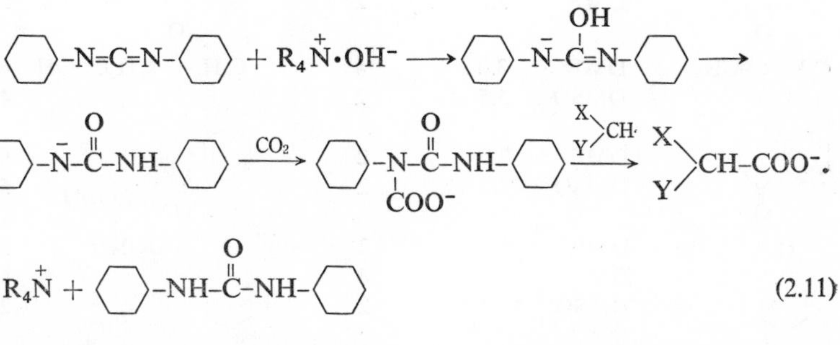

$$(2.11)$$

This carboxylation is not as simple as it appears in Eq. (2.11), however. The reaction may actually proceed via the following mechanism (Eq. 2.12), since a small amount of methyl cyclohexylcarbamate is produced as a reaction product.

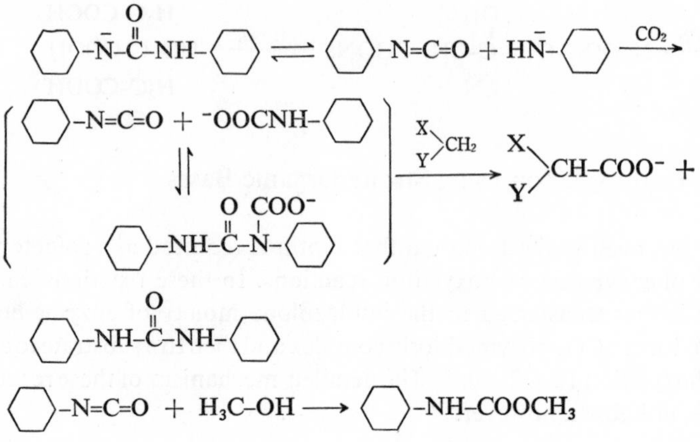

$$\langle\rangle-\overset{-}{N}-\overset{\overset{O}{\parallel}}{C}-NH-\langle\rangle \;\rightleftharpoons\; \langle\rangle-N=C=O + H\overset{-}{N}-\langle\rangle \;\xrightarrow{CO_2}\;$$

$$\left[\begin{array}{c}\langle\rangle-N=C=O + {}^{-}OOCNH-\langle\rangle \\ \Updownarrow \\ \overset{O\;\;COO^{-}}{\langle\rangle-NH-\overset{\parallel}{C}-\overset{\mid}{N}-\langle\rangle}\end{array}\right] \xrightarrow{\;\overset{X}{\underset{Y}{>}}CH_2\;} \overset{X}{\underset{Y}{>}}CH-COO^{-} +$$

$$\langle\rangle-NH-\overset{\overset{O}{\parallel}}{C}-NH-\langle\rangle$$

$$\langle\rangle-N=C=O + H_3C-OH \longrightarrow \langle\rangle-NH-COOCH_3 \qquad (2.12)$$

The formation of methyl cyclohexylcarbamate shows that methanol is present in this reaction mixture. A small amount of methanol remains in this reaction system because Triton B can be prepared from the solution by evaporation of the methanol *in vacuo* at room temperature. As a result, there is a partial exhaustion of the effective complex base during carboxylation. Results of carboxylations using this reagent system are summarized in Table 2.3.[49]

Table 2.3.   Carboxylation of active hydrogen compounds with DCC-Triton B-CO₂.

Substrate	Solvent	Triton B / Substrate	Reaction time (h)	Reaction product	Yield (%)
	DMF	3.6	2		56
$C_6H_5-\overset{\overset{O}{\parallel}}{C}-CH_3$	DMF	3.6	4	$C_6H_5-\overset{\overset{O}{\parallel}}{C}-CH_2COOH$	66
	DMSO	3.5	2		47
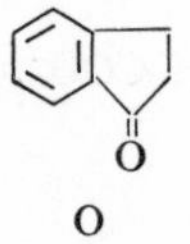 (indanone)	DMF	3.0	2	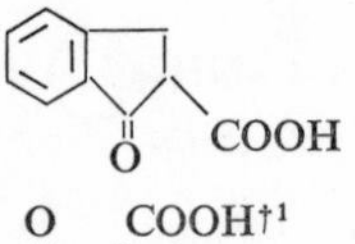 (indanone-COOH)	66
	DMSO	3.0	2		58
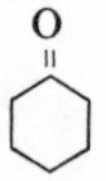 (cyclohexanone)	DMF	4.0	2	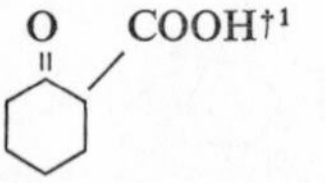 (cyclohexanone-COOH†1)	37
	DMF	2.9	4		40
	DMSO	3.0	2		21
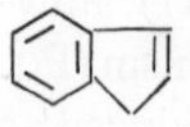 (indanone)	DMF	3.6	2	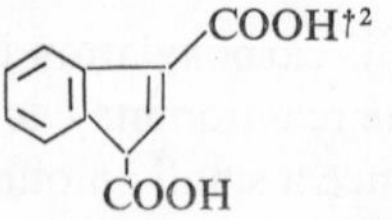 (indanone-COOH†2, COOH)	71
	DMF	3.6	4		77
	DMSO	3.5	2		64

†1 Contains a small amount of 2,6-dicarboxylic acid.
†2 Contains a small amount of monocarboxylic acid.

Equation (2.12) shows that cyclohexylcarbamate ion and cyclohexylisocyanate in DMF should be able to carboxylate active hydrogen compounds. If a mixture of cyclohexylamine and an equimolar amount of 1,5-diazabicyclo[5.4.0]-5-undecene(DBU) in DMF is saturated with carbon dioxide, and cyclohexylisocyanate and indene are added to the mixture in that order at room temperature, indene-1,3-dicarboxylic acid is produced in 75% yield (Eq. 2.13).[50)]

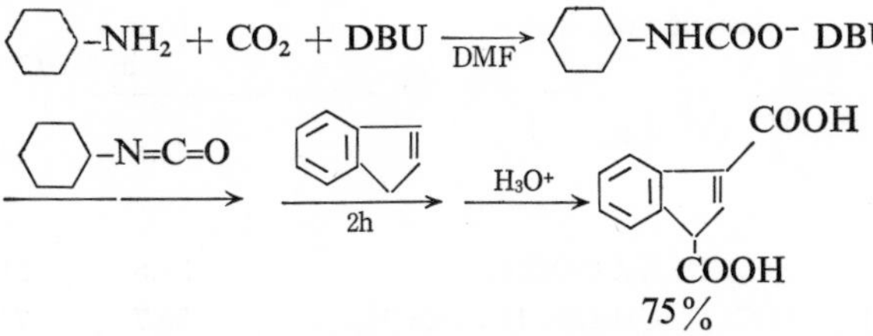

(2.13)

No carboxylation takes place when phenylisocyanate is used instead of cyclohexylisocyanate. The details of this carboxylation will be reported in the near future.

The same type of reaction is shown in the following example (Eq. 2.14).[51)] This reaction is especially effective for the monocarboxylation of indene.

$$\begin{matrix}X\\Y\end{matrix}\!\!>\!CH_2 + CO_2 + K_2CO_3 + C_6H_5NH\text{-}\overset{\displaystyle O}{\overset{\|}{C}}\text{-}NH\text{-}C_6H_5 \xrightarrow[\text{solv.}]{} \xrightarrow{H_3O^+}$$

$$\begin{matrix}X\\Y\end{matrix}\!\!>\!CH\text{-}COOH \qquad (2.14)$$

N-Lithiated ethyleneurea is not a good reagent for the carboxylation of active methylene compounds because it is insoluble in most organic solvents.[48)] The magnesium derivative of ethyleneurea is soluble in dipolar aprotic solvents, however, and it takes up carbon dioxide and carboxylates active methylene compounds (Eq. 2.15).[52)] The results of such reactions are summarized in Table 2.4.

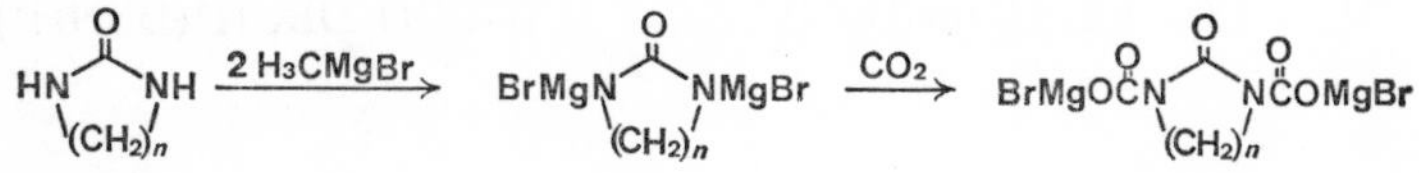

a; n=2

b; n=3

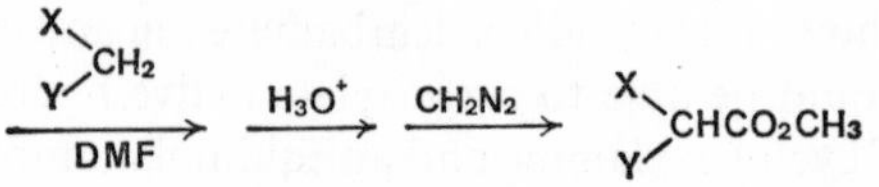

$$\underset{Y}{\overset{X}{>}}CH_2 \xrightarrow[DMF]{} \xrightarrow{H_3O^+} \xrightarrow{CH_2N_2} \underset{Y}{\overset{X}{>}}CHCO_2CH_3 \qquad (2.15)$$

Table 2.4.   Carobxylation of active hydrogen compounds with magnesium-ureid-CO_2 complex.

Active hydrogen compound	Reaction product	Yield (%) from a in Eq. (2.15)	from b
	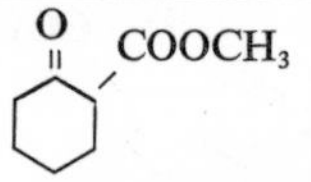 COOCH₃	44.8	74.3
$C_6H_5COCH_3$	$C_6H_5COCH_2COOCH_3$	39.6	51.7
$(H_3C)_2CHCH_2COCH_3$	$(H_3C)_2CHCH_2COCH_2COOCH_3$	36.7	73.3
$H_3C)_2C{=}CHCOCH_3$	$(H_3C)_2C{=}CHCOCH_2COOCH_3$ (90%) $(H_3C)_2C{=}CH(COOCH_3)COCH_3$ (10%)	30.7	66.6
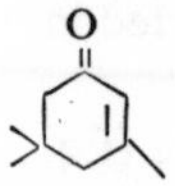	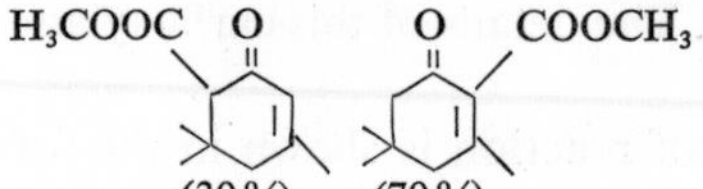 H_3COOC O O COOCH₃ (30%) (70%)	0	43.8
$H_3CCH_2NO_2$	$H_3CCH_2(NO_2)COOCH_3$	10.2	20.3
$H_3CCH_2CH_2NO_2$	$H_3CCH_2CH(NO_2)COOCH_3$	4.1	48.0
$(H_3C)_2CHCH_2CH_2NO_2$	$(H_3C)_2CHCH_2CH(NO_2)COOCH_3$	7.8	24.0

Reaction temperature 110°C; reaction time 3 h.

The carboxylations using urea derivatives provide an important model for biological reactions rather than a reagent for organic syntheses.

B.   Carboxylation using Amidine and Guanidine Derivatives

In the course of studies on carboxylation using urea derivatives and bases, it was found that either 1,5-diazabicyclo[5.4.0]-5-undecene (DBU) (Eq. 2.16)[53] or 1,5-diazabicyclo[4.3.0]-5-nonene (DBN) (Eq. 2.17)[54] could be used as an effective reagent for the carboxylation of active hydrogen compounds under a carbon dioxide atmosphere at room temperature.

$$\underset{Y}{\overset{X}{>}}CH_2 + CO_2 + DBU \text{ (or DBN)} \longrightarrow \underset{Y}{\overset{X}{>}}CH{-}COO^-DBUH^+(DBNH^+)$$

$$\xrightarrow{H_3O^+} \underset{Y}{\overset{X}{>}}CH{-}COOH \qquad (2.16)$$

$$R-C\equiv CH + CO_2 + DBU(\text{or } DNN) \longrightarrow R-C\equiv C-COO^-$$

$$DBUH^+(DBNH^+) \xrightarrow{\text{H}_3\text{O}^+} R-C\equiv C-COOH \qquad (2.17)$$

In this reaction DBU is preferable to DBN as a carboxylating reagent because DBN forms insoluble precipitates with carbon dioxide even in DMF or DMSO, while DBU does not form precipitates under the same conditions and reacts readily. In addition to DBU and DBN, pentamethylguanidine is useful for the carboxylation of indene, cyclopentadiene, fluorene, and aralkylcyanides.[55] Tetramethylguanidine[56] and tetramethylethylenediamine[57] are also slightly effective for the carboxylation, but triethylenediamine, triethylamine, and pyridine are not.[57]

The results of this carboxylation with DBU and DBN are summarized in Tables 2.5–2.7.

Table 2.5. Carboxylation of active hydrogen compounds with DBU and CO_2 (5 kg/cm²).

Substrate	Reaction solvent	Reaction time (h)	Reaction product	Yield (%)
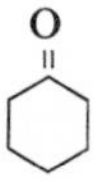	DMSO	3	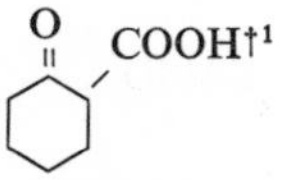 COOH[†1]	63 (52)[†2]
	DMSO	6		73
	DMSO	24		77
	none	24		90
	$(C_2H_5)_2O$	4		50
$C_6H_5-\overset{O}{\overset{\|}{C}}-CH_3$	DMSO	3	$C_6H_5-\overset{O}{\overset{\|}{C}}-CH_2COOH$	48 (41)
	none	25		72
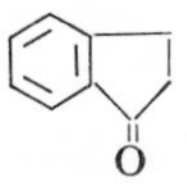	DMSO	3	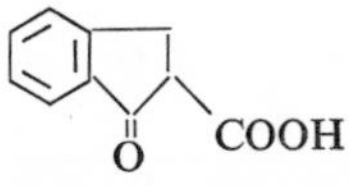 COOH	74
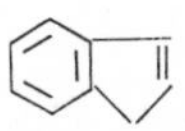	DMSO	1	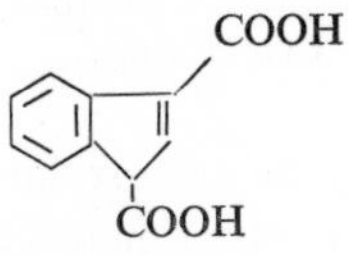 COOH / COOH	87
	none	1		95
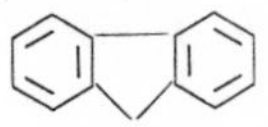	DMSO	18	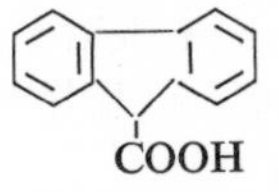 COOH	49 (30)
$C_6H_5-C\equiv CH$	none	48	$C_6H_5-C\equiv C-COOH$	91
$C_6H_5-\overset{OH}{\underset{\|}{CH}}-C\equiv CH$	DMSO	165	$C_6H_5-\overset{OH}{\underset{\|}{CH}}-C\equiv C-COOH$	22

†1 Contains about 7% of 2,6-dicarboxylic acid.
†2 ( ) Yields under CO_2 (1 atm).

Table 2.6. Carboxylation of active hydrogen compounds with DBU and CO_2 (50 kg/cm²).

Substrate	Reaction solvent	Reaction time (h)	Reaction product	Yield (%)
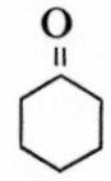 cyclohexanone (O)	DMSO	3	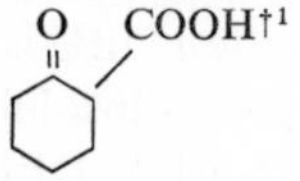 2-oxocyclohexanecarboxylic acid (O, COOH[†1])	90 (52)[†2]
C_6H_5-C(=O)-CH_3	DMSO	3	C_6H_5-C(=O)-CH_2COOH	84 (41)
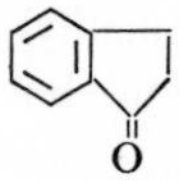 indanone	none	3	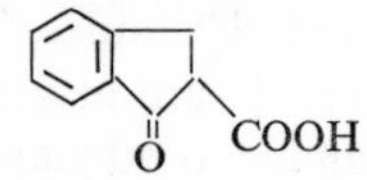 indanone-COOH	86
C_6H_5-C≡CH	DMSO	6	C_6H_5-C≡C-COOH	82
C_6H_5-CH(OH)-C≡CH	none	165	C_6H_5-CH(OH)-C≡C-COOH	68
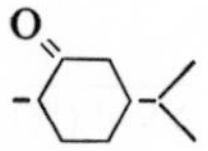 isopropylidene cyclohexanone	DMSO	6	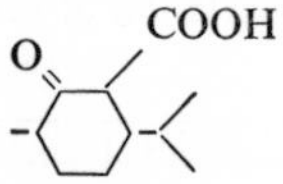 isopropylidene cyclohexanone-COOH	58
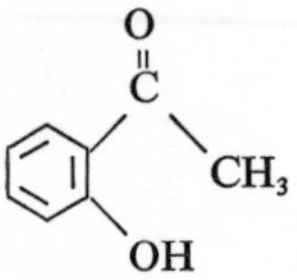 2-hydroxyacetophenone (C=O, CH_3, OH)	DMSO	6	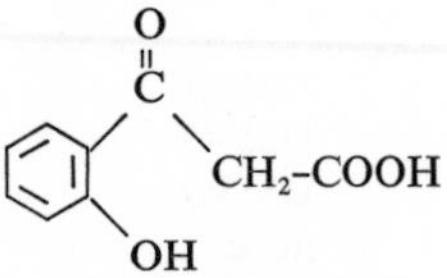 (C=O, CH_2-COOH, OH)	85
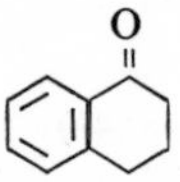 tetralone (O)	DMSO	6	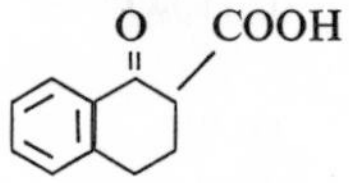 tetralone (O, COOH)	91
$C_6H_5CH=CH$-C(=O)-CH_3	DMSO	6	$C_6H_5CH=CH$-C(=O)-CH_2-COOH	82

[†1] Contains about 7% of 2,6-dicarboxylic acid.
[†2] ( ) Yields under CO_2 (1 atm).

At present, the carboxylation of active hydrogen compounds with DBU or DBN is the best method for the preparation of the corresponding carboxylic acids, both in yield and convenience of procedure.[58] Details of this reaction are described below.

General procedure A mixture of substrate (5–10 mmol), DBU or DBN (20–25 mmol), and solvent (0–10 ml) is placed in an autoclave or glass reaction vessel at −40° to 60°C, then carbon dioxide is charged into the reaction mixture at a pressure of 1–100 kg/cm². After 1–6 h, carbon dioxide is excluded as quickly as possible, and the reaction mixture is poured onto a mxiture of 100 ml of concentrated hydrochloric acid and 100 g of ice which is covered by a layer of 100 ml of ether. After stirring, the ether layer is removed,

Table 2.7. Carboxylation of active hydrgen compounds with DBN and CO_2 (5 or 50 kg/cm^2).

substrate	Reaction solvent	CO_2 (kg/cm^2)	Reaction time (h)	Reaction product	Yield (%)
O	DMSO	5	3	O COOH[1]	56 (63)[2]
	THF	5	3		11
	DMSO	50	3		46 (90)
	THF	50	3		6
O ‖ C_6H_5-C-CH_3	DMSO	5	3	O ‖ C_6H_5-C-CH_2- COOH	32 (48)
	DMF	5	48		12
	DMSO	50	3		23 (84)
O	DMSO	5	3	O COOH	49 (74)
	DMSO	5	1	COOH / COOH	88 (87)
C_6H_5-C≡CH	none	5	30	C_6H_5-C≡C-COOH	3 (78)
OH | C_6H_5-CH-C≡CH	DMSO	50	52	OH | C_6H_5-CH-C≡C-COOH	5 (57)

[1] Contains about 7% of 2,6-dicarboxylic acid.
[2] ( ) Yields when using DBU.

and the aqueous solution is extracted twice with 50 ml portions of ether. The combined ether extracts are extracted again three times with 50 ml portions of saturated aqueous solution of sodium hydrogen carbonate. After the addition of ice (100 g), the combined aqueous extracts are carefully acidified with concentrated hydrochloric acid and extracted three times with 70 ml portions of ether. The combined ether extracts are dried over anhydrous magnesium sulfate and evaporated by rotary evaporator. The residue is a carboxylated product and is nearly pure without recrystallization. The yield is generally high ($\sim$95%).

a) Effect of the carbon dioxide pressure

This carboxylation is affected by the pressure of the carbon dioxide, as shown in Tables 2.5–2.7. For example, phenylacetylene is not carboxylated under one atmosphere pressure of carbon dioxide, but it is carboxylated to give phenylpropiolic acid in satisfactory yield under a carbon dioxide pressure of 5 kg/cm^2 or 50 kg/cm^2. Tables 2.5–2.7 show that most substrates, except indene, give carboxylated products in much higher yields under higher pressures of carbon dioxide. Indene is only slightly affected by the pressure of carbon dioxide.

b) Effect of the solvents

In general, the use of dipolar aprotic solvents gives good results in this carboxylation. Fluorene is not carboxylated in DMSO under a pressure of 50 kg/cm² of carbon dioxide, but it is carboxylated in benzene under the same conditions. Fluorene dissolved in DBU-DMSO immediately precipitates out of the reaction medium on introduction of carbon dioxide at 50 kg/cm², but fluorene is carboxylated without precipitation under a lower pressure of carbon dioxide (5 kg/cm²), even in DMSO. Substrates other than fluorene can be carboxylated easily in DMSO or DMF. Indene, cyclopentadiene, and phenylacetonitrile can also be carboxylated in benzene, ether, tetrahydrofuran, etc. with good yields.

c) Effect of the reaction temperature

This carboxylation is affected by the reaction temperature, and the result is remarkable at some temperatures. At a higher reaction temperature, less carbon dioxide is dissolved in the reaction medium, but the reaction proceeds more rapidly. These factors balance each other in the results of the reaction. The results of these effects on several substrates are shown in Fig. 2.1.[59]

Phenylacetonitrile is carboxylated to give phenylcyanoacetic acid in 90% yield at −40°C for 3 h, but at a higher reaction temperature, no phenylcyanoacetic acid is produced. In the case of acetophenone, the maximum yield is obtained with reaction temperatures in the vicinity of

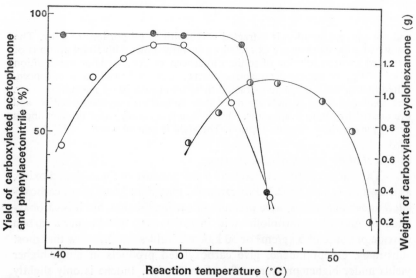

Fig. 2.1   Effect of reaction temperature on the carboxylation of active hydrogen compounds with DBU and CO₂. Substrate, 10 mmol; DBU, 25 mmol; DMF, 5 ml; CO₂, 30–88 kg/cm². —●— Phenylacetonitrile, —○— acetophenone, —◑— cyclohexanone.

$-10°$ to $5°$ C. The best reaction temperature for cyclohexanone is about $30°$ C. Thus, the best reaction temperature is individual for each substrate. The low yields of carboxylated products make it necessary to adjust the reaction temperature.

The effect of the reaction temperature on the carboxylation of phenylacetonitrile is illustrated in more detail in Eq. (2.18).

$$C_6H_5CH_2CN + CO_2 + DBU \xrightarrow[\text{DMF}]{28°C,\ 3h} C_6H_5\underset{\underset{CN}{|}}{C}HCOOH\ (25\%)$$

$$C_6H_5CH_2CN + CO_2 + DBU \xrightarrow[\text{DMF}]{0°C,\ 3h} C_6H_5\underset{\underset{CN}{|}}{C}HCOOH\ (93\%) \qquad (2.18)$$

$$C_6H_5CH_2CN + CO_2 + DBU \xrightarrow[\text{DMF}]{0°C,\ 3h} \xrightarrow{30°C,\ 3h} C_6H_5\underset{\underset{CN}{|}}{C}HCOOH\ (59\%)$$

$$C_6H_5CH_2CN + CO_2 + DBU \xrightarrow[\text{DMF}]{0°C,\ 3h} \xrightarrow{30°C,\ 3h} \xrightarrow{0°C,\ 3h} C_6H_5\underset{\underset{CN}{|}}{C}HCOOH\ (81\%)$$

Another example is acetophenone, with which the same pattern of results is obtained by the same procedure, indicating that the reversible fixation of carbon dioxide at the active methylene group of phenylacetonitrile or acetophenone occurs when the reaction temperature is changed. It has recently been reported that several transition metal complexes can effect the reversible fixation of carbon dioxide.[60] One of them, copper (I)-cyanoacetate-phosphine complex, reacts with cyclohexanone to produce cyclohexanone-2-carboxylic acid (Eq. 2.19).[61]

$$NCCH_2CO_2Cu \cdot (PBu_3)_x \rightleftharpoons NCCH_2Cu \cdot (PBu_3)_x + CO_2 \xrightarrow[\text{ii) } H_2C=CHCH_2Br]{\text{i)}}$$

$$(2.19)$$

d) Effect of the reaction time

In general, this carboxylation proceeds at a fair rate with phenylacetonitrile and acetophenone at low temperatures ($-40°$ to $0°$ C). The relationships between the yields of the carboxylated products and the reaction time are shown in Fig. 2.2.

After one hour, the carboxylated products are obtained in the best yields ($\sim90\%$). Phenylacetonitrile, in particular, is carboxylated to cyanophenylacetic acid (about 60%) within several minutes. Indene and cyclopentadiene are also carboxylated rapidly. On the other hand, the

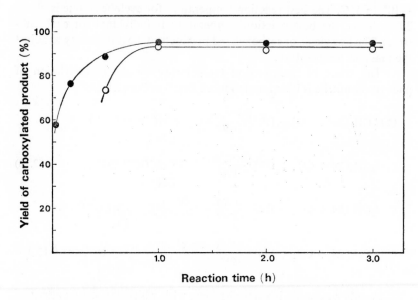

Fig. 2.2   Effect of reaction time on the carboxylation of acetophenone (—○—)and phenylacetonitrile (—●—).
Substrate, 10 mmol; DBU, 25 mmol; DMF, 5 ml; CO_2, 40–45 kg/cm², reaction temp. 0–5°C.

carboxylation of terminal acetylenic derivatives requires a longer reaction time. For example, 3-hydroxy-3-phenylpropyne is less reactive and is carboxylated in 68% yield only after 165 h.

e) Effect of DBU

In the carboxylation of acetophenone and phenylacetylene, higher yields of carboxylated products are obtained when higher concentrations of DBU are employed, as shown in Tables 2.5–2.7. Figure 2.3 shows the relationships between the yields of carboxylated products and the amounts of DBU used in the carboxylation of indene, cyclopentadiene, and *o*-hydroxyacetophenone.[62]

Mixtures of mono-, di-, and tricarboxylic acids are obtained in the carboxylation of indene and cyclopentadiene.[63] Therefore, the total yield of carboxylated products can be used as a measure of the amount of fixed carbon dioxide. The most striking characteristic of these reactions is the occurrence of points at which the amount of the carboxylated product reaches a maximum relative to the amount of DBU used. This phenomenon is seen in the carboxylation of substrates which give carboxylated products bearing two or more anions by dissociation.

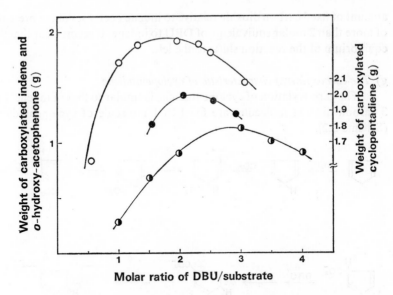

Fig. 2.3 Effect of DBU on the carboxylation of cyclopentadiene, indene, and *o*-hydro-
xyacetophenone, —●— cyclopentadiene, —○— indene, —◐— hydroxyacetophenone.

f) *The carboxylation characteristic of indene*

   The reaction mechanism in the carboxylation of indene is shown in
Scheme 2.1.

Scheme (2.1)

   As shown in Fig. 2.3, indene gives a maximum fixation of carbon
dioxide (over 90%, as dicarboxylic acid) in the presence of 1.5–1.6 molar
equivalents of DBU to indene. Thus, the reaction mixture would contain
approximately equimolar amounts of substances 5 and 6 in Scheme 2.1. The

amount of fixed carbon dioxide relative to indene decreases in the presence of more than 2 molar equivalents of DBU to indene. It seems clear that the equilibrium of the reaction shifts to the left.

g) *The carboxylation characteristic of cyclopentadiene*

The carboxylation of cyclopentadiene is similar to that of indene, but 3 carbon dioxide molecules are fixed to 1 molecule of cyclopentadiene (Scheme 2.2).

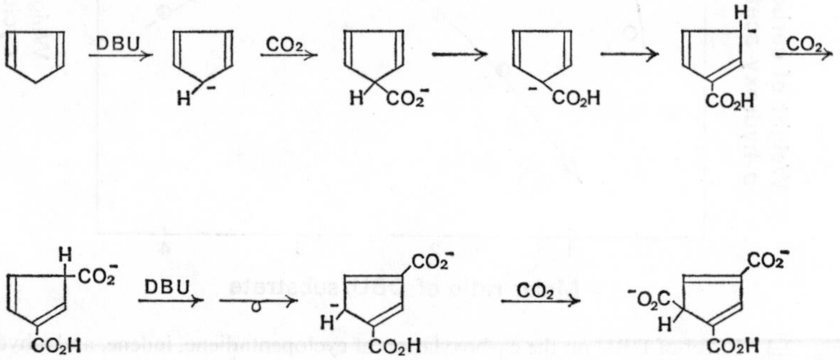

Scheme (2.2)

The carboxylation of cyclopentadiene is carried out at $0°-5°C$, and the acidification of the reaction mixture should be carefully performed below $13°C$. Under such reaction conditions, the carboxylation of cyclopentadiene using 2 molar equivalents of DBU gives tricarboxylated cyclopentadiene in over 90% yield. When the temperature is over $13°C$, the

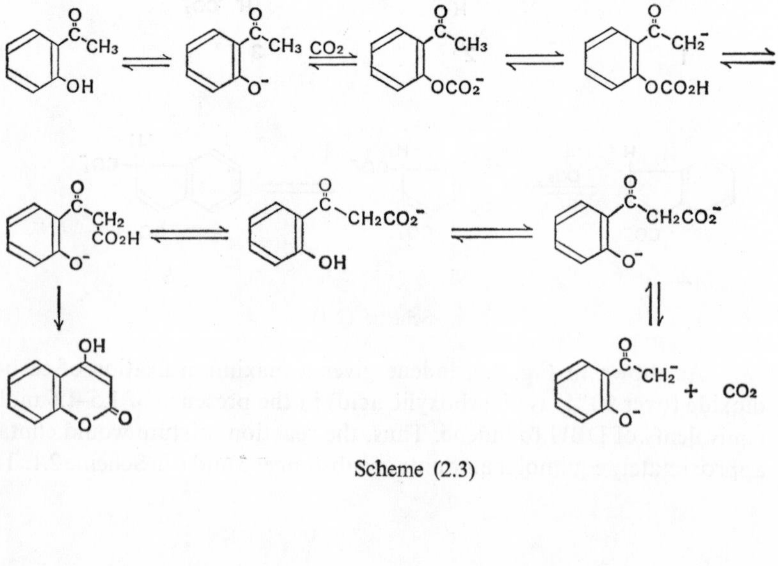

Scheme (2.3)

carboxylated product is a mixture of di- and tricarboxylic acids or only dicarboxylic acid, which gives Thiele's acid (tricyclo[5.2.1.0]deca-3,8-diene-4,9-dicarboxylic acid) by recrystallization from methanol. This carboxylation is the first example of a polycarboxylation of cyclopentadiene.[64]

In the carboxylation of *o*-hydroxylacetophenone (Scheme 2.3), it is unclear why the use of 3–4 molar equivalents of DBU gives the maximum amount of carboxylated product.[65]

h) Decarboxylation of β-oxocarboxylic acid and cyanoacetic acid derivatives

The reversibility of the carboxylation obtained by changing the reaction temperature (Eq. 2.20) is described in *c* (p. 19).

$$\begin{matrix} X \\ Y \end{matrix}\!\!\!>\!\!CH_2 + CO_2 + DBU \underset{\text{at high temp.}}{\overset{\text{at low temp.}}{\rightleftharpoons}} \begin{matrix} X \\ Y \end{matrix}\!\!\!>\!\!CH-COO^- \cdot DBUH^+ \qquad (2.20)$$

The reversibility of the carboxylation that results from charging and discharging carbon dioxide has also been investigated.[66] Forty kg/cm^2 carbon dioxide was charged into a mixture of 10 mmol of phenylacetonitrile and 25 mmol of DBU in 5 ml of DMF at $0°C$.

$$C_6H_5CH_2CN + DBU \xrightarrow[0°C]{CO_2,\ 3h} C_6H_5\underset{CN}{\overset{|}{C}}HCOOH\ (93\%)$$

$$C_6H_5CH_2CN + DBU \xrightarrow[0°C]{CO_2,\ 3h} \xrightarrow[0°C]{3h} C_6H_5\underset{CN}{\overset{|}{C}}HCOOH\ (10\%)$$

$$C_6H_5CH_2CN + DBU \xrightarrow[0°C]{CO_2,\ 3h} \xrightarrow[28°C]{3h} C_6H_5\underset{CN}{\overset{|}{C}}HCOOH\ (3\%)$$

$$C_6H_5CH_2CN + DBU \xrightarrow[0°C]{CO_2,\ 3h} \xrightarrow[0°C]{H_2O(30ml),\ 3h} C_6H_5\underset{CN}{\overset{|}{C}}HCOOH\ (92\%)$$

(2.21)

The reaction system was allowed to stand for 3h at $0°C$, carbon dioxide was then discharged from the reaction system, which was allowed to stand again for 3 h. The results are shown in Eq. (2.21). Similar results are obtained in the carboxylation of acetophenone, indicators that cyanoacetic acid and β-oxo acids are rapidly decarboxylated in the presence of DBU in aprotic solvents even at $0°C$. But decarboxylation is prevented by the addition of water to the reaction system.

$$C_6H_5\underset{\underset{CN}{|}}{C}HCOOH \xrightarrow[\text{DMF, 3h}]{\text{DBU, 28°C}} C_6H_5\underset{\underset{CN}{|}}{C}HCOOH \text{ (17\% recovered)} \qquad (2.22)$$

$$NC-CH_2COOH \xrightarrow[\text{DMF, 3h}]{\text{DBU, 28°C}} NC-CH_2COOH \text{ (20\% recovered)} \qquad (2.23)$$

$$\underset{}{R-\overset{\overset{O}{\|}}{C}-CH_2COO^-} \rightleftharpoons R-\overset{\overset{O}{\|}}{C}-CH_2^- + CO_2 \qquad (2.24)$$

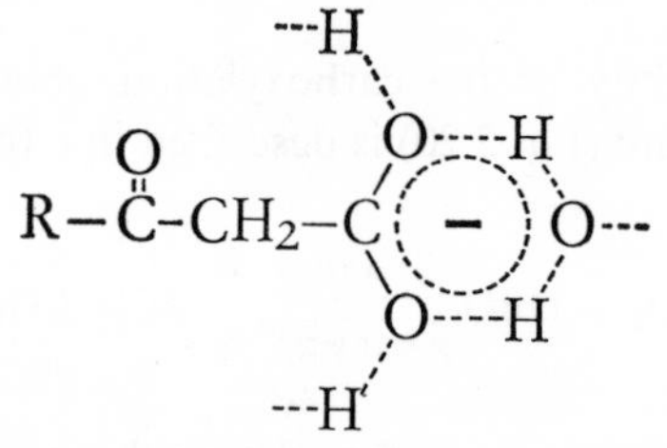

Fig. 2.4   Stabilization of β-ketocarboxylate ion by hydration.

The carboxylate anion produced by the reaction is stabilized by hydrogen bonding with water at room temperature. At elevated temperatures, β-oxo acids are gradually decarboxylated to produce the corresponding ketones. The decarboxylation of benzoylacetic acid in DMF on addition of triethylamine or p-toluenesulfonic acid was investigated,[67] and the results are shown in Fig. 2.5.

The rate of decarboxylation of β-oxo acids in an aqueous solution is represented by the following equation. However, k' is much

$$v = k\,[\beta\text{-oxo acid}] + k'\,[\beta\text{-oxo acid anion}] \qquad k' < k$$

larger than k in an aprotic solvent.

Cyanoacetic acid is more stable than the β-oxo acids, a difference which is important in the synthetic reaction.

The carboxylic acids obtained from active hydrogen compounds are easily decarboxylated by the addition of a small amount of tertiary amines. The active hydrogen compounds do not react appreciably with carbon dioxide in the presence of tertiary amines such as triethylamine, but by the addition of anhydrous magnesium chloride or other bivalent metal ion to the reaction system, the corresponding carboxylic acids are obtained in satisfactory yields as their magnesium or other metal chelates (Eq. 2.25; Tables 2.8 and 2.9).[68]

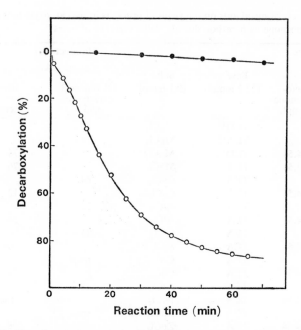

Fig. 2.5 Decarboxylation of benzoylacetic acid.
—○— Reaction conditions: C$_6$H$_5$COCH$_2$COOH, 2.44 mmol; DMF, 3.617 g; triethylamine, 1.67 mmol; temp., 14.5–14.6°C.
—●— Reaction conditions: C$_6$H$_5$COCH$_2$COOH, 3.14 mmol; DMF, 10.150 g; p-TosOH, 5.71 mmol; temp., 16°C.

$$\text{R-}\overset{\text{O}}{\overset{\|}{\text{C}}}\text{-CH}_2\text{R}' + \text{tertiary amine} \longrightarrow \text{R-}\overset{\text{O}}{\overset{\|}{\text{C}}}\text{-}\bar{\text{C}}\text{HR}' \xrightarrow{\text{CO}_2} \text{R-}\overset{\text{O}}{\overset{\|}{\text{C}}}\text{-CH-CO}_2^- \underset{\text{R}'}{}$$

(2.25)

$$\xrightarrow{\text{Mg}^{2+}} \quad \text{R-C} \overset{\text{R}'}{\underset{\text{O}}{}} \overset{\text{C}}{\underset{\text{O}}{}} \text{C=O} \underset{\text{Mg}}{}$$

The decarboxylation of β-oxo acids is accelerated by the addition of bases, which must accelerate the carboxylation of active hydrogen compounds, but only the strong bases actually effect the carboxylation in the absence of metal ions.

i) *Organic syntheses using the DBU-CO$_2$ carboxylation system.*

β-Oxo acids are decarboxylated easily and cannot be stored, but they can be converted into stable and useful derivatives. For example, they are useful as carbanion sources in the cross-aldol condensation (Eq. 2.26).

Table 2.8.   Effects of metal ions on the yields of products in the carboxylations of indene and acetophenone with carbon dioxide in the presence of bases[1].

Active methylene compound		Base (27.2 mmol)	Salt (9.1 mmol)	Product (%)	
Indene (mmol)	Aceto-phenone (mmol)			1H-Indene-3-carboxylic acid	Benzoyl-acetic acid
4.54		TED[2]	MgCl$_2$	78	
4.54		TEA[3]	MgCl$_2$	79	
	4.54	TED	MgCl$_2$		74
	4.54	TEA	MgCl$_2$		76
4.54		TEA	CaCl$_2$	64	
	4.54	TEA	CaCl$_2$		12
4.54		TEA	AlCl$_3$	22	
	4.54	TEA	AlCl$_3$		24
4.54		TEA	SrCl$_2$	11	
	4.54	TEA	SrCl$_2$		trace
4.54		TEA	CoCl$_2$	trace	
	4.54	TEA	CoCl$_2$		11
4.54		TEA	NiCl$_2$	9	
	4.54	TEA	NiCl$_2$		trace
4.54		TEA	ZnCl$_2$	trace	
	4.54	TEA	ZnCl$_2$		3
4.54		TEA	FeCl$_3$ MnCl$_2$ CuCl$_2$ BaCl$_2$ PdCl$_2$ AgCl[4] NaCl[4] KCl	0	
	4.54	TEA	CsCl		0

[1] The reaction was carried out at room temperature under 5 kg/cm^2 pressure of carbon
dioxide and in 10 ml of DMF.
[2] TED: Triethylenediamine.
[3] TEA: Triethylamine.
[4] Heterogeneous reaction.
Source: ref. 68. Reproduced by kind permission of the Chemical Society of Japan.

$$R-\overset{O}{\overset{\|}{C}}-\underset{R'}{\overset{|}{CH}}COOH + R''CHO \xrightarrow{\text{base}} R-\overset{O}{\overset{\|}{C}}-\underset{R'}{\overset{|}{CH}}\overset{OH}{\overset{|}{CH}}R'' \qquad (2.26)$$

Preparation   of   β-Oxo acids obtained by carboxylation with the DBU–CO$_2$
β-hydroxy acids[69] system can be reduced with sodium borohydride to give
β-hydroxy acids without separation (Eq. 2.27). The results are shown in Table
2.10.

$$R-\overset{O}{\overset{\|}{C}}-\underset{R''}{\overset{|}{CH}}R' \xrightarrow{\text{DBU, CO}_2} R-\overset{O}{\overset{\|}{C}}-\underset{R''}{\overset{R'}{\overset{|}{C}}}-COO^+ \xrightarrow[\text{aq.NaOH}]{\text{NaBH}_4} R-\underset{H}{\overset{HO}{\overset{|}{C}}}-\underset{R''}{\overset{R'}{\overset{|}{C}}}-COOH \qquad (2.27)$$

Table 2.9. Carboxylation of active methylene compounds with base-magnesium chloride-carbon dioxide mixtures[1].

Active methylene compound (4.54 mmol)	Base (27.2 mmol)	MgCl$_2$ (mmol)	Time (h)	Product	Yield[2] (%)
Indene	TED[3]	9.1	17	1 *H*-Indene-3-carboxylic acid	78
	TEA[4]	9.1	17	1 *H*-Indene-3-carboxylic acid	79
Acetophenone	TED	9.1	17	Benzoylacetic acid	74
	TEA	9.1	17	Benzoylacetic acid	76
p-Bromoacetophenone	TEA	9.1	17	*p*-Bromobenzoylacetic acid	75
Cyclohexanone	TEA	9.1	17	2-Oxocyclohexan-1-carboxylic acid	74
1-Indanone	TEA	9.1	17	1-Oxoindan-2-carboxylic acid	72
Benzylideneacetone	TEA	9.1	17	3-Oxo-5-phenyl-4-pentenoic acid	85
o-Nitroacetophenone	TEA	9.1	17	*o*-Nitrobenzoylacetic acid	65
β-Ionone	TEA	9.1	17	3-Oxo-5-(2,6,6-trimethyl-1-cyclohexenyl)-4-pentenoic acid	66
o-Acetylphenol	TEA	9.1	17	*o*-Hydroxybenzoylacetic acid	58
Dihydroresorcinol	TEA	9.1	17	2, 6-Dioxocyclohexane-1-carboxylic acid	54
Fluorene	TEA	9.1	64	Fluorene-9-carboxylic acid	40
s-Benzyl thioacetate	TEA	9.1	72	(Benzylthiocarbonyl)acetic acid	40
Nitromethane	TEA	9.1	17	Nitroacetic acid	37

[1] The reaction was carried out at room temperature under 5 kg/cm^2 pressure of carbon dioxide and in 10 ml of DMF.
[2] The yield was calculated on the basis of the substrate used.
[3] TED: Triethylenediamine.
[4] TEA: Triethylamine.
Source: ref. 68. Reproduced by kind permission of the Chemical Society of Japan.

Carboxylic acids from indene Indene can be carboxylated by several methods to give indene-1, 3-dicarboxylic acid, which is easily decarboxylated to produce indene-3-carboxylic acid. Esterification of indene-1, 3-dicarboxylic acid with diazomethane gives 1,3-dimethoxycarbonyl-1-methyl indene and an unidentified cycloadduct of diazomethane.[63]

Indene-1,3-dicarboxylic acid is hydrogenated with palladized charcoal in acetic acid-acetone (5:1) at atmospheric pressure (Eq. 2.28), and it is decarboxylated in acetone at room temperature to indene-3-carboxylic acid on addition of palladized charcoal or active charcoal. The decarboxylation is retarded completely by the addition of acetic acid to the reaction system.[65]

Carboxylic acids from cyclopentadiene The carboxylation of cyclopentadiene gives different products depending on the reaction conditions, as described in (*g*) above. The positions of the double bonds in the cyclopentadiene ring in the resulting tricarboxylic and dicarboxylic acids have not been determined. The unidentified products are shown in brackets in Eq. (2.29). The relative positions among the carboxyl groups are also illustrated.[63]

Table 2.10.   Synthesis of β-hydroxycarboxylic acids from ketones.

Substrate	Product	Yield(%)
C₆H₅-C(=O)-CH₃	C₆H₅-CH(OH)-CH₂-COOH	83
(p)Cl-C₆H₅-C(=O)-CH₃	(p)Cl-C₆H₅-CH(OH)-CH₂-COOH	80
(p)H₃C-C₆H₅-C(=O)-CH₃	(p)H₃C-C₆H₅-CH(OH)-CH₂-COOH	80
(p)Br-C₆H₅-C(=O)-CH₃	(p)Br-C₆H₅-CH(OH)-CH₂-COOH	75
(p)O₂N-C₆H₅-C(=O)-CH₃	(p)O₂N-C₆H₅-CH(OH)-CH₂-COOH	28
C₆H₅CH=CH-C(=O)-CH₃	C₆H₅-CH=CH-CH(OH)-CH₂-COOH	79
(indanone structure)	*(hydroxy-acetic acid indane structure)*	78
(tetralone structure)	*(hydroxy-carboxylic tetralin structure)*	92
H₃C-C₆H₃(CH₃)-C(=O)-CH₃	H₃C-C₆H₃(CH₃)-CH(OH)-CH₂-COOH	70
(menthone-type cyclohexanone structure)	*(hydroxy-carboxylic cyclohexane structure)*	70
(dimethyl cyclohexane ketone structure)	*(dimethyl cyclohexane hydroxy-carboxylic structure)*	26
H₃CO-C₆H₅-C(=O)-CH₃	H₃CO-C₆H₅-CH=CH-COOH	71

Unlike indene-1,3-dicarboxylic acid, the tricarboxylic and dicarboxylic acids of cyclopentadiene in acetone at room temperature do not evolve carbon dioxide on the addition of palladized or active charcoal.

Synthesis of 4-hydroxycumarine When *o*-hydroxybenzoylacetic acid is produced **from** *o*-hydroxyacetophenone by the carboxylation of *o*-hydroxyacetophenone using DBU and carbon dioxide, the reaction product contains a small amount of 4-hydroxycumarine (Eq. 2.30).[65]

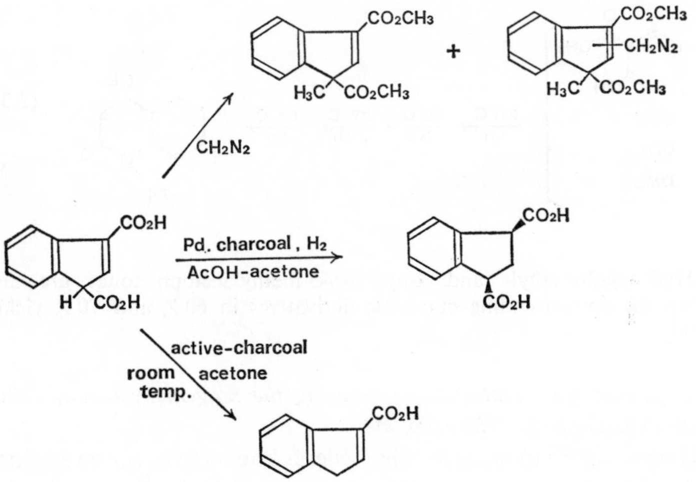

(2.28)

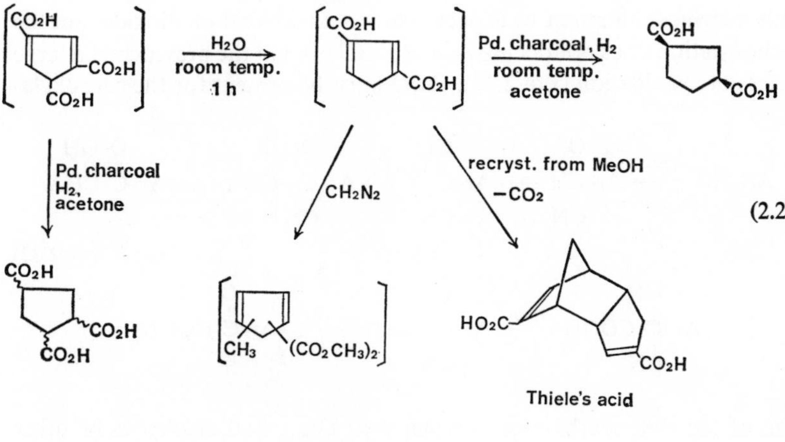

(2.29)

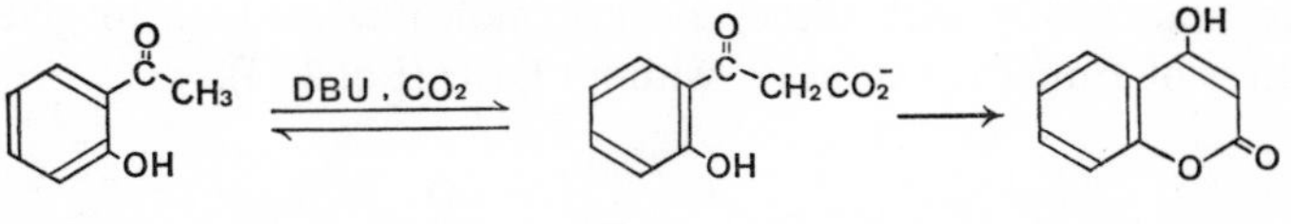

Thiele's acid

(2.30)

The yield of 4-hydroxycumarine increases with an extension of the reaction time or an elevation of the reaction temperature. At high temperatures, the total yield of the reaction products decreases, but the relative amount of 4-hydroxycumarine in the reaction mixture increases. *o*-Hydroxybenzoylacetic acid is decarboxylated at elevated temperatures (40° ~ 50° C), but it also cyclizes to give 4-hydroxy-cumarine under the reaction conditions (Eq. 2.31).[70]

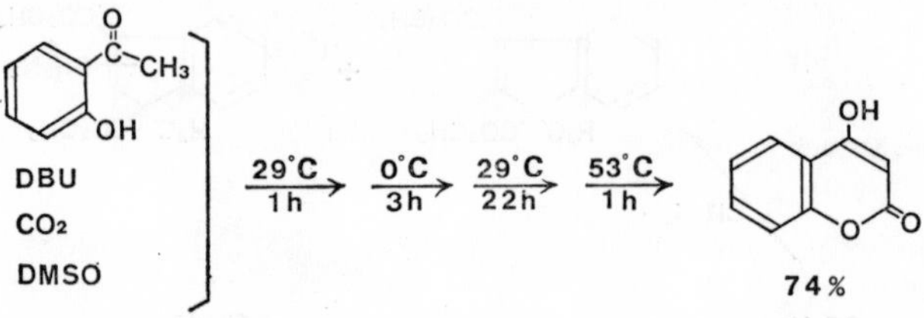

(2.31)

74%

2-Hydroxy-3-methyl- and 2-hydroxy-5-methylacetophenones are also cyclized to the corresponding cumarine derivatives in 60% and 70% yields, respectively.

j) Carboxylation of aromatic aldehydes using the reagent system of DBU, potassium cyanide, and carbon dioxide

"Umpolung"[71] in aromatic aldehydes is known to occur on addition of cyanide ion in benzoine condensation, in which the cyanocarbanion serves as an intermediate that attacks electron-deficient carbon atoms. This intermediate might be expected to react with carbon dioxide, but the carboxylation does not occur without DBU even in the presence of a large excess of cyanide ion (Eq. 2.32). DBU must be present for the carboxyla-

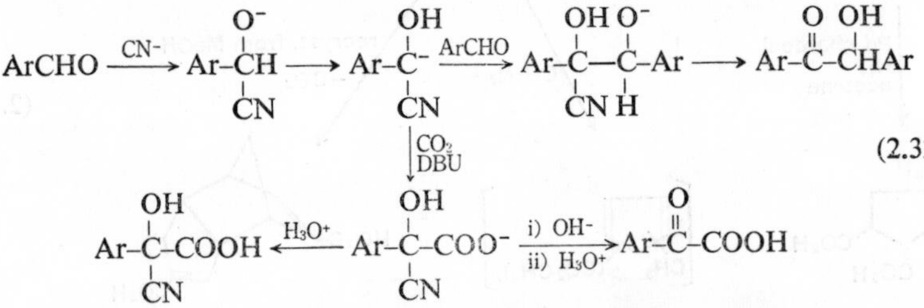

(2.32)

tion of the cyanocarbanion intermediate. The initial aldehydes or other electrophiles, such as acrylonitrile, alkyl acrylates, and α,β-unsaturated ketones, react easily with the cyanocarbanion intermediate to give the derivatives of aromatic ketones[72] without DBU (Eq. 2.33).

$$ArCHO + H_2C=CH-CN \xrightarrow{CN^-} Ar-\overset{O}{\overset{\|}{C}}CH_2CH_2CN \qquad (2.33)$$

It is unclear why carbon dioxide requires DBU to react with the cyanocarbanion intermediate while other electrophiles do not. Perhaps *N*-carboxylated DBU serves as an activator of the carbon dioxide in the carboxylation. This speculation arises from the observation that using

DBN instead of DBU has no effect at all on the carboxylation. A reaction system consisting of 10 mmol of aromatic aldehydes, 1.5 g of potassium cyanide, and 25 mmol of DBN in 10 ml of DMF solidified as soon as carbon dioxide was introduced. The same phenomenon was described in connection with the carboxylation of active hydrogen compounds with DBN and carbon dioxide in (B) (p. 15). There is a high planarity in the molecular structure of DBN which may cause the construction of a crystalline complex through the interaction of carbon dioxide, potassium cyanide, substrate, and solvent in the carboxylation of aromatic aldehydes (Fig. 2.6). In contrast, a DBU molecule, consisting of a seven-membered

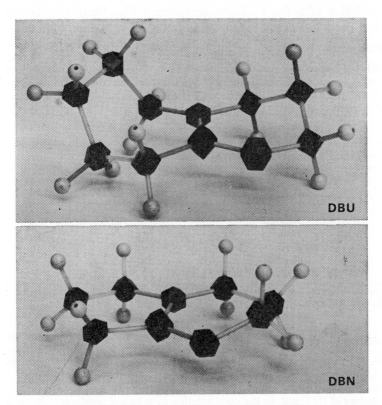

Fig. 2.6   Molecular structures of DBU and DBN.

ring fused to a six-membered ring (Eq. 2.34), has a bulging or puckered structure which disturbs the formation of a crystalline complex under the same reaction conditions, and so permits the carboxylation of aromatic aldehydes. This carboxylation of aromatic aldehydes is as effective as the

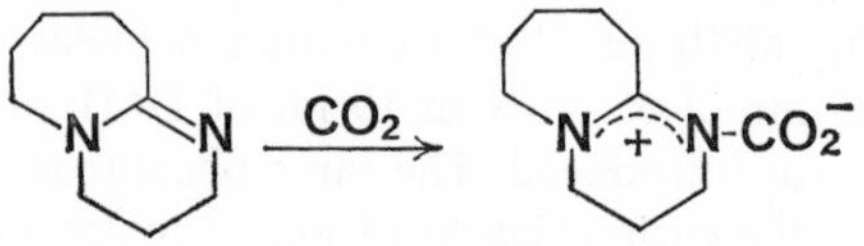

$$\text{(2.34)}$$

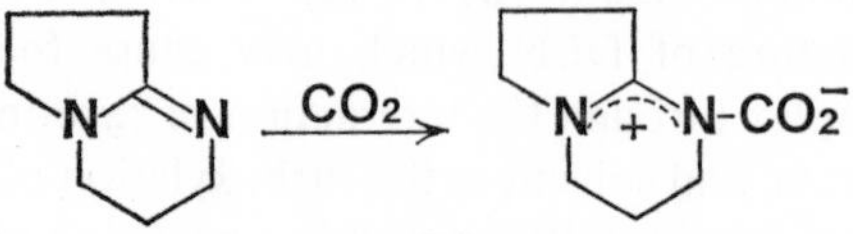

$$\text{(2.35)}$$

carboxylation of active hydrogen compounds with DBU and carbon dioxide. Potassium cyanide is the most available cyanide ion source.[73]

Reaction solvent The solvent for this carboxylation should dissolve the mixture of potassium cyanide, substrate, DBU, and carbon dioxide. DMF is the best solvent that meets this condition. Diglyme is an unsatisfactory solvent because it is difficult to separate from the reaction product. The carboxylation is retarded completely if benzene or HMPA is used as a solvent; the latter result is unexpected. Table 2.11 summarizes these findings.

Table 2.11.   Solvent effect on the carboxylation of benzaldehyde with KCN–DBU–CO_2.

Solvent	Yield (%)
DMF	84–93
DMSO	72
Diglyme	43
Dioxane	18
THF	8
Benzene	0
HMPA	0

Reaction conditions: Benzaldehyde, 10 mmol; KCN, 1.5 g; DBU, 4.0 g; CO_2, 5 kg/cm²; DMF, 10 ml; reaction time, 18 h; reaction temp., 10°–15°C.

Cyanide ion source Sodium cyanide and potassium cyanide are useful as sources of cyanide ion. The latter is more effective for the carboxylation. In this reaction, Since potassium cyanide becomes incorporated into the reaction products, it must be present in more than 1 molar equivalent. Use of 2.3 molar equivalents of potassium cyanide gives the maximum yield of the product.

The amount of DBU and the yield of product In the carboxylation of aromatic aldehydes, the yield of the products increases with an increase in the amount of DBU, as shown in Fig. 2.7.

Reaction temperature The maximum yield (~90%) of the carboxylated product is obtained the range of about 10°–15°C, when the reaction is carried out in as shown in Fig. 2.8. At elevated reaction temperatures (50°–60°C), benzoine condensation occurs rather than carboxylation. The results are summarized in Table 2.12.

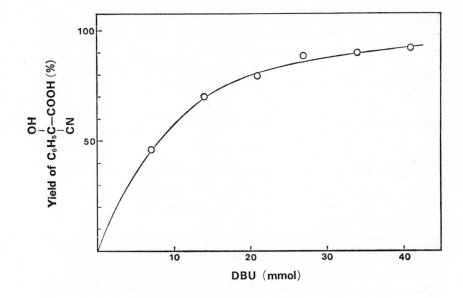

Fig. 2.7 Effect of DBU on the carboxylation of benzaldehyde.
Benzaldehyde, 10 mmol; KCN, 23 mmol; DMF, 10 ml; CO_2, 5 kg/cm².

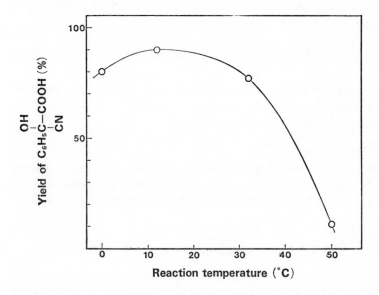

Fig. 2.8 Effect of reaction temperature on the carboxylation of benzaldehyde.
Benzaldehyde, 10 mmol; KCN, 23 mmol; DBU, 27 mmol; DMF, 10 ml; CO_2, 5 kg/cm².

Table 2.12. Carboxylation of aromatic aldehydes with KCN–DBU–CO_2.

Substrate	Reaction product	Yield (%)
C_6H_5CHO	$\underset{\overset{\mid}{OH}}{\overset{\overset{CN}{\mid}}{C_6H_5-C-COOH}}$	84–93
$(p)Cl-C_6H_4CHO$	$\underset{\overset{\mid}{OH}}{\overset{\overset{CN}{\mid}}{(p)Cl-C_6H_4-C-COOH}}$	70
$(o)Cl-C_6H_4CHO$	$\underset{\overset{\mid}{OH}}{\overset{\overset{CN}{\mid}}{(o)Cl-C_6H_4-C-COOH}}$	80
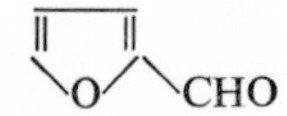 furan-CHO	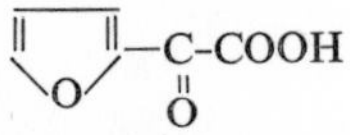 furan–C(=O)–COOH	50

Reaction conditions: Substrate, 10 mmol; KCN, 1.5 g; DBU, 4.0 g; CO_2, 5 kg/cm²; DMF, 10 ml; reaction time, 3–18 h; reaction temp., 10°–15°C.

2.1.4. The carboxylation of active hydrogen compounds with potassium carbonate and carbon dioxide

Under extreme conditions, anhydrous potassium acetate will react with carbon dioxide in the presence of anhydrous potassium carbonate to give potassium malonate[74,75] (Eq. 2.36).

$$H_3CCOOK \xrightarrow[Fe,\ 300°C,\ 3h]{K_2CO_3,\ CO_2(500\,atm)} \xrightarrow{H_3O^+} HOOC-CH_2-COOH\ 75\% \quad (2.36)$$

This carboxylation cannot be employed with other active hydrogen compounds because most of the resulting carboxylic acids are easily decarboxylated under the reaction conditions.

A. Carboxylation in DMSO

It has been found that acetophenone is carboxylated in the presence of finely powdered anhydrous potassium carbonate in DMSO under a pressure of 50–60 kg/cm² carbon dioxide at room temperature (Eq. 2.37).[76]

$$\underset{C_6H_5-\overset{\overset{O}{\mid\mid}}{C}-CH_3}{} + K_2CO_3 \xrightarrow[DMSO,\ room\ temp.]{CO_2\ (50kg/cm^2)} \underset{C_6H_5-\overset{\overset{O}{\mid\mid}}{C}-CH_2COOK}{} + KHCO_3 \quad (2.37)$$

The apparatus is a 400 ml pressure ball mill which contains steel balls and which is equipped with a pressure gauge, as shown in Fig. 2.9. The ball mill is rotated at a rate of 140 rpm in all the experiments.

Fig. 2.9   Reaction vessel contains steel balls.

Anhydrous rubidium carbonate and cesium carbonate also facilitate carboxylation, but anhydrous lithium carbonate and sodium carbonate have no effect; anhydrous tripotassium phosphate reacts only slightly. Anhydrous potassium carbonate is the most effective reagent, and DMSO is the best solvent. The carboxylation of acetophenone as a substrate model for active hydrogen compounds has been investigated in detail. The reaction requires more than 70 h for the preparation of the carboxylated product in 60% yield, as shown in Fig. 2.10.

Potassium phenoxide is an effective catalyst, but a small amount of potassium phenoxide is ineffective without anhydrous potassium carbonate (Eq. 2.38). Neither anhydrous sodium carbonate and phenol nor anhydrous lithium carbonate and a phenol reagent system is effective for this carboxylation.

$$K_2CO_3 \quad C_6H_5OH \quad C_6H_5\text{-}\overset{\overset{O}{\|}}{C}\text{-}CH_2CO_2K$$

$$KHCO_3 \quad C_6H_5OK \quad C_6H_5\text{-}\overset{\overset{O}{\|}}{C}\text{-}CH_3 \quad + \quad CO_2$$

(2.38)

Acetophenone is carboxylated to benzoylacetic acid in more than 80% yield after 15 h at room temperature when 0.34 molar equivalent of phenol is added to substrate. The addition of a small amount of benzyl-

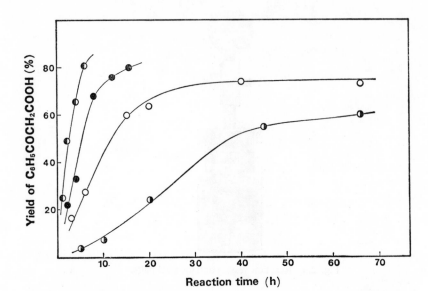

Fig. 2.10 Effect of phenol on the carboxylation of acetophenone. Reaction conditions: C₆H₅COCH₃, 25 mmol; K₂CO₃, 87 mmol; reaction temp. room temp. CO₂, ~ 60 kg/cm²; DMSO, 50 ml.
—◐—, Without phenol; —●—, Addition of phenol (0.8 g); —○—, Addition of PhOK (0.5 g); —◑—, Addition of phenol (0.8 g) and Et₃NCH₂CH₆H₅·Cl⁻ (1.0 g)

triethylammonium chloride promotes this carboxyation further. The effect of the addition of several phenols on the product yields is summarized in Table 2.13.

The results of this carboxylation with other substrates are given in Table 2.14.

It might be expected that acetylphenols would be easily carboxylated by this carboxylation without phenols, but *m*-acetylphenol and *p*-acetylphenol are carboxylated only slightly; *o*-acetylphenol is carboxylated to give *o*-hydroxybenzoylacetic acid. The reaction product contains a small

Table 2.13. Catalytic effect of phenols.

Phenol	Yield (%)
C_6H_5OK	86
C_6H_5OH	80
$(o)H_3C-C_6H_5OH$	55
$(m)H_3C-C_6H_5OH$	83
$(p)H_3C-C_6H_5OH$	86
$(p)H_5C_2C_6H_5OH$	85

Table 2.14. Carboxylation of active hydrogen compounds with K_2CO_3–CO_2 in DMSO.

Substrate	Reaction product	Yield (%)
C_6H_5–$\overset{O}{\overset{\|}{C}}$–$CH_3$	C_6H_5–$\overset{O}{\overset{\|}{C}}$–$CH_2COOH$	80
$(p)H_3C$–C_6H_5–$\overset{O}{\overset{\|}{C}}$–$CH_3$	$(p)H_3C$–C_6H_4–$\overset{O}{\overset{\|}{C}}$–$CH_2COOH$	94
$(p)Cl$–C_6H_4–$\overset{O}{\overset{\|}{C}}$–$CH_3$	$(p)Cl$–C_6H_4–$\overset{O}{\overset{\|}{C}}$–$CH_2COOH$	86
$(p)Br$–C_6H_4–$\overset{O}{\overset{\|}{C}}$–$CH_3$	$(p)Br$–C_6H_5–$\overset{O}{\overset{\|}{C}}$–$CH_2COOH$	89
$(p)H_3CO$–C_6H_4–$\overset{O}{\overset{\|}{C}}$–$CH_3$	$(p)H_3CO$–C_6H_4–$\overset{O}{\overset{\|}{C}}$–$CH_2COOH$	80
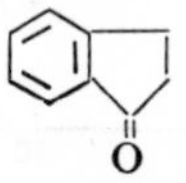 (indanone structure)	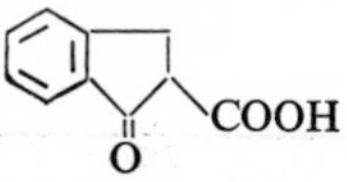 (indanone-COOH structure)	45
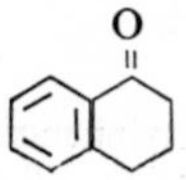 (tetralone structure)	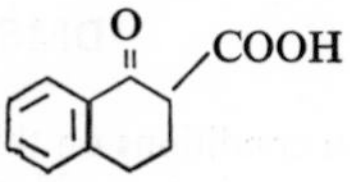 (tetralone-COOH structure)	53
$C_6H_5CH{=}CH$–$\overset{O}{\overset{\|}{C}}$–$CH_3$	$C_6H_5CH{=}CH$–$\overset{O}{\overset{\|}{C}}$–$CH_2COOH$	43
 (cyclohexanone structure)	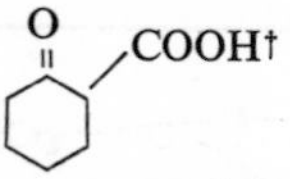 (cyclohexanone-COOH† structure)	35

† Contains a small amount of 2,6-dicarboxylic acid

amount of 4-hydroxycumarine, which increases with an increase in the reaction time.[70]

The yield of the carboxylated product rises with a decrease in the amount of DMSO added, and it reaches 76% without DMSO (Fig. 2.11). Scheme 2.3 shows that the carboxylated product is produced in high yield even under heterogeneous reaction conditions. Carbon dioxide fixed on the phenolate anion would lie close to the methyl group, from which the carbon dioxide would abstract a proton and then migrate to the generated carbanion.

The carboxylation of acetophenone was carried out in the presence of phenol without solvent to confirm the intramolecular carboxylation from the orthohydroxyl group to the methyl group in *o*-hydroxyacetophenone. However, acetophenone was appreciably carboxylated in these nearly heterogeneous reaction conditions.

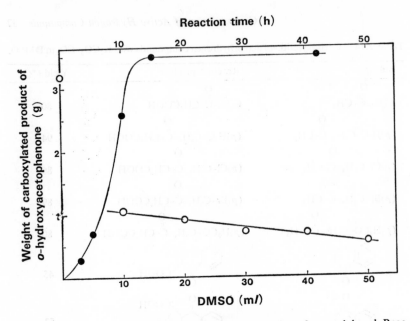

Fig. 2.11 Effect of reaction conditions on the carboxylation of *o*-acetylphenol. Reaction conditions: *o*-Hydroxyacetophenone, 25 mmol; K_2CO_3, 87 mmol; CO_2, -60 kg/cm². —○—, DMSO—Yield; —●—, Reaction time—Yield (without DMSO)

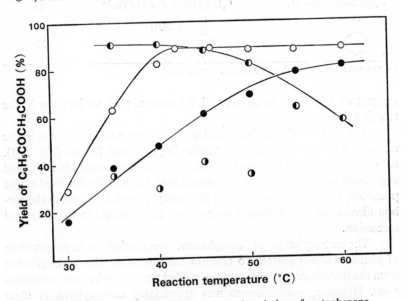

Fig. 2.12 Effect of reaction conditions on the carboxylation of acetophenone.

—●— ⎫ Acetophenone, 25 mmol   —◐— ⎫ Acetophenone, 25 mmol
—○— Phenol ⎬ K_2CO_3, 217 mmol   —◑— Phenol ⎬ K_2CO_3, 87 mmol
  (0.8 g) ⎭ CO_2, 77–87 kg/cm²   (0.8 g) ⎭ DMSO, 50 ml
                                                    CO_2, 77–87 kg/cm²

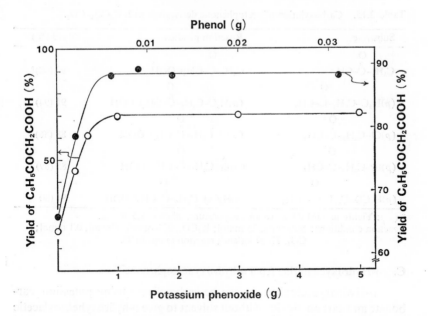

Fig. 2.13 Effect of additional phenol.
Reactions conditions: $C_6H_5COCH_3$, 25 mmol; reaction time, 15 h;
—○— K_2CO_3, 87 mmol; DMSO, 50 ml, reaction temp., room temp., CO_2, 62–65 kg/cm², PhOk, 0 ~ 5g
—●— K_2CO_3, 217 mmol; reaction temp., 50°C, CO_2, 77–87 kg/cm²; PhOH 0 ~ 0.032g

B.   Carboxylation without solvent[77]

The carboxylation of acetophenone with potassium carbonate and carbon dioxide depends on the reaction temperature, as shown in Fig. 2.12. The yield of benzoylacetic acid increases with a rise in the reaction temperature in the range of 30–60° C, but the reaction product is colored at a reaction temperature over 55° C. It is best to carry out the carboxylation at about 50° C. The catalytic effect of phenol without solvent is much larger than that in DMSO, as shown in Fig. 2.13. For instance, the addition of a slight amount of phenol (0.4 mol% for acetophenone) is sufficient to produce the effect. All the phenols which have a catalytic effect in DMSO also have a catalytic effect without solvent (Eq. 2.39).

$$C_6H_5\overset{O}{\overset{\|}{C}}\text{-}CH_3 + K_2CO_3 \xrightarrow[\text{PhOH (cat. amount), 3h}]{CO_2(\sim 80 kg/cm^2),\ 50°C} C_6H_5\overset{O}{\overset{\|}{C}}CH_2COOH\sim90\% \quad (2.39)$$

The carboxylation without solvent can also be employed with other substrates, as shown in Table 2.15.

Table 2.15.   Carboxylation of acetophenone derivatives with K_2CO_3–CO_2.

Substrate	Reaction product	Yield (%)
C_6H_5–$\overset{O}{\overset{\|}{C}}$–$CH_3$	C_6H_5–$\overset{O}{\overset{\|}{C}}$–$CH_2COOH$	90 (80)†
$(p)H_3C$–C_6H_5–$\overset{O}{\overset{\|}{C}}$–$CH_3$	$(p)H_3C$–C_6H_5–$\overset{O}{\overset{\|}{C}}$–$CH_2COOH$	91 (94)
$(p)Cl$–C_6H_5–$\overset{O}{\overset{\|}{C}}$–$CH_3$	$(p)Cl$–C_6H_5–$\overset{O}{\overset{\|}{C}}$–$CH_2COOH$	88 (86)
$(p)Br$–C_6H_5–$\overset{O}{\overset{\|}{C}}$–$CH_3$	$(p)Br$–C_6H_5–$\overset{O}{\overset{\|}{C}}$–$CH_2COOH$	85 (89)
$(p)H_3CO$–C_6H_5–$\overset{O}{\overset{\|}{C}}$—$CH_3$	$(p)H_3CO$–C_6H_5–$\overset{O}{\overset{\|}{C}}$–$CH_2COOH$	84 (80

† ( ): Yields in DMSO at room temperature, phenol 8.5 mmol.
Reaction conditions: Substrate, 25 mmol; K_2CO_3, 271 mmol; phenol, 0.1 mmol; CO_2, 77–87 kg/cm²; reaction temp. 50°C.

C.   4-Hydroxycumarine from o-hydroxyacetophenone[70]

o-Hydroxyacetophenone can be carboxylated using potassium carbonate and carbon dioxide without solvent to give o-hydroxybenzoylacetic acid with a satisfactory yield. The carboxylated product contains a small amount of 4-hydroxycumarine which increases when the carboxylated pro-

Table 2.16.   Synthesis of cumarines from o-hydroxyacetophenone derivatives.

Substrate	Reaction conditions	Reaction product	Yield (%)
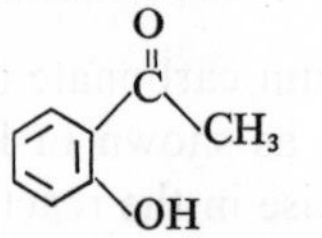	65°C, 15 h	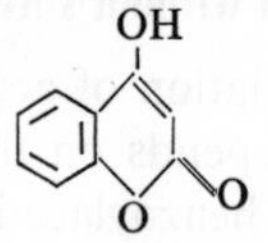	93
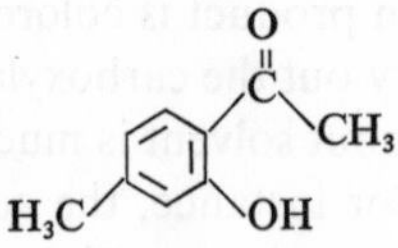	87°C, 15 h	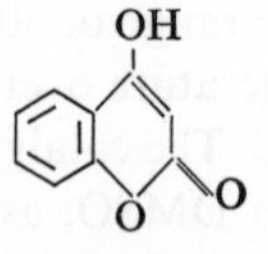	73
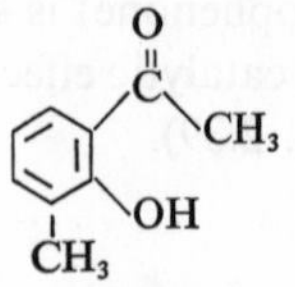	85°C, 15 h	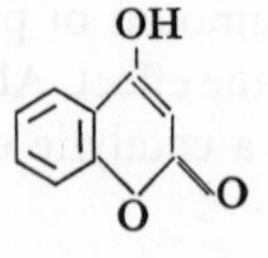	89
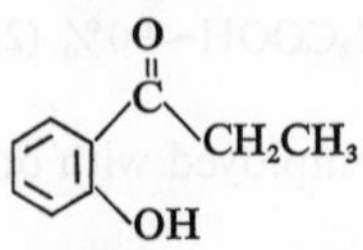	85°C, 15 h	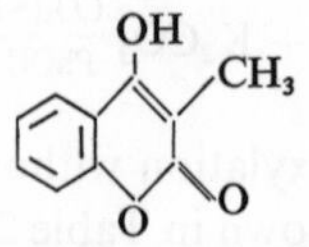	94

duct is heated. This reaction (Eq. 2.40) offers a convenient preparation for

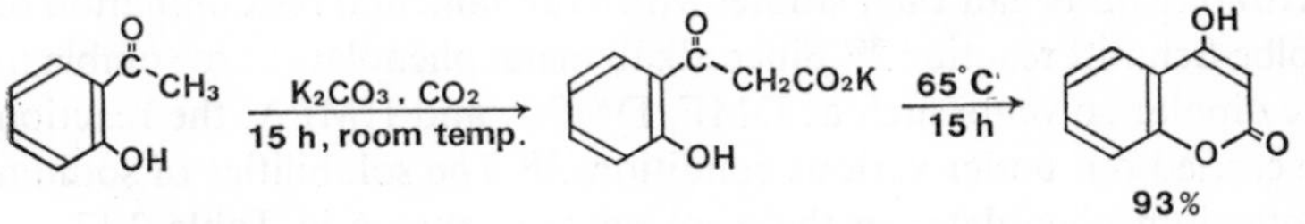

(2.40)

93%

4-hydroxycumarine derivatives. The results of this reaction with several other substrates are summarized in Table 2.16.

2.2. THE KOLBE-SCHMITT REACTION AND RELATED REACTIONS

One century ago, the reaction of alkali metal phenolates with carbon dioxide at an elevated temperature was discovered by Kolbe and Lautemann[78] and further developed by Schmitt and Burkard.[79] Sodium or potassium phenolate reacts with carbon dioxide under high pressure at an elevated temperature to produce salicylic acid and *p*-hydroxybenzoic acid (Eq. 2.41, 2.42). The Kolbe-Schmitt reaction is sensitive to differences

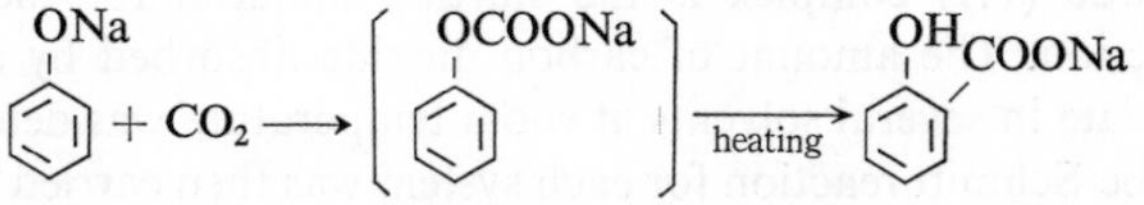

(2.41)

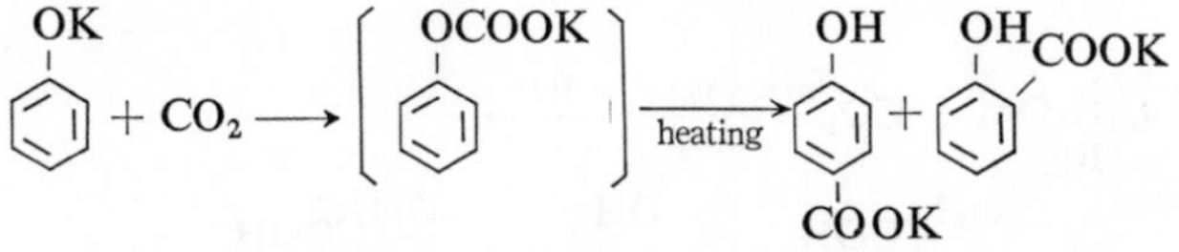

(2.42)

in the reaction temperature, the pressure of the carbon dioxide, and the amounts of bases or phenols, however.

It is difficult to determine the details at the reaction mechanism because the reaction is carried out in a heterogeneous system. Possible reaction mechanisms have been proposed by several investigators.[80-85] Despite this gap in our knowledge, it is evident that salicylic acid is obtained from sodium phenolate, and a mixture of *p*-hydroxybenzoic acid and salicylic acid is obtained from potassium phenolate.

2.2.1   The Kolbe-Schmitt Reaction in Solvents

Hirao *et al.* began their studies with a fundamental reexamination of the Kolbe-Schmitt reaction.[86] Since alkali metal phenolates are soluble in aprotic dipolar solvents such as DMF, DMSO, and HMPA, the reaction can be carried out under various conditions.[87] The solubilities of sodium and potassium phenolates in these solvents are shown in Table 2.17.

Table 2.17.   Solubility of C_6H_5OM in organic solvents.

Solvent	C_6H_5OM	Solubility (g C_6H_5OM/100 g solvent)		
		20°C	50°C	70°C
DMF	C_6H_5OK	22.7	31.1	35.3
	C_6H_5ONa	25.5	27.1	36.8
HMPA	C_6H_5OK	18.9	22.0	22.9
	C_6H_5ONa	8.5	11.7	16.1
DMSO	C_6H_5OK	76.4	88.3	108.3
	C_6H_5ONa	1.3	1.5	1.5

Sodium phenolate is slightly soluble in DMSO, but a solution of sodium phenolate suspended in DMSO becomes transparent upon addition of carbon dioxide. It is thought that an alkali metal phenolate and carbon dioxide (1:1) complex is the starting material for the Kolbe-Schmitt reaction. The amount of carbon dioxide absorbed by an alkali metal phenolate in several solvents at room temperature was determined, and the Kolbe-Schmitt reaction for each system was then carried out. The results are shown in Table 2.18.

$$C_6H_5OM + CO_2 \xrightarrow[\text{heat}]{\text{solv. (70g)}} \xrightarrow{H_3O^+}$$
$$\quad\; 10g$$

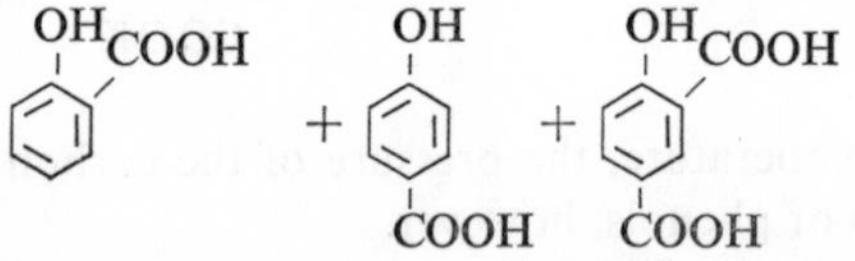

(2.43)

The yield of total acids increases with a rise in the reaction temperature, but the proportion of *p*-hydroxybenzoic acid in the total acids decreases. *p*-Hydroxybenzoic acid is obtained in 60% and 80% yields (conversion yields) from sodium phenoxide and potassium phenoxide, respectively. The yield of 4-hydroxyisophthalic acid from sodium phenolate increases suddenly over 200°C, but it is little affected by temperature in the same reaction with potassium phenolate. This large difference between sodium and potassium salts echoes the difference between sodium

Table 2.18. Carboxylation of sodium or potassium phenoxide in solvent.

$C_6H_5 \cdot$ OM M	Solvent	Initial CO_2 pressure	Reaction temp. (°C)	Total yield of (%)	(p) HO– $C_6H_5 \cdot$ COOH (%)	(o)HO– $C_6H_5 \cdot$ COOH (%)	4–OIP† (%)
K	DMF	1 atm	100	24.2	88	12	0
K	DMF	1	150	30.8	83	16	1
K	HMPA	1	140	13.7	81	14	5
K	HMPA	1	190	34.4	78	18	4
K	DMSO	1	100	22.0	94	2	4
K	DMSO	1	140	11.5	90	4	6
K	DMF	5 kg/cm²	140	33.5	88	9	3
K	DMF	5	180	41.1	83	13	4
K	HMPA	5	140	28.2	87	11	2
K	HMPA	5	230	46.9	55	31	14
K	DMSO	5	100	23.9	94	2	4
K	DMSO	5	140	25.8	85	6	9
K	DMSO	5	180	6.7	49	29	22
Na	DMF	1 atm	100	10.0	66	32	2
Na	DMF	1	140	22.0	63	36	1
Na	HMPA	1	140	11.8	76	19	5
Na	HMPA	1	190	18.5	24	65	11
Na	DMSO	1	100	18.5	83	11	6
Na	DMSO	1	140	5.9	75	17	8
Na	DMSO	1	170	1.7	11	73	16
Na	DMF	5 kg/cm²	140	36.1	70	27	3
Na	DMF	5	180	45.4	53	35	12
Na	HMPA	5	140	40.7	86	7	7
Na	HMPA	5	230	57.1	3	81	16
Na	DMSO	5	100	22.7	88	8	4
Na	DMSO	5	140	31.1	81	13	6
Na	DMSO	5	180	10.9	52	32	16

Reaction time, 30 min.
† 4–OIP: 4-hydroxyisophthalic acid

and potassium carbonates seen in the carboxylation of active hydrogen compounds (2.1.4). Hirao et al.[88] studied the Kolbe-Schmitt reaction of potassium and sodium phenolates in 41 solvents and found that the solvents fell into three groups.

(A) Those solvents which give the same results as the Kolbe-Schmitt reaction of potassium phenolate without solvent. They are aprotic solvents insoluble in water. The high boiling points of these solvents afford a high yield of total acids and p-hydroxybenzoic acid.

The Kolbe-Schmitt reaction of potassium phenolate in light oil or kerosene has been investigated in detail by Hirao and Kito.[89]

(B) Those solvents which give results like these of DMF. They include

DMF, dimethylacetamide, HMPA, DMSO, acetonitrile, and *N*-methyl-pyrrolidone. The carboxylation occurs at a relatively low reaction temperature, but no *p*-hydroxybenzoic acid is obtained at a high reaction temperature.

(C) Other solvents, including protic solvents, ethylenechloride, acetone, and ethylmethylketone.

There are many patents for preparations of *p*-hydroxybenzoic acid. The solvents used in these preparations include diphenylether,[89] diphenyl,[90] kerosene,[91,92] liquid hydrocarbon,[93,94] C_{5-20} aliphatic higher alcohols,[95] and aromatic hydrocarbons,[96] and all produce a high yield of *p*-hydroxybenzoic acid. Figure 2.14 illustrates one such process.

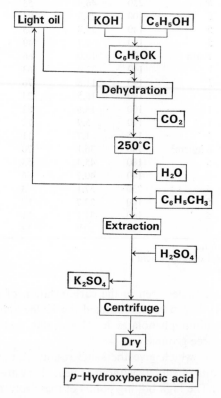

Fig. 2.14 Flow sheet for the continuous production of *p*-hydroxybenzoic acid used by the Ueno pharmaceutical Co.

C_6H_5OK (200 kg/h in 820 kg of Ph_2O) $\left.\begin{array}{c}\end{array}\right\}$ $\xrightarrow{230°}$ Reaction mixture 1055 kg/h
CO_2 (5 kg/cm^2)

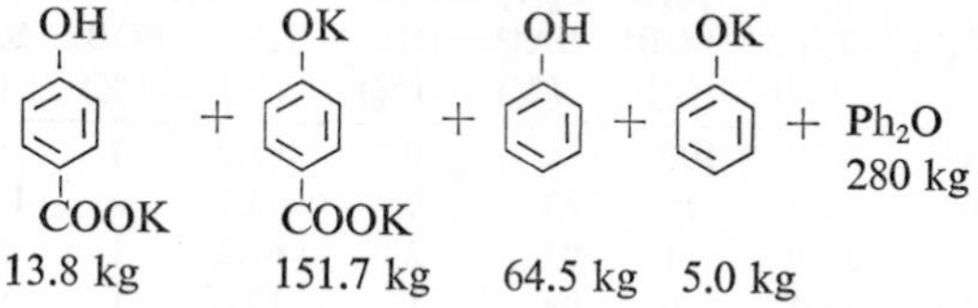

OH   OK   OH   OK

$+$ $+$ $+$ $+$ Ph_2O (2.44)
280 kg

COOK   COOK
13.8 kg   151.7 kg   64.5 kg   5.0 kg

2.2.2. Thermostability of Alkali Metal Salts of Salicylic Acid and p-Hydroxybenzoic Acid[97]

Potassium phenoxide preferentially produces p-hydroxybenzoic acid at 220° C in DMF under a carbon dioxide pressure of 5 kg/cm^2 for 1 h, but the product yield decreases rapidly with an increase in reaction time, because of the decarboxylation or rearrangement of p-hydroxybenzoic acid under the reaction conditions. Hirao *et al.* examined the thermostability of sodium or potassium salts of salicylic acid and p-hydroxybenzoic acid in detail. The disodium salt of salicylic acid and the dipotassium salt of p-hydroxybenzoic acid are relatively stable when heated in DMF. The monosodium, monopotassium, and dipotassium salts of salicylic acid, and the mono-sodium, disodium, and monopotassium salts of p-hydroxybenzoic acid are all unstable, undergoing decarboxylation or rearrangement under the reaction conditions. The salts of salicyclic acid yield very little p-hydroxy-benzoic acid (Eq. 2.45), but the salts of p-hydroxybenzoic acid do give salicylic acid (Eq. 2.46). These results are very useful for the regioselective carboxylation of phenol and its derivatives. They are summarized in Tables 2.19 and 2.20.

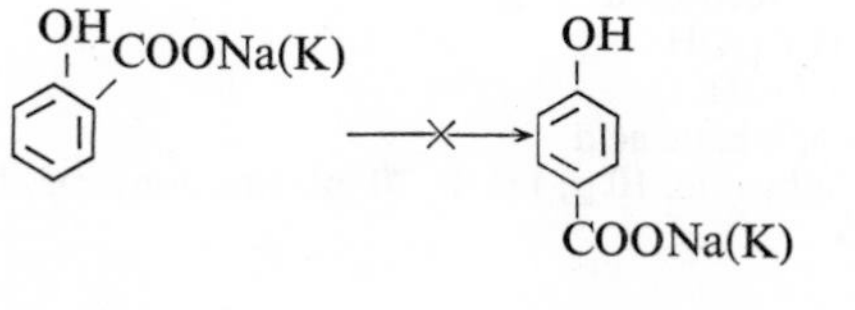

OH COONa(K)   OH

$\longrightarrow\!\!\times\!\!\longrightarrow$ (2.45)

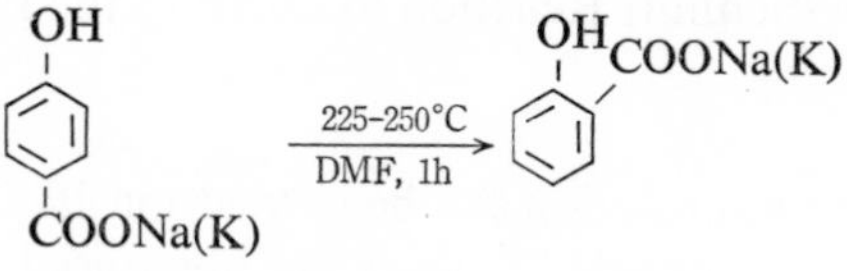

COONa(K)

OH   OH COONa(K)

$\xrightarrow[\text{DMF, 1h}]{225-250°C}$ (2.46)

COONa(K)

Table 2.19. Thermostability of hydroxybenzoic acid salts in DMF.

Salt	P[1]	T (°C)	(p)HO–C6H5COOH				(o)HO–C6H5COOH			
			T.Y.[3] (%)	POB[4] (%)	SA[5] (%)	OIP[6] (%)	T.Y. (%)	POB (%)	SA (%)	OIP (%)
Monosodium salt	V[2]	140	86.9	100	0	0	95.2	1	99	0
		200	45.0	10	83	7	93.2	0	100	0
		225	22.0	1	97	2	67.2	1	99	1
		250	17.4	0	99	1	17.6	1	97	2
	O	140	89.2	100	0	0	96.8	0	100	0
		200	55.0	2	73	25	92.7	0	100	0
		225	50.3	0	86	14	58.6	1	97	2
		250	19.9		1		18.8	0	99	1
	5	140	92.3	100	0	0	97.6	1	98	1
		200	48.7	2	80	18	92.7	0	100	0
		225	52.6	0	70	30	73.0	0	100	0
		250	54.9		1		57.5	1	94	5
Disodium salt	V	140	90.1	96	0	4	98.2	1	99	0
		200	91.0	96	0	4	100.0	1	99	0
		225	7.5	9	89	2	99.1	0	100	0
		250	3.0		2		98.9	0	98	2
	O	140	73.7	98	0	2	97.6	0	100	0
		200	7.3	83	15	2	98.4	0	100	0
		225	6.1	2	82	16	92.0	0	98	2
		250	3.3		2		58.4	0	88	12
	5	140	91.0	99	0	1	97.6	0	100	0
		200	6.9	78	3	19	99.5	0	100	0
		225	2.4				103.0	0	95	5
		250	3.2		2		79.6	0	78	22

[1] P: Initial pressure in reaction vessel (kg/cm^2)
[2] V: 3–5 mm Hg
[3] T.Y.: Total yield of recovered acids
[4] POB: (p) HO–C6H5COOH
[5] SA: (o)HO–C6H5COOH
[6] OIP: 4-hydroxyisophthalic acid
Reaction conditions: Substrate, 10 g; DMF, 70 ml; reaction time, 1.0 h.

2.2.3. The Kolbe-Schmitt Reaction in DMF in the Presence of Carbonates

The Kolbe-Schmitt reaction can be carried out in DMF in the presence of alkali metal carbonates[98] which are often used for the conventional heterogeneous Kolbe-Schmitt reaction. Potassium carbonate is the most effective base (Eq. 2.47).

The proportion of salicylic acid in the products increases at higher reaction temperatures. The molar ratio of p-hydroxybenzoic acid in the

Table 2.20.   Thermostability of hydroxybenzoic acid salts in DMF.

Salt	p†¹	T (°C)	$(p)HO-C_6H_5COOH$ T.Y.†³ (%)	POB†⁴ (%)	SA†⁵ (%)	OIP†⁶ (%)	$(o)HO-C_6H_5COOH$ T.Y. (%)	POB (%)	SA (%)	OIP (%)
Monopotassium salt	V†²	140	79.1	98	0	2	98.3	1	98	1
		200	67.9	99	1	0	92.4	0	100	0
		225	37.6	29	69	12	68.4	0	98	2
		250	19.2	2	95	3	18.6	0	99	1
	O	140	84.2	98	0	2	96.7	1	99	0
		200	62.5	99	0	1	86.8	0	100	0
		225	33.8	73	27	0	52.1	3	95	2
		250	25.5	3	96	1	29.0	0	99	1
	5	140	84.8	99	0	1	98.8	0	100	0
		200	64.4	96	0	4	93.1	0	100	0
		225	52.8	7	88	5	88.9	0	100	0
		250	47.2	1	98	1	42.8	0	99	1
Dipotassium salt	V	140	100.0	99	1	0	95.3	1	99	0
		200	95.3	96	1	3	67.0	2	96	2
		225	90.7	97	1	2	57.4			
		250	85.1	94	4	2	25.8		1	
	O	140	98.6	97	1	2	96.2	1	98	1
		200	99.4	95	1	4	94.0	5	74	21
		225	86.7	98	1	1	77.5			
		250	75.2	69	29	2	52.2		1	
	5	140	100.0	95	5	0	96.7	1	97	2
		200	100.0	97	1	2	98.0		1	
		225	91.5	95	3	2				
		250	90.1	83	13	4	100.0		1	

†¹ P: Initial pressure in reaction vessel (kg/cm²)
†² V: 3–5 mm Hg
†² T.Y.: Total yield of recovered acids
†⁴ POB: $(p)HO-C_6H_5COOH$
†⁵ SA: $(o)HO-C_6H_5COOH$
†⁶ OIP: 4-hydroxyisophthalic acid
Reaction conditions: Substrate, 10 g; DME, 70 ml; reaction time, 1.0 h.

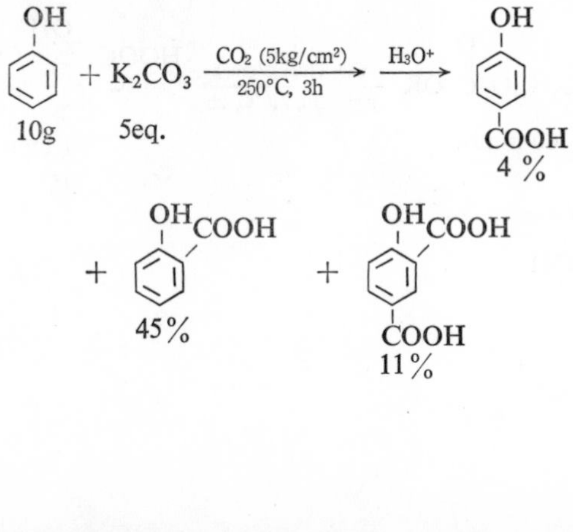

$$(2.47)$$

products increases, but the yield of total acids decreases, with lower reaction temperatures.

2.2.4.   The Carboxylation of Phenol with Alkyl Carbonates

Jones[99] reported that phenol reacts with alkali metal ethyl carbonates to give p-hydroxybenzoic acid and salicylic acid (Eq. 2.48). Hirao *et al.*[100] also examined this reaction, and they reported that potassium phenolate reacts with potassium alkyl carbonates in a 1:1 ratio to give p-hydroxybenzoic acid in a high yield under a carbon dioxide atmosphere at 220° C (Eq. 2.49).

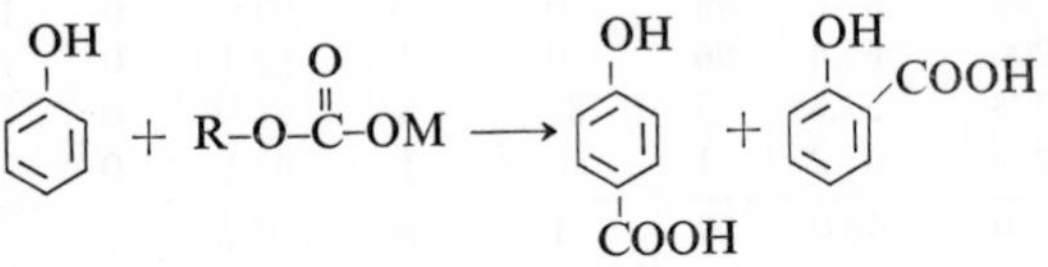

$$(2.48)$$

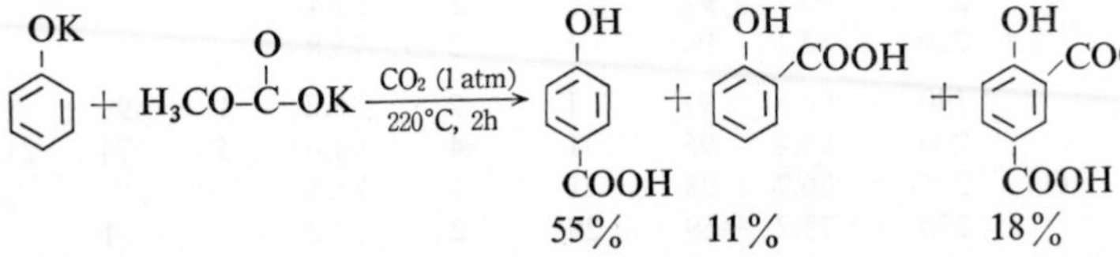

$$55\% \quad 11\% \qquad 18\% \quad (2.49)$$

This carboxylation was carried out in several solvents[101] As an example, the reaction in light oil is shown in (Eq. 2.50). Satisfactory results are

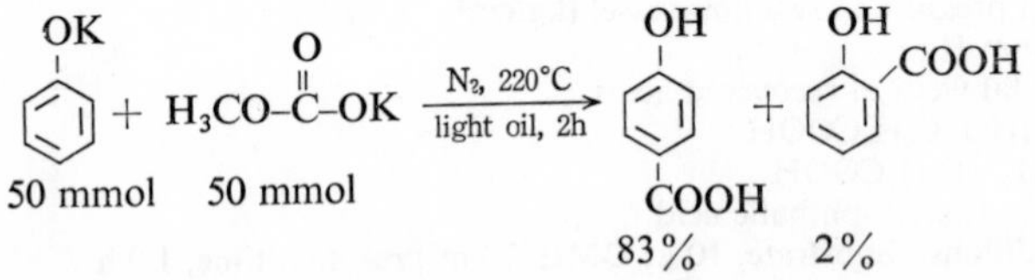

$$83\% \quad 2\% \qquad (2.50)$$

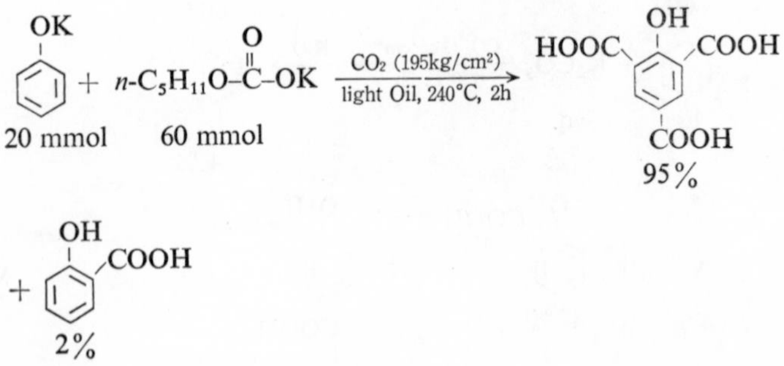

$$(2.51)$$

obtained, as shown in Table 2.21. In hydrocarbon solvents *p*-hydroxy-benzoic acid is obtained, but a good yield of hydroxytrimesic acid results if a large excess of potassium alkyl carbonates is used (Eq. 2.51).[102]

Table 2. 21. Solvent effect on carboxylation of potassium phenolate with potassium methyl carbonate.

Solvent	Total yield of acids (%)	$(p)HO-C_6H_4\cdot$ COOH (%)	$(o)HO-C_6H_4\cdot$ COOH (%)	4–OIP† (%)
Light oil	85	98	2	0
$(o)Cl-C_6H_4Cl$	83	94	2	4
$(p)Cl-C_6H_4Cl$	81	96	0	4
$(C_6H_5)_2O$	77	93	4	3
Kerosene	73	99	0	1
$(C_6H_5)_2$	73	93	4	3
C_6H_5Cl	73	87	5	8
C_6H_5OCH	71	85	4	11
Methylcyclohexane	70	93	1	6
$(C_2H_5)_2O$	70	89	5	6
$(C_2H_5)_3N$	69	98	1	1
N-Methyl-2-pyrrolidone	68	87	9	4
No solvent	68	84	11	5
$C_6H_5N(CH_3)_2$	65	75	5	20
DMF	6	65	27	8

† 4-Hydroxyisophthalic acid
Reaction conditions: C_6H_5OK, 50 mmol; CH_3OCOOK, 50 mmol; solvent, 40 g; reaction temp., 220°C; reaction time, 2 h; in N_2.

2.2.5. Carboxylation of the Alkali Fusion Product of Benzene-Sulfonates

In the synthesis of phenol by alkali fusion of benzenesulfonic acid with potassium hydroxide, the reaction mixture contains a large excess of potassium hydroxide. Alcohol is added to the reaction mixture and water is removed by adiotropic distillation, then carbon dioxide is introduced to the reaction mixture to give potassium alkyl carbonate without the sepa-

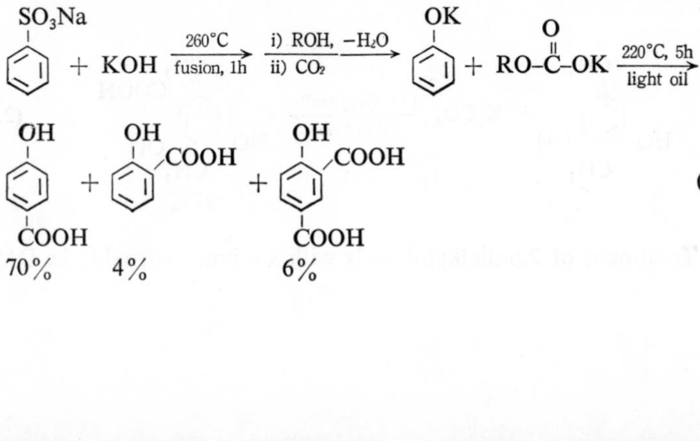

(2.52)

ration of phenol. Light oil is added to the mixture of potassium phenoxide, potassium alkyl carbonate, and other reaction products. This reaction mixture is then heated at 220° C to give p-hydroxybenzoic acid (Eq. 2.52). Using sodium hydroxide instead of potassium hydroxide gives a good yield of salicylic acid (Eq. 2.53).[103]

$$
\begin{array}{c}
\underset{\text{0.1 mol}}{\text{SO}_3\text{Na}} + \underset{\text{0.3 mol}}{\text{NaOH}} \xrightarrow[\text{0. 25h}]{\text{300°C}} \xrightarrow[\text{ii) CO}_2]{\text{i) ROH(}-\text{H}_2\text{O)}} \underset{\text{ONa}}{\bigcirc} + \text{RO–C–ONa}
\end{array}
$$

$$
\xrightarrow[\text{light oil}]{\text{220°C, 2h}} \quad \underset{\text{COOH}}{\overset{\text{OH}}{\bigcirc}} \;+\; \underset{}{\overset{\text{OH}}{\bigcirc}}\!\text{COOH} \;+\; \underset{\text{COOH}}{\overset{\text{OH}}{\bigcirc}}\!\text{COOH} \qquad (2.53)
$$

$$
\quad\quad\quad\quad 4\% \quad\quad 54\% \quad\quad\quad 2\%
$$

2.2.6. The Kolbe-Schmitt Reaction with Phenol Derivatives

The carboxylation of hydroquinone dipotassium salt in the presence of 0.02–0.7 mol of water gives 2,5-dihydroxybenzoic acid as a main product and 2,5-dihydroxyterephthalic acid as a by-product (Eq. 2.54). The same substrate can also be carboxylated with alkali metal alkyl carbonates in organic solvents.[104,105]

$$
\underset{\text{OK}}{\overset{\text{OK}}{\bigcirc}} + \text{K}_2\text{CO}_3 \xrightarrow[\text{H}_2\text{O, 210°C, 3h}]{\text{CO}_2\ (50\text{kg/cm}^2)} \underset{\text{HO}}{\overset{\text{COOH}}{\bigcirc}}\!\!\text{OH} \;+\; \underset{\text{HO}\quad\ \text{COOH}}{\overset{\text{COOH}}{\bigcirc}}\!\!\text{OH} \qquad (2.54)
$$

$$
\quad\quad\quad\quad\quad\quad\quad\quad \text{major} \quad\quad\quad\quad \text{minor}
$$

2,5-Dimethylresorcinol can be carboxylated with carbon dioxide at high pressure when heated in the presence of alkali metal carbonates (Eq. 2.55).[106]

$$
\underset{\underset{\text{CH}_3}{\text{HO}\quad\text{OH}}}{\overset{\text{CH}_3}{\bigcirc}} + \text{K}_2\text{CO}_3 \xrightarrow[\text{150°C, 5. 5h}]{\text{CO}_2\ (50\text{kg/cm}^2)} \underset{\underset{\text{CH}_3}{\text{HO}\quad\text{OH}}}{\overset{\text{CH}_3}{\bigcirc}}\!\!\text{COOH} \qquad (2.55)
$$

$$
\quad\quad\quad\quad\quad\quad\quad\quad\quad\quad\quad\quad\quad\quad 95\%
$$

Treatment of 2,6-dialkylphenols with sodium hydroxide in DMF, then with carbon dioxide at 70°–80°C, gives 3,5-dialkyl-4-hydroxybenzoic acid (Eq. 2.56).[107]

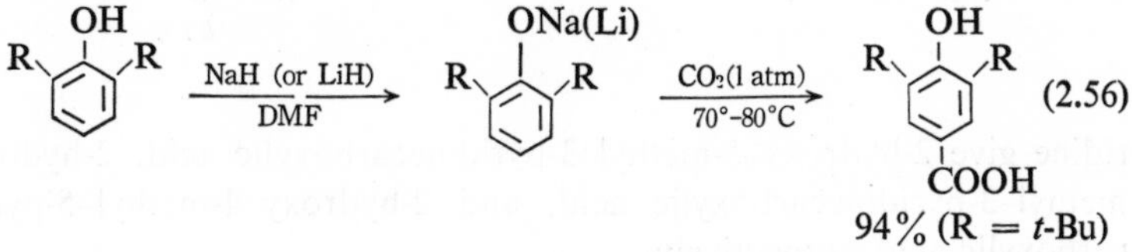

$$94\% \ (R = t\text{-Bu}) \tag{2.56}$$

The carboxylation of meta-aminophenol in the presence of potassium carbonate gives 4-amino-2-hydroxybenzoic acid with a good yield (Eq. 2.57).[108]

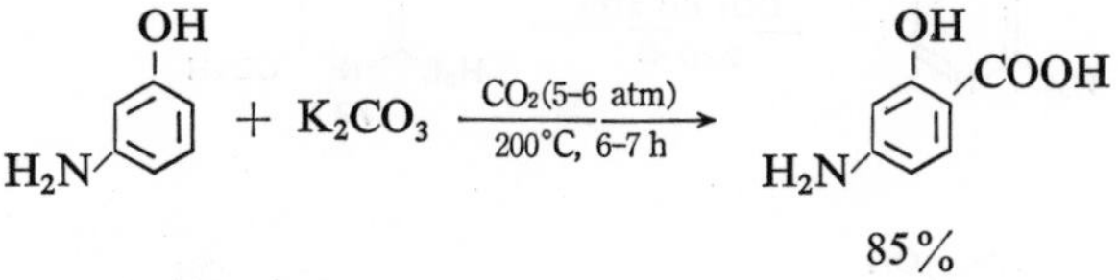

$$85\% \tag{2.57}$$

p-(Hydroxyalkoxy)benzoic acids can be prepared by carboxylation with carbon dioxide (Eq. 2.58).[109]

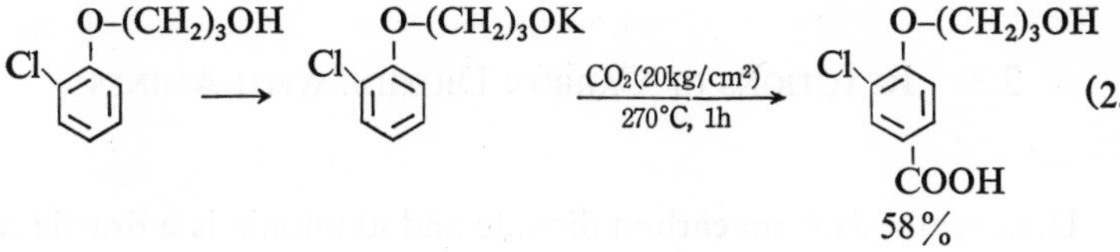

$$58\% \tag{2.58}$$

2-Hydroxynaphthalene-3-carboxylic acid is prepared via the carboxylation of 2-sodiumnaphthoxide with carbon dioxide at 180°–260°C (Eq. 2.59).[110]

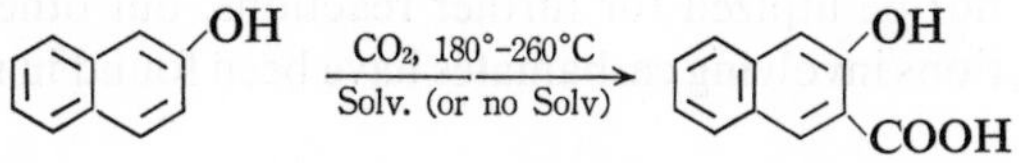

$$\tag{2.59}$$

Treatment of a mixture of 2-hydroxy-3-methylpyridine and potassium carbonate with 55 atm of carbon dioxide for 9 h gives 2-hydroxy-3-methyl-5-pyridinecarboxylic acid (Eq. 2.60).[111] Carboxylation of the sodium and potassium salts of 2-hydroxy-3-methylpyridine gives only 49.5% and 53% yields of the corresponding acid, respectively. Similarly, 2-hydroxy–5-methyl–, 2-hydroxy–6-methyl–, and 2-hydroxy–4-methylpy-

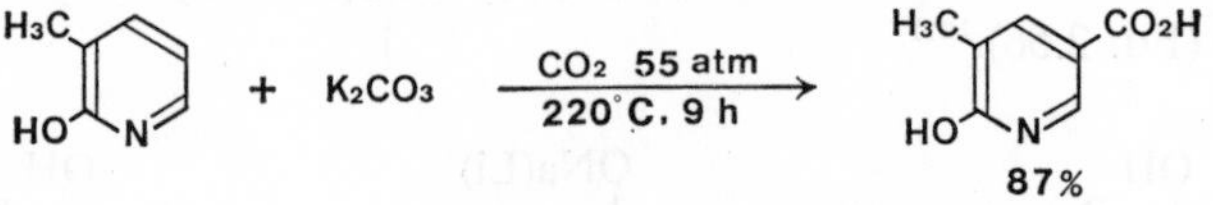

$$(2.60)$$

ridine give 2-hydroxy–5-methyl–3-pyridinecarboxylic acid, 2-hydroxy–6-methyl–3-pyridinecarboxylic acid, and 2-hydroxy–4-methyl–5-pyridinecarboxylic acid, respectively.

Heating 3-hydroxypyridine derivatives with carbon dioxide (60 atm) at 220°C produces the corresponding carboxylic acids.[112]

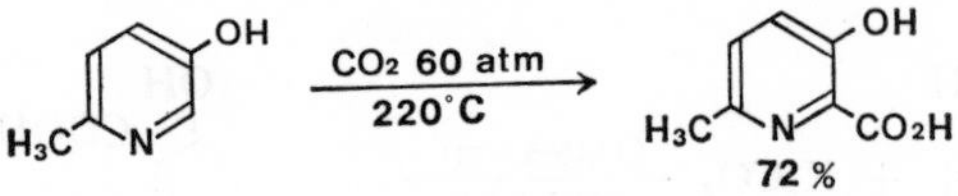

$$(2.61)$$

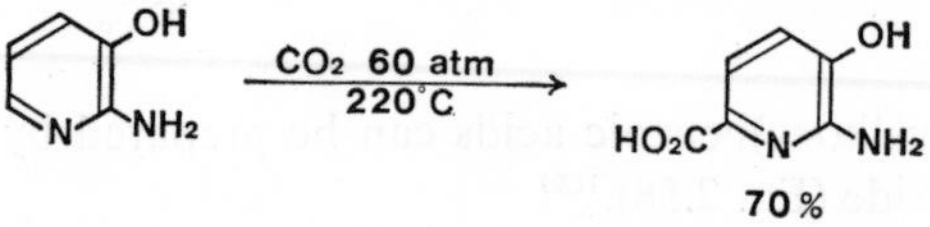

$$(2.62)$$

2.3. REACTIONS OF CARBON DIOXIDE WITH AMINES

Urea synthesis from carbon dioxide and ammonia is a drastic reaction requiring a high reaction temperature and high pressure, and the conditions for the synthesis of urea derivatives from carbon dioxide and primary or secondary amines are also extreme. The reaction of primary and secondary amines with carbon dioxide to give carbamates, however, occurs at room temperature. Carbon dioxide incorporated into amines is labile and can not be utilized for further reactions, but other novel and interesting reactions involving carbamates have been found in recent years.

2.3.1. The Synthesis of Urea Derivatives

Yamazaki et al.[113,114] reported an interesting synthesis of ureas from carbon dioxide and various amines with a number of phosphorus compounds in the presence of tertiary amines under mild conditions (Eq. 2.63; Scheme 2.4).

$$CO_2 + R\,NH_2 + HOP(O\,C_6H_5)_2 \xrightarrow[40°C]{Py}$$

$$\overset{\overset{\text{O}}{\|}}{R\text{–NHCNHR}} + C_6H_5OH + (HO)_2POC_6H_5 \qquad (2.63)$$

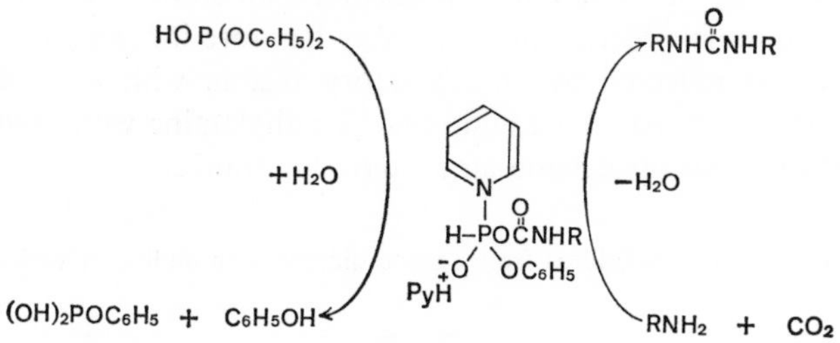

(Scheme 2.4)

Table 2.22. Symmetrical ureas from carbon dioxide and amines by reaction with diphenyl phosphite in pyridine[†1]

Amine	Yield (%) of ureas[†2†]
Aniline	85
o-Toluidine	74
Isopropylamine	23
Cyclohexylamine	27
N-Methylaniline	0
Diphenylamine	0
Diethylamine	0

[†1] $HO–P(OC_6H_5)_2$ = Amine = 50 mmoles in 40ml of pyridine; temp. = 40°C; time = 4h.
[†2] Based on the amine used.

Results of the preparation of several symmetrical ureas with diphenyl phosphite in pyridine are given in Table 2.22.

General procedure   Gaseous carbon dioxide at 40°C is passed for 4 h into a mixture of diphenyl phosphite (50 mmol) and aniline (50 mmol) in 40 ml of pyridine. The resulting mixture is evaporated under reduced pressure to a syrup. On treatment of the syrup with 20 ml of 50% aqueous ethanol, N,N'-diphenylurea is obtained in 85% yield.

Primary aromatic amines such as aniline and o-toluidine give higher yields of ureas than do aliphatic amines such as isopropylamine and cyclohexylamine. Secondary amines failed to yield ureas. Since aliphatic amines are more basic than pyridine, the unsatisfactory results obtained with these compounds may be due to the formation of pyridine-insoluble and unreactive ammonium salts. The lack of reaction with secondary amines cannot be accounted for by steric hindrance.

Tertiary amines facilitated the reaction of carbon dioxide with aniline, as shown in Table 2.23. The yield of N,N'-diphenylurea is affected by both the basicity of the tertiary amines and their steric hindrance around the nitrogen atoms. Steric hindrance may be more influential in pyridine derivatives, as reflected by unsatisfactory results with α-picoline, 2,4-lutidine, and (surprisingly) 2,6-lutidine. Triethylamine with high basicity gives favorable results despite large steric hindrance.

Table 2.23. N,N'-Diphenylurea from carbon dioxide with aniline in various tertiary amines

Tertiary amine	Yield (%)[1]	
	pK_a	The urea[2]
Imidazole	7.12	93
γ-Picoline	6.02	88
β-Picoline	5.52	84
Pyridine	5.23	85
2,4-Lutidine	6.99	74
α-Picoline	5.97	60
2,6-Lutidine	6.99	32
Triethylamine	10.87	70
None	—	4[3]

[1] Based on aniline used.
[2] $HO-P(OC_6H_5)_2$ = aniline = 50 mmoles in 40 ml of tertiary amines; imidazole = mmoles in 40 ml of DMF; temp. = 40°C; time = 4h.
[3] The reaction was carried out in 40 ml of DMF.

The reaction is favored also by the presence of phosphorus compounds other than diphenyl phosphite, as shown in Table 2.24. Thus, triaryl phosphite, diphenyl ethyl phosphonite, phenyl diphenylphosphinite, and diphenyl n-butylphosphonate are effective in the reaction between carbon dioxide and amines. On the other hand, their alkyl esters failed to yield urea. Both triphenyl and triethyl phosphate failed to promote the reaction.

Both phosphorus compounds and tertiary amines are found to be stoichiometrically involved in the reactions of carbon dioxide with aniline. A similar dependence of yield on the amount of tertiary amine (imidazole) is observed in these reactions. The effect of the reaction temperature on the yield of urea in the reaction with aniline (2 moles) when using diphenylphosphite (1 mole) in pyridine is shown in Fig 2.15. Better results are obtained at higher reaction temperatures despite the diminished concentration of carbon dioxide in the reaction mixture at these temperatures. This

Table 2.24. *N,N'*-Diphenylurea from carbon dioxide and aniline by reaction with various phosphorus compounds[1]

Phosphorus compound	Yield (%)[2]
HO—P(O—⟨○⟩)₂	94
P(O—⟨○⟩)₃	68
P(O—⟨○⟩—Cl)₃	87
P(O—⟨○⟩)₃ CH₃	o- 21 m- 50 p- 40
Et–P(O—⟨○⟩)₂	90
Et–P(OEt)₂	0
(⟨○⟩—)₂—P—O—⟨○⟩	71
Bu—P(O)(O—⟨○⟩)₂	15

[1] Phosphorus compounds = aniline = 50 mmoles in 40 ml of DMF containing imidazole (100 mmoles); temp. = 40°C ; time = 4h.
[2] Based on aniline used.

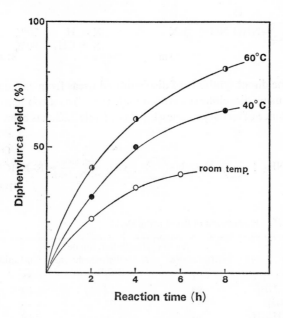

Fig. 2.15 Effect of reaction conditions on diphenyl urea formation. Aniline, 50 mmol; diphenyl phosphite, 25 mmol; pyridine, 40 ml.
(Source: ref. 106. Reproduced by kind permission of Pergamon Press)

urea synthesis can be applied to the synthesis of a polymer-containing urea unit. These results are described in the next chapter in detail.

Yamazaki *et al.* also reported the synthesis of trisubstituted ureas from hexaalkylphosphorus triamides, primary amines, and carbon dioxide (Eq. 2.64). This reaction proceeds almost quantitatively.[115]

$$
\begin{array}{c}
P{-}\!\!\begin{array}{c} N(R^1)_2 \\ N(R^1)_2 \\ N(R^1)_2 \end{array}
\xrightarrow[\text{ii) } R^2NH_2, \, 60°C, \, 3h]{\text{i) } CO_2, \, 60°C, \, 3h}
\begin{array}{c} R^1 \\ \diagdown \\ R^1 \diagup \end{array}\!\!N{-}\!\overset{\displaystyle O}{\overset{\|}{C}}{-}NH{-}R^2
\end{array}
\qquad (2.64)
$$

$$ R^1 = \; CH_3, \, C_2H_5 $$

$$ R^2 = -CH_2C_6H_5, \; -\!\!\big\langle\,\big\rangle, \; -C_6H_5 $$

N-Trimethylsilylaniline derivatives take up carbon dioxide in the presence of several transition metal halides at room temperature and yield carbamate derivatives which produce urea derivatives on heating in a carbon dioxide atmosphere (Eq. 2.65).[116]

$$
X{-}\!\!\big\langle\,\big\rangle{-}NHSi(CH_3)_3 \xrightarrow[\text{THF, room, temp.}]{CO_2, \, FeCl_3} X{-}\!\!\big\langle\,\big\rangle{-}NHCO_2Si(CH_3)_3 \xrightarrow[\text{4-9h}]{140\text{-}145°C}
$$

$$
X{-}\!\!\big\langle\,\big\rangle{-}NH\overset{\displaystyle O}{\overset{\|}{C}}{-}NH{-}\!\!\big\langle\,\big\rangle{-}X
$$

$X = H$	70%
$X = CH_3$	90%

(2.65)

The direct synthesis of disubstituted ureas from carbon dioxide and amines using dicyclohexylcarbodiimide (DCC) as a condensing reagent in the presence of tertiary amines (Eq. 2.66, Table 2.25, has been described.[117]

$$
R{-}NH_2 + \big\langle\,\big\rangle{-}N{=}C{=}N{-}\big\langle\,\big\rangle + CO_2 \xrightarrow[-78°C, \, 1\text{-}16h]{\text{tert. amine}} R{-}NH\overset{\displaystyle O}{\overset{\|}{C}}NHR
$$

$$ 31\text{-}98\% $$

(2.66)

Table 2.25.   Preparation of ureas using DCC.

Amine	Yield (%) with tert. amine:			
	Triethylamine	*N*-Methylmorpholine	2,6-Lutidine	Pyridine
$(C_6H_5)_2CHNH_2$	91	81	85	91
c-$C_6H_{11}NH_2$	98	75	89	91
$C_6H_5NH_2$	—	31	—	—
$C_6H_5CH_2NH_2$	—	—	80	—
i-$C_3H_7NH_2$	48	—	—	—
n-$C_3H_7NH_2$	—	—	—	58

In this reaction, tertiary amines play an important role, and no urea is obtained if tertiary amines are absent. When 1-ethyl-3-(3-dimethylamino-propyl)-carbodiimide is used as a condensing reagent, low yields of ureas are obtained (30–40%) after 1–2 days. However, the reactions proceed smoothly in an autoclave under pressure of carbon dioxide at room temperature to yield 95–100% ureas (Eq. 2.67).

$$RNH_2 + C_2H_5N=C=N(CH_2)_3N(CH_3)_2 + \text{tert. amine} + CO_2(60kg/cm^2)$$

$$\xrightarrow[1h]{\text{room temp.}} \underset{95 \sim 100\%}{R-NHCNHR}^{\overset{O}{\overset{\|}{}}} \qquad (2.67)$$

The mechanism of the formation of ureas is shown in Scheme 2.5; evidence for path A includes a broad band that appeared at 2200 cm^{-1} in the infrared spectrum of a crude reaction mixture.

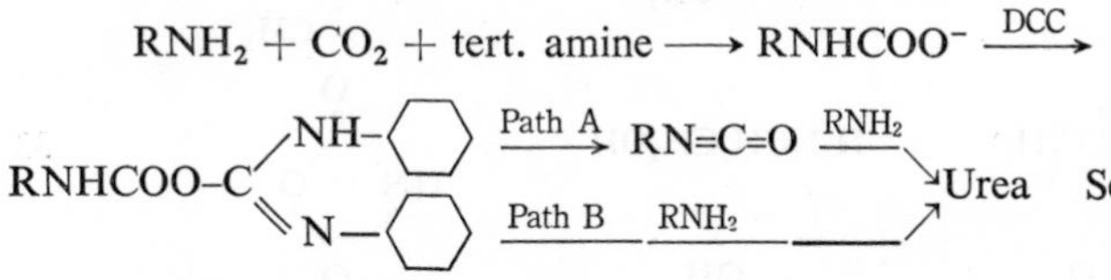

$$RNH_2 + CO_2 + \text{tert. amine} \longrightarrow RNHCOO^- \xrightarrow{DCC}$$

Scheme (2.5)

2-Oxazoline derivatives are obtained from β-hydroxyamine derivatives and carbon dioxide in the presence of tetraphenyl pyrophosphate or phosphinic acid anhydride, which has a P–O–P bond (Eq. 2.68).[118]

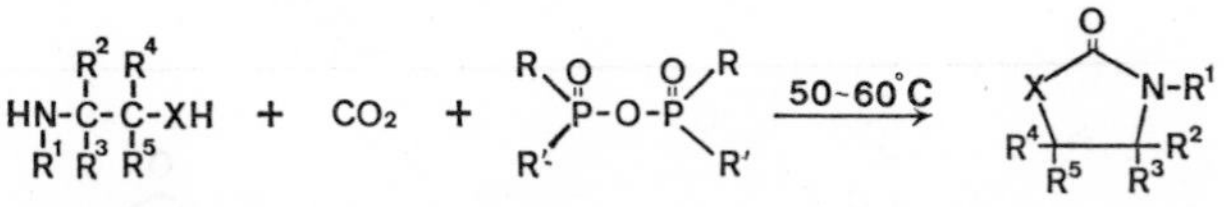

$$(2.68)$$

The results of such reactions are summarized in Table 2.26.

2.3.2.   Synthesis of Carbamic Esters

The use of carbamic acids would appear to offer certain advantages for studying carbon dioxide incorporation into organic compounds under mild conditions.

The reaction of α-haloacylophenones with carbon dioxide and aliphatic amines in methanol produces oxazolidone derivatives (Eq. 2.69, Table 2.27).[119]

Table 2.26. Preparation of cyclic carbamates.

Condensing reagent	Amino compound	Reaction product	Yield (%)
$(C_6H_5O)_2\overset{O}{\underset{\parallel}{P}}-O-\overset{O}{\underset{\parallel}{P}}(OC_6H_5)_2$	$H_2NCH_2CH_2OH$	(oxazolidin-2-one ring) HN–O, C=O	32
$(C_6H_5O)_2\overset{O}{\underset{\parallel}{P}}-O-\overset{O}{\underset{\parallel}{P}}(OC_6H_5)_2$	$HN-\underset{\underset{C_6H_5}{\mid}}{\overset{\overset{CH_3}{\mid}}{C}}-CH-OH$, CH_3	H_3C-N ring, H_3C, C_6H_5, C=O	90
$(C_6H_5O)_2\overset{O}{\underset{\parallel}{P}}-O-\overset{O}{\underset{\parallel}{P}}(OC_6H_5)_2$	$H_2NCH_2CH_2SH$	HN–S ring, C=O	30
$(C_6H_5O)_2\overset{O}{\underset{\parallel}{P}}-O-\overset{O}{\underset{\parallel}{P}}(OC_6H_5)_2$	$H_2NCH_2\underset{\underset{CH_3}{\mid}}{CHOH}$	HN–O ring, CH_3, C=O	32
$(C_6H_5)_2\overset{O}{\underset{\parallel}{P}}-O-\overset{O}{\underset{\parallel}{P}}(C_6H_5)$	$H_2NCH_2CH_2OH$	HN–O ring, C=O	33
$(C_6H_5O)_2\overset{O}{\underset{\parallel}{P}}-O-\overset{O}{\underset{\parallel}{P}}(OC_6H_5)_2$	benzene ring with OH and NH$_2$	benzoxazolone =O, N–H	62
$(C_6H_5O)_2\overset{O}{\underset{\parallel}{P}}-O-\overset{O}{\underset{\parallel}{P}}(OC_6H_5)_2$	$H_2NCH_2CH_2NH$	HN–NH ring, C=O	32

$$R\text{-}NH_2 + CO_2 + CH_3OH \xrightarrow[]{C_6H_5\text{-}\overset{O}{\underset{\parallel}{C}}\text{-}\overset{X}{\underset{\mid}{C}}(R')CH_3} \xrightarrow[\text{heat,10-12 h}]{CO_2(\text{bubble})} \quad R\text{-}N \ldots O,\ C_6H_5,\ CH_3,\ OH,\ R' \quad (2.69)$$

When α-bromopropiophenone is employed, a mixture of stereoisomers of 3-alkyl-4-hydroxy-5-methyl-4-phenyloxazolidone-2 is obtained. The reaction of α-bromoacetophenone or α-chloroacetophenone does not yield the corresponding oxazolidone derivatives. The same results are obtained with aromatic or secondary aliphatic amines. A reasonable mechanism for these reactions is shown in Eq. (2.70).

When the oxiranes are treated under similar conditions, the corresponding oxazolidones are obtained in much better yields. Chloromethyl-

Table 2.27.   Formation of oxazolidone derivatives

Oxazolidones	R[1]	R'	Yield(%)
1	Methyl	Me	60
2	Ethyl	Me	45
3	*i*-Propyl	Me	50
4	Cyclohexyl	Me	65
5	Benzyl	Me	40
6	Cetyl	Me	40
7	*i*-Propyl	H	40[2]

[1] Derived from amines (RNH_2).
[2] Yield of a mixture.

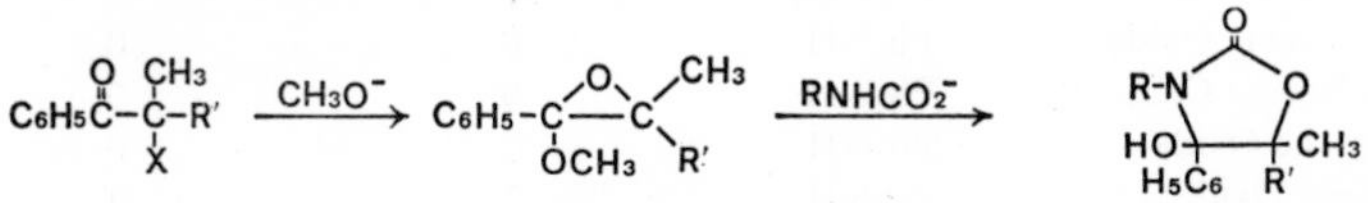

$$\text{(2.70)}$$

oxirane and phenyloxirane can be reacted with carbamate to give cyclic carbamates and hydroxyphenethyl carbamates in 2–20% yields (Eq. 2.71 and 2.72).

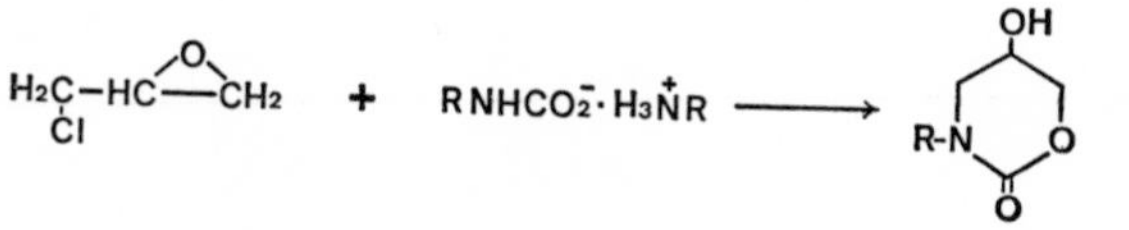

$$\text{(2.71)}$$

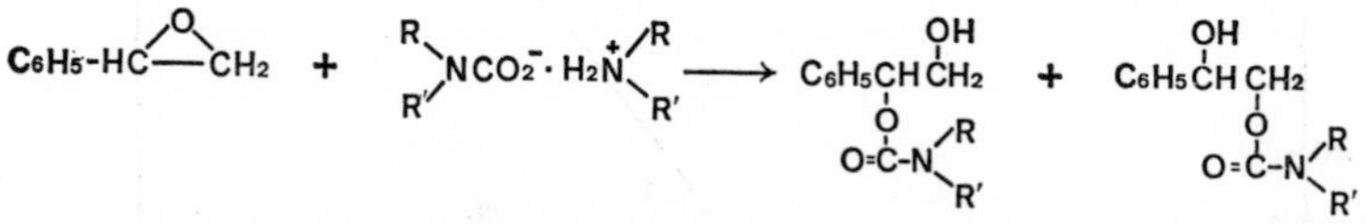

$$\text{(2.72)}$$

Yoshida and Inoue[120,121] reported a novel and simple method for the direct preparation of monocarbamic esters of 1,2-diol from carbon dioxide, epoxide, and amine (Eq. 2.73, Table 2.28).

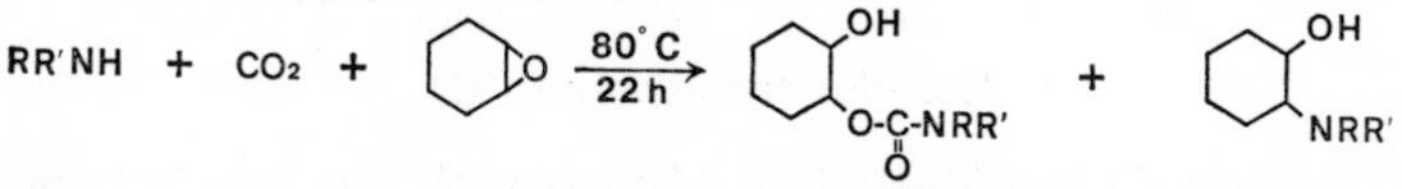

$$\text{(2.73)}$$

Figure 2.16 shows that a reaction temperature of 80° C is preferable for the synthesis of carbamic ester under these conditions. Carbamic ester is obtained from other epoxides such as epoxyethane and 1,2-epoxypropane, together with a considerable amount of amino alcohol or polymerized product. In the reactions of carbamate with α-bromoacylophenone and

Table 2.28.   Reaction of Carbon Dioxide, Epoxide, and Amine.[†1]

Epoxide	Amine	Product(%)	
		Carbamate	Aminoalcohol
Cyclohexene Oxide	Me$_2$NH	42	40
Cyclohexene Oxide	Et$_2$NH	62	5
Cyclohexene Oxide	nPrNH$_2$	54	5
Propylene Oxide	Me$_2$NH	16	2
Propylene Oxide	Et$_2$NH	24	42
Propylene Oxide	nPrNH$_2$	20	21
Ethylene Oxide[†2]	Me$_2$NH	6	15
Ethylene Oxide[†2]	Et$_2$NH	21	15
Ethylene Oxide[†2]	nPrNH$_2$	15	6
Cyclohexene Oxide	Ph$_2$NH	0	0
Cyclohexene Oxide[†3]	Ph$_2$NH	0	0
Propylene Oxide	Ph$_2$NH	0	0
Ethylene Oxide	Ph$_2$NH	0	0

[†1] CO$_2$, 50 kg/cm^2; amine and epoxide, 0.2 mol each; 80°C; 22 h.
[†] Oligomer of ethylene oxide was formed.
[†3] 70 h.
Source: *The Report of Asahi Glass Foundation for Industrial Technology*, **32**, 91 (1978).
Reproduced by kind permission of the Asahi Glass Foundation.

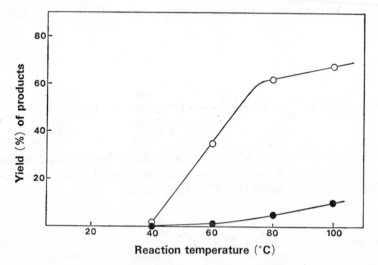

Fig. 2.16   Reaction of CO$_2$, (C$_2$H$_5$)$_2$NH and 1,2-epoxycyclohexane. CO$_2$, 50 kg/cm^2, 22 h.; ○, carbmate; ●, Amino alchohol.
Source: Reproduced by kind permission of the Chemical Society of Japan.

with epoxide, carbamate derivatives from secondary amines gave no corresponding products with the former, but yielded carbamates with the latter.

Yoshida and Inoue[122] reported that 2-hydroxydimethyl carbamate was synthesized quantitatively by the reaction of carbon dioxide, tetrakis (dimethyamido)titanium(IV), and 1,2-epoxycyclohexane (Eq. 2.74, Table 2.29).

$$Ti(NMe_2)_4 \ + \ CO_2 \ + \quad \underset{O}{\bigcirc} \quad \xrightarrow[220\,h]{80\degree C} \xrightarrow{H_2O} \quad \underset{O\overset{O}{\underset{\parallel}{C}}NMe_2}{\overset{OH}{\bigcirc}} \qquad (2.74)$$

Table 2.29. Reaction of tetrakis (dimethylamido) titanium, carbon dioxide, and 1,2-epoxycyclohexane[†1]

Run	Ti(NMe₂)₄ (g)	Epoxide (g)	Solvent	Temp. (°C)	Time (h)	(1)[†2] (g)	(%)[†3]	(2)[†4] (g)
1	0.4	9.7	CH₂Cl₂	35	360	0.18	(13)	trace
2	1.0	9.7	THF	35	290	0.17	(5.0)	0
3	1.0	9.7	CH₂Cl₂	80	150	0.07	(2)	0.034
4	1.0	9.7	Benzene	80	210	1.7	(50)	0.027
5	1.0	9.7	THF	80	240	2.5	(74)	0.025
6	1.0	9.7	THF	80	23	0.30	(8.8)	0
7	1.0	39.0	—	80	220	3.4	(100)	0.007
8	1.8	3.1	THF	80	190	0.94	(15)	0

[†1] CO_2, 50 kg/cm²; solvent 40 ml.   [†2] 2-Hydroxycyclohexyl *N,N*-dimethy carbamate.
[†3] Yield based on four Ti-N bonds.   [†4] Epoxide-CO_2 copolymer.
Source: *Bull. Chem. Soc. Japan*, **51**, 559 (1978). Reproduced by kind permission of the Chemical Society of Japan.

General procedure   A mixture of tetrakis (dimethylamido) titanium (IV) (1 g, 4.5 mmol) and 1,2-epoxycyclohexane (10 ml, 99 mmol) in benzene (40 ml) is placed in a 200 ml autoclave and constantly stirred at 80°C under carbon dixide pressure (50 kg/cm²). After 210 h of reaction, a few drops of water are added to the reaction mixture, which is then filtered to remove insoluble materials, and the filtrate is concentrated under reduced pressure at room temperature. To the filtrate concentrate is added about 100 ml of petroleum ether with stirring. The mixture is seeded with a small amount of pure crystalline carbamate and cooled in a refrigerator overnight, during which time white needles separate out. The crystals are filtered off and dried at room temperature *in vacuo*.

White, needle-like crystals of *trans*-2-hydroxycyclohexyl dimethyl carbamate, mp 52°–53° C are obtained with a 50% yield (1.7 g).

This reaction is the first example of a fixation of carbon dioxide by a reaction between a metal carbamate and an epoxide. The mechanism of this reaction is thought to proceed as shown on Eq. (2.75). Since tetrakis

$$\text{Ti(NMe}_2)_4 \;+\; \text{CO}_2 \longrightarrow \text{Ti(OCONMe}_2)_4 \xrightarrow{} \text{Ti}\!\left[\text{O}\!\!\bigcirc\!\!\text{O}\overset{O}{\overset{\|}{\text{C}}}\text{NMe}_2\right]_4 \quad (2.75)$$

(dimethylamido)titanium(IV) reacts with carbon dioxide to form its tetra-carbamate derivative,[123, 124] and since tetrakis(dimethylamido)titanium(IV) does not react with epoxyethane and 1,2-epoxypropane,[124] tetrakis(di-methylamido)titanium(IV) is expected to react preferentially with carbon dioxide even in the presence of epoxide. It was found that 1,2-epoxycyclo-hexane does not react with tetrakis(dimethylamido)titanium(IV) at room temperature.

A novel direct synthesis of carbamic ester from carbon dioxide, vinyl ether, and amine has been described (Eq. 2.76).[125] The amines used

$$R_2NH + CO_2 + H_2C{=}CHOCH_2CH_3 \xrightarrow[70h]{35\text{-}100°C} R_2NCOOCH{-}OCH_2CH_3$$
$$50 \text{ kg/cm}^2 \qquad\qquad\qquad\qquad\qquad \underset{\textstyle CH_3}{\big|} \qquad (2.76)$$

are dimethyl- and diethylamine, as shown in Table 2.30, and carbamic ester is obtained with a 0.06–11% yield. In these reactions, neither the regional isomer, 2-ethoxyethyl carbamate, nor the simple adduct of amine to vinyl ether is formed. The vinyl ether polymer cannot be detected, either. These findings are advantageous for the application of this novel reaction to or-ganic syntheses.

Table 2.30.   The formation of 1-ethoxyethyl carbamate (I) from carbon dioxide, ethyl vinyl ether (EVE), and amine.[†1]

Amine	Amine/EVE mole ratio	Temp. (°C)	Yield of (1)[†2] (%)
Me₂NH	1	80	7.2
Et₂NH	1	80	3.7
Et₂NH	2	80	5.7
Et₂NH	10	80	11.0
Et₂NH	1	35	0.06
Et₂NH	1	100	11.0

[†1] CO_2, 50 kg/cm². Time, 70 h.   [†2] On the basis of EVE.
Source: Reproduced by kind permission of the Chemical Society of Japan.

Since the sole product is 1-ethoxyethyl carbamate, and not isomeric 2-ethoxyethyl ester, the reaction is thought to proceed by the electrophilic addition of carbamic acid, which is formed by the reversible reaction of carbon dioxide and amine, to vinyl ether (Eq. 2.77); Eq. (2.78) is the rate-determining step.

$$R_2NH + CO_2 \rightleftharpoons R_2NCOO^- + H^+ \tag{2.77}$$

$$H_2C=CHOC_2H_5 \xrightarrow{H^+} [H_3C\overset{+}{C}HOC_2H_5] \xrightarrow{R_2NCOO^-} R_2NCOOCHCH_3 \tag{2.78}$$
$$\underset{OC_2H_5}{\mid}$$

2,3-Dihydrofuran also gives carbamic ester with a 5.8% yield. This reaction is also very appealing as a new technique for the synthesis of carbamic ester by direct fixation of carbon dioxide under moderate conditions. There are many interesting reactions between amines and carbon dioxide that use an organometallic compound as a reagent or catalyst. For example, transition metal amide complexes react with carbon dioxide to produce transition metal carbamato complexes, and certain transition metal complexes can act as catalysts in the reaction between carbon dioxide and amine that produces formamides. These reactions are described in detail in the following chapter.

2.4. MISCELLANEOUS

2.4.1. Electroreduction of Carbon Dioxide

Carbon dioxide can be reduced electrochemically. Formic acid can be produced quantitatively by the electrochemical reduction of low concentrations of carbon dioxide,[126] but the reduction of high concentrations of carbon dioxide gives a mixture of the higher carboxylic acids. Details of these carboxylations remain unclear, and new developments in this field are expected in the future.

$$CO_2 + e \quad \ldots\ldots\ldots\ldots\ldots COO^-$$
$$COO^- + H^+ \quad \ldots\ldots\ldots\ldots HCOO$$
$$HCOO + e \quad \ldots\ldots\ldots\ldots HCOO^-$$
$$HCOO^- + H^+ \ldots\ldots\ldots\ldots HCOOH$$
$$COO^- \rightarrow \rightarrow \rightarrow HOCH_2COOH + RCOOH$$

2.4.2. Electrocarboxylation of Organic Compounds

Unsaturated compounds (aromatic compounds and some olefins) electrochemically produce anion radicals which are carboxylated with carbon dioxide. For instance, the electrocarboxylation of naphthalene produces mono- and dicarboxylic acid of dihydronaphthalene (Eq. 2.79).[127,128]

Osa and Shinzaki[129] reported that 1- and 2-cyanonaphthalene are

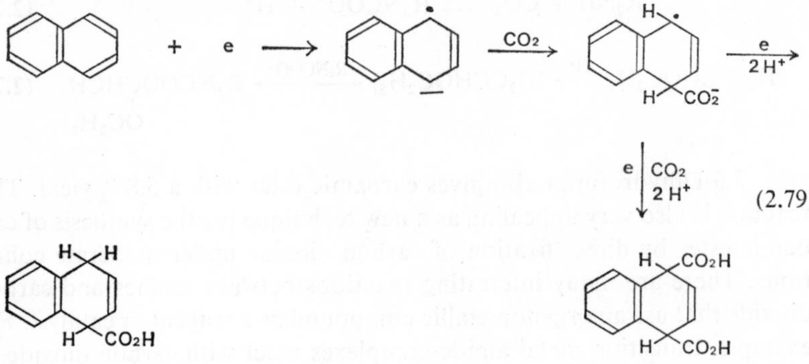

$$(2.79)$$

electrocarboxylated to produce the corresponding tetrahydroderivatives, as shown in Eqs. (2.80) and (2.81).

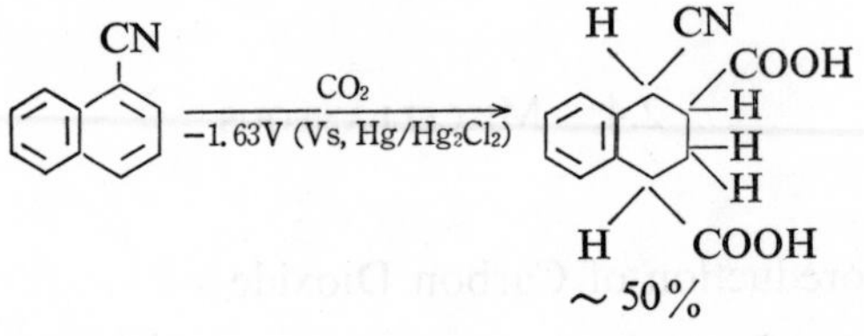

$$(2.80)$$

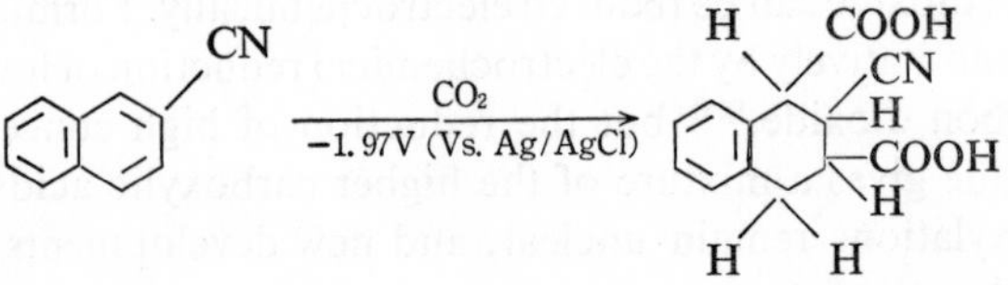

$$\sim 50\%$$

$$(2.81)$$

There have been several studies in which carbon dioxide was used as a trapping agent to demonstrate the presence of anionic intermediates in anhydrous media.[130]

Tyssee and Baizer[131] described the electrochemical behavior of activated olefins in the presence of dissolved carbon dioxide in both non-aqueous and partially aqueous systems, and found that under these conditions the olefins are either mono- or dicarboxylated. Under nonaqueous conditions, the reaction pathway is as shown in Eq. (2.82). For example, in the presence of carbon dioxide, dimethyl maleate is reduced to its anion radical [$E_{1/2}^1 = -1.53$ V], which reacts with carbon dioxide to form a carboxylate radical (5). The radical (5) is stable against further reduction at this potential, and subsequent reduction of (5) to the carboxylate carbanion (6) does not occur until -1.84 V. Reaction of (6)

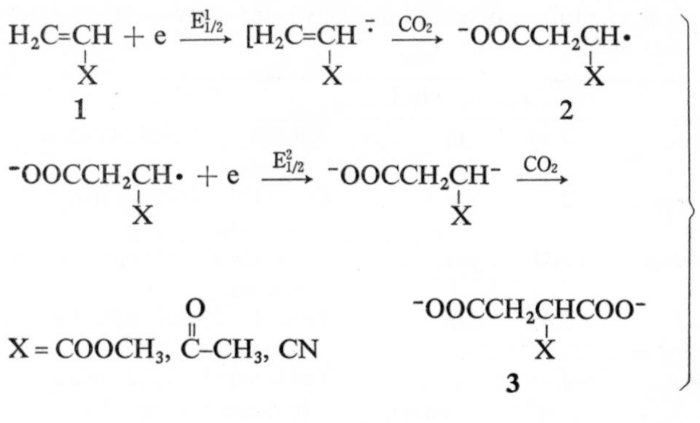

$$H_2C=CH + e \xrightarrow{E^1_{1/2}} [H_2C=CH \overset{\cdot}{} \xrightarrow{CO_2} {}^-OOCCH_2CH\cdot$$
$$\underset{X}{} \qquad \underset{X}{} \qquad \underset{X}{}$$
$$1 \qquad\qquad\qquad 2$$

$${}^-OOCCH_2CH\cdot + e \xrightarrow{E^2_{1/2}} {}^-OOCCH_2CH^- \xrightarrow{CO_2}$$
$$\underset{X}{} \qquad\qquad \underset{X}{}$$

$$\hspace{6cm} {}^-OOCCH_2CHCOO^-$$
$$X = COOCH_3, \overset{O}{\overset{\|}{C}}-CH_3, CN \hspace{2cm} \underset{X}{}$$
$$\hspace{7cm} 3$$

$$(2.82)$$

with carbon dioxide gives the ethanetetracarboxylate derivative (7) (Eq. 2.83).

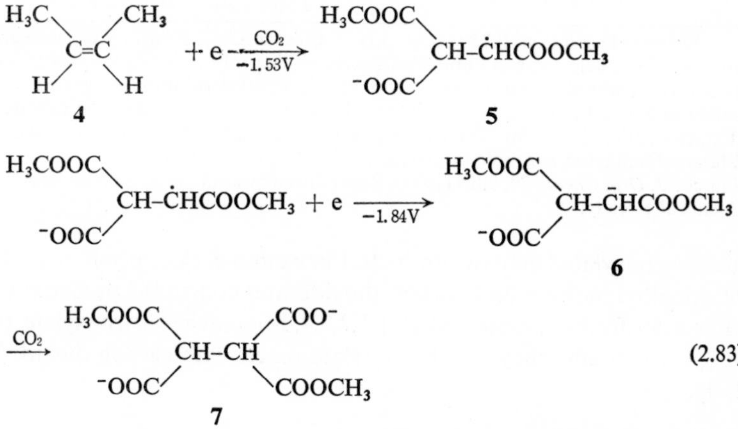

$$(2.83)$$

The results of bulk electrolyses under anhydrous conditions are presented in Table 2.31.

Electrocarboxylation of activated olefins produces monocarboxylated products by replacing the anhydrous solvent-electrolyte mixture with a carbon dioxide-saturated solution of aqueous tetraethylammonium bicarbonate (Eq. 2.84).

$$H_2C=CHCN + CO_2 + H_2O \longrightarrow NCCH_2CH_2COOH + 1/2\ O_2 \quad (2.84)$$

Carboxylative dimerization of olefins has been reported by Wawzonek and Gundersen.[132] Tyssee and Baizer[131] have reported details of an interesting electrocarboxylation which, in the presence of excess olefin,

Table 2.31. Bulk electrolyses of activated olefins in the presence of dissolved carbon dioxide.[†1]

Olefin	$-E^1/_2$(sec)	Control[†2]	Product[†3]	Yield (ce)[†4]
Methyl acrylate	2.10	cpe	Trimethyl 1,1,2-ethanetricar-boxylate	61
Methyl methacrylate	2.27	cie	Trimethyl 1,2,2-propanetricar-boxylate	42
Methyl crotonate	2.41	cie	Trimethyl 1,1,2-propanetricar-boxylate	38
Methyl trans-β-methoxyacrylate	2.67	cie	Trimethyl 2-methoxy-1,1,2-ethanetricarboxylate	10
Dimethyl maleate	1.84	cpe (second wave)	Tetramethyl 1,1,2,2-ethane-tetracarboxylate	31
Acrylonitrile	2.14	cpe	Dimethyl 2-cyanosuccinate	41
Methacrylonitrile	2.31	cie	Dimethyl 2-methyl-2-cyano-succinate	28
Methyl vinyl ketone	1.91	cpe	Methyl levulinate	22
			Dimethyl 2-acetylsuccinate	16

[†1] Electrolyte solution, 0.25 M $(C_2H_5)_4N + OTs^-$ in CH_3CN with CO_2 saturation (1 atm) at a Hg pool. [†2]cpe = controlled potential electrolysis, cie = controlled current electrolysis (cathode potential = − 2.15 V). [†3]Catholyte solution treated with methyl iodide to convert tetraethylammonium carboxylates to methyl esters for isolation and characterization (ref. 128). [†4]Yields expressed as current efficiencies (ce), assuming a 2-faraday reduction per mole of product.
Source: J. Org. Chem., 39, 2819 (1974). Reproduced by kind permission of the American Chemical Society.

gives carboxylated dimeric products. For example, electrolysis of a solution of dimethyl maleate and carbon dioxide at a controlled potential yields, after esterification, hexamethyl 1,1,2,3,4,4-butanehexacarboxylate (8) in 46% current efficiency. Methyl acrylate reacts with carbon dioxide similarly.

$$
2 \quad
\begin{array}{c}
H_3COOC \\
\diagdown \\
\quad C=C \\
\diagup \quad \diagdown \\
H \qquad H
\end{array}
\begin{array}{c}
COOCH_3 \\
\diagup \\
\\
\end{array}
\; + \; 2e \; + \; 2CO_2 \quad \xrightarrow[\text{ii) CH}_3I]{\text{i) E}=-1.65V}
$$

$$
\begin{array}{c}
H_3COOC \\
\diagdown \\
\quad CH-CH-CH-CH \\
\diagup \qquad | \qquad \diagdown \\
H_3COOC \qquad COOCH_3 \quad COOCH_3
\end{array}
\begin{array}{c}
COOCH_3 \\
\diagup \\
\\
\\
\end{array}
\qquad (2.85)
$$

8

$$
2H_2C=CHCN + 2e + 2CO_2 \longrightarrow NCCHCH_2CH_2CHCN \qquad (2.86)
$$
$$
\qquad\qquad\qquad\qquad\qquad\qquad\quad | \qquad\qquad |
$$
$$
\qquad\qquad\qquad\qquad\qquad\qquad COO^- \qquad COO^-
$$

Puglisi and Bard[133] have characterized the reaction mechanism of dimerization.

Upon reduction, the bisactivated olefins may be expected to undergo a variety of electrocarboxylative reactions. Tyssee and Baizer[134] described pertinent reactions with these derivatives.

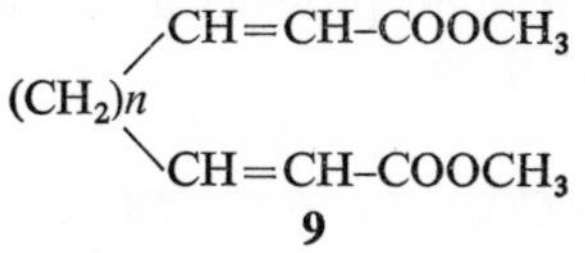

$$
(CH_2)n \begin{cases} CH{=}CH{-}COOCH_3 \\ \\ CH{=}CH{-}COOCH_3 \end{cases}
$$
$$
9
$$

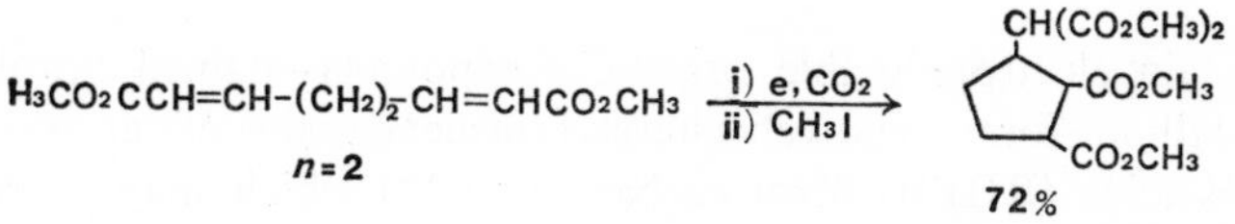

$$
H_3CO_2C\,CH{=}CH{-}(CH_2)_2{-}CH{=}CHCO_2CH_3 \xrightarrow[\text{ii) CH}_3\text{I}]{\text{i) e, CO}_2} \quad \text{(2.87)}
$$
$$
n=2
$$
CH(CO_2CH_3)_2, CO_2CH_3, CO_2CH_3 — 72%

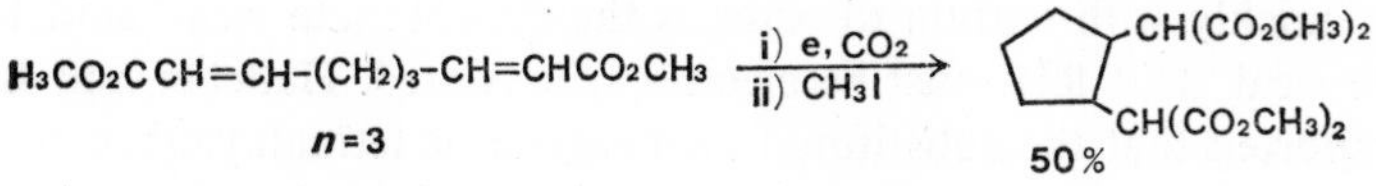

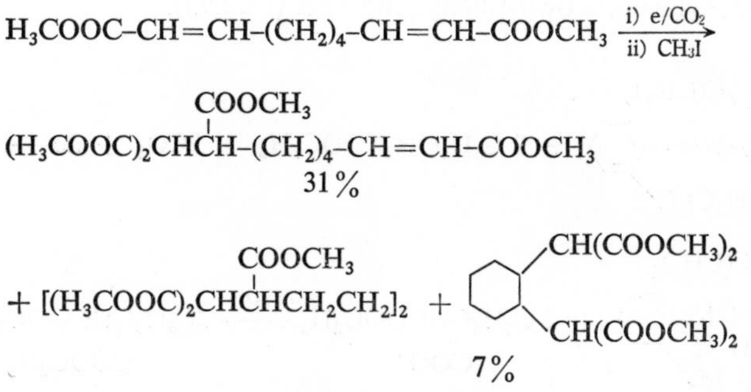

$$
H_3CO_2C\,CH{=}CH{-}(CH_2)_3{-}CH{=}CHCO_2CH_3 \xrightarrow[\text{ii) CH}_3\text{I}]{\text{i) e, CO}_2} \quad \text{(2.88)}
$$
$$
n=3
$$
CH(CO_2CH_3)_2, CH(CO_2CH_3)_2 — 50%

$$
H_3COOC{-}CH{=}CH{-}(CH_2)_4{-}CH{=}CH{-}COOCH_3 \xrightarrow[\text{ii) CH}_3\text{I}]{\text{i) e/CO}_2}
$$

$$
\overset{\displaystyle COOCH_3}{(H_3COOC)_2CHCH}{-}(CH_2)_4{-}CH{=}CH{-}COOCH_3
$$
$$
31\%
$$

$$
+\ [(H_3COOC)_2CH\overset{\displaystyle COOCH_3}{CHCH}_2CH_2]_2\ +\ \quad \text{(2.89)}
$$
CH(COOCH_3)_2, CH(COOCH_3)_2 — 7%

In the electrocarboxylation of allyl chloride, neither 3-butenoic nor crotonic acid is obtained in more than trace amounts. The major coupling product is allyl crotonate (Eq. 2.90).[135]

$$
H_2C{=}CHCH_2Cl + CO_2 \longrightarrow [H_2C{=}CHCH_2COO^- \rightleftharpoons H_3C{-}CH{=}CH{-}COO^-]R_4\overset{+}{N}
$$

$$
H_3CCH{=}CH{-}COO^- + H_2C{=}CH{-}CH_2Cl \longrightarrow H_3CCH{=}CH{-}COOCH_2CH{=}CH_2
$$

$$
\text{(2.90)}
$$

Reduction of benzylchloride in the presence of carbon dioxide at −2.37 V leads directly to benzyl phenylacetate with a good yield (Eq. 2.91). This one-step synthesis of the ester from carbon dioxide is appealing as a synthetic reaction.

$$2\ C_6H_5CH_2Cl \xrightarrow[CO_2]{2e} C_6H_5CH_2COOCH_2C_6H_5 + 2\ Cl^- \qquad (2.91)$$

Electrocarboxylation of perfluoro-*n*-hexyl iodide ($C_6F_{13}I$), which is one of the special activated halides, provides a very easy route for the synthesis of R_fCOOH (Eq. 2.92).[136]

$$C_6F_{13}I + CO_2 \xrightarrow{-0.9V} \underset{\sim 90\%}{C_6F_{13}COOH} \qquad (2.92)$$

Certain electroreducible organic compounds—named "probases" —at the cathode form negatively charged species capable of deprotonating weak CH-acids (RH) to form carbanions (R^-) which may react with carbon dioxide. If the negatively charged species from the probases do not react as quickly with carbon dioxide as the deprotonate weak acid RH does, the weak acid RH must be electrocarboxylated. Hallcher and Baizer[137] reported that the substituted azobenzene or tetraalkylethenetetracarboxylates are useful as probases in the carboxylation of weak acids such as ethyl phenylacetate or 9-phenylfluorene. (Eq2.93).

(2.93)

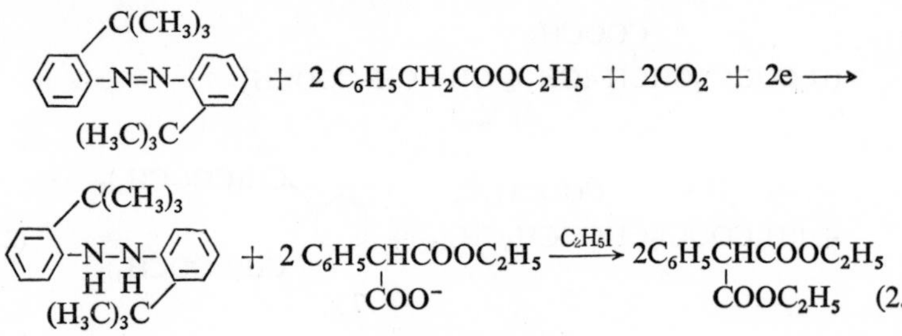

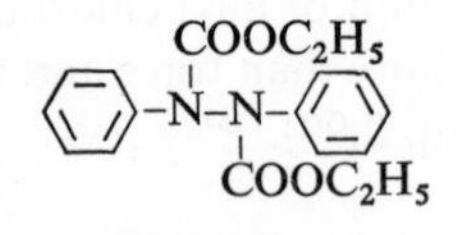

Azobenzene gives only the ethyl carbamate of hydrazobenzene[138] under the same reaction conditions (Eq. 2.94, 2.95). These carboxylated products are obtained in satisfactory yields.

$$(ROOC)_2C{=}C(COOR)_2 + 2\ C_6H_5CH_2COOC_2H_5 + 2\ CO_2 + 2e \longrightarrow$$

$$(ROOC)_2CHCH(COOR)_2 + 2\ C_6H_5\underset{\underset{COO^-}{|}}{CH}COOC_2H_5 \qquad (2.94)$$

$$R = C_2H_5,\ n\text{-}C_4H_9$$

$$(C_2H_5OOC)_2C = C(COOC_2H_5)_2 + 2\ \underset{\underset{H_5C_6}{}}{\overset{}{\bigcirc}}\!\!-\!\!\underset{\overset{}{H}}{\overset{}{\bigcirc}} + 2CO_2 + 2e$$

$$\longrightarrow (C_2H_5OOC)_2CHCH(COOC_2H_5)_2 + 2\ \underset{\underset{H_5C_6\quad COO^-}{}}{\overset{}{\bigcirc}}\!\!-\!\!\overset{}{\bigcirc} \qquad (2.95)$$

2.4.3. Cyanocarboxylation of Activated Olefins

Cyanide ion, especially in the form of a tetraalkylammonium salt, is a very strong nucleophile. Its nucleophilic addition to polarized double bonds is well known and plays an important role in various reactions. White discovered a novel reaction between carboxylated tetraethylammonium cyanide and a number of electrophiles.[139] Tetraalkylammonium cyanide in acetonitrile or dimethylformamide solution takes up about 1 molar equivalent of carbon dioxide. In analogy with the reaction of cyanide with carbon disulfide in which ($NC\text{-}CS_2^-$ is formed), it is assumed that the cyanoformate is formed as shown in Eq. (2.96).

$$\underset{\mathbf{1}}{[Et_4N^+]CN^-} + CO_2 \rightleftharpoons [Et_4N^+]NC\text{-}COO^- \qquad (2.96)$$

Whereas the cyanodithioformate ion reacts with methyl iodide to give corresponding methyl ester, reactions between cyanide (1) saturated with carbon dioxide and alkyl halides lead to the liberation of carbon dioxide and the formation of the alkyl cyanide. This can be anticipated if the carboxylation is reversible because the cyanoformate ion should be a relatively poor nucleophile compared to the cyanide ion. Reactions between activated olefins (2) and carboxylates (1) give products (3) with cyanide ions and carbon dioxide added to the olefins (Eq. 2.97, 2.98). The treatment of salt solutions ($3_{a-c\ and\ e}$), prepared in acetonitrile with methyl iodide produced good yields of the corresponding esters ($4_{a-c\ and\ e}$).

$$\underset{\mathbf{2}}{R^1CH = \underset{\underset{R^2}{|}}{C}\text{-}X} + [Et_2N^+]NC\text{-}COO^- \longrightarrow \underset{\mathbf{3a\text{-}e}}{[Et_4N^+]NC\text{-}\underset{\underset{R^1}{|}}{C}\overset{\overset{X}{|}}{H}\underset{\underset{R^2}{|}}{C}\text{-}COO^-} \qquad (2.97)$$

$$3 + H_3CY \longrightarrow \underset{\overset{|}{R^1}\ \overset{|}{R^2}}{NCCH\overset{\overset{X}{|}}{C}\text{-}COOCH_3} + Y^- \qquad (2.98)$$

$$\underset{4}{}$$

a : $R^1 = H$, $R^2 = H$, $X = COOCH_3$
b : $R^1 = H$, $R^2 = CH_3$, $X = COOCH_3$
c : $R^1 = H$, $R^2 = CH_3$, $X = COOCH_3$
d : $R^1 = H$, $R^2 = H$, $X = COOCH_3$
e : $R^1 = H$, $R^2 = H$, $X = CN$

However, (3d) gives a mixture of the expected product [i.e., (4d)] and (4e). Further treatment with dimethyl sulfate gives pure (4d) in high yield, and with diethyl sulfate, the corresponding ethyl ester (5).

$$NC\text{-}CH_2\text{-}\underset{\overset{|}{CN}}{CH}\text{-}COOC_2H_5$$

$$5$$

The results of the over-all conversion of activated olefins (2) into esters (4) and (5) are summarized in Table 2.32.

Table 2.32.   1,2-Cyanocarboxylation of activated olefins.

Substrate	Reaction time (h)†	Alkylating agent	Reaction product	Yield (%)
$H_2C=CH\text{-}COOCH_3$	2	CH_3I	$NC\text{-}CH_2\text{-}CH(COOCH_3)_2$	77
$H_2C=\underset{\overset{|}{CH_3}}{C}\text{-}COOCH_3$	20	CH_3I	$NC\text{-}CH_2\text{-}\underset{\overset{|}{CH_3}}{C}(COOCH_3)_2$	82
$H_3C=CH\text{-}CH\text{-}COOCH_3$	20	CH_3I	$NC\text{-}CH\text{-}CH(COOCH_3)_2$	68
$H_2C=CH\text{-}CN$	2	$(CH_3)_2SO_4$	$NC\text{-}CH_2\text{-}\underset{\overset{|}{CN}}{\overset{\overset{CH_3}{|}}{CH}}\text{-}COOCH_2$	78
$H_2C=\underset{\overset{|}{CH_3}}{C}\text{-}CN$	2	$(C_2H_5)_2SO_4$	$NC\text{-}CH_2\text{-}\underset{\overset{|}{CN}}{C}\text{-}COOC_2H_5$	75
	20	CH_3I	$NC\text{-}CH_2\text{-}\underset{\overset{|}{CN}}{\overset{\overset{CH_3}{|}}{C}}\text{-}COOCH_3$	85

† Time allowed for the formation of the intermediate salt

General procedure   Carbon dioxide is bubbled through a suspension of tetra-ethylammonium cyanide (15.6 g, 0.1 mol) in acetonitrile (100 ml) for 1 h. Methacrylonitrile (6.7 g, 0.1 mol) is added and the flask is stoppered and set aside for 20 h. The mixture is then cooled to 5°C and methyl iodide (15.6 g, 0.11 mol) is added. The flask is allowed to warm up to room temperature (about 1 h). The tetraethylammonium iodide is filtered off and washed with acetonitrile. The combined filtrate and washings are evaporated to dryness. The residue is stirred with ether (approx. 200 ml) and the additional tetraethylammonium iodide is filtered off. The residue is washed with ether and the combined filtrate and washings are evaporated and distilled. The fraction with bp 78° – 79°C at 0.03 mm Hg is collected as a colorless liquid, 2,3-dicyano-2-methylpropionate (12.9 g, 85%).

2.4.4. Friedel-Crafts Reaction using Carbon Dioxide

Carbon dioxide can be looked upon as the acid anhydride of carbonic acid. Friedel and Crafts[140, 141] reported that a small amount of benzoic acid forms when dry carbon dioxide is passed through a mixture of aluminum chloride and benzene at the boiling temperature. Although there have been a number of reports of investigations of the Friedel-Crafts reaction, it is difficult to perform the procedure and obtain a good yield without side reactions. High reaction temperatures and high pressure are generally needed, and only reactive substrates and Olah (such as phenols) combine readily. These reactions are reviewed by Olah[142].

The reaction of toluene with carbon dioxide in the presence of aluminum chloride gives p-methylbenzoic acid in 20–35% yields with side reaction products (Eq. 2.99).[143, 144]

$$C_6H_5CH_3 + CO_2 \xrightarrow[50°-120°]{AlCl_3} p\text{-}H_3C\text{-}C_6H_4COOH + H_3C\text{-}C_6H_4\overset{O}{\overset{\|}{C}}C_6H_4CH_3$$

$$+ \text{ polymers} \tag{2.99}$$

The carboxylation of ferrocene[145] with carbon dioxide by the Friedel-Crafts reaction proceeds smoothly under mild conditions to give a single product, ferrocene monocarboxylic acid, in satisfactory yields (Eq. 2.100).

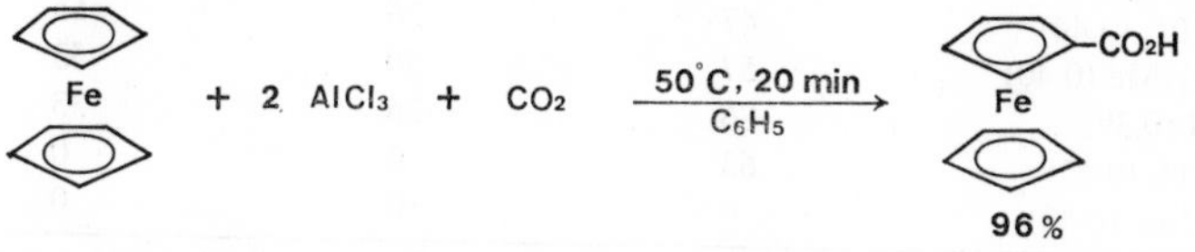

$$\tag{2.100}$$

Two equivalents of aluminum chloride are acquired in the formation of ferrocenemonocarboxylic acid via the complex $(C_5H_5)_2Fe \cdot (AlCl_3)_2 \cdot CO_2$. The details of this reaction are not clear, but it is certain that at least one half-equivalent of ferrocene is recovered, and another half-equivalent of ferrocene takes part directly in the reaction with carbon dioxide.

2.4.5. Photocarboxylation of Aromatic Hydrocarbons

Tazuke and Ozawa[146] have described an interesting carboxylation of anion radical-cation radical pairs photochemically formed in solution (Eq. 2.101).

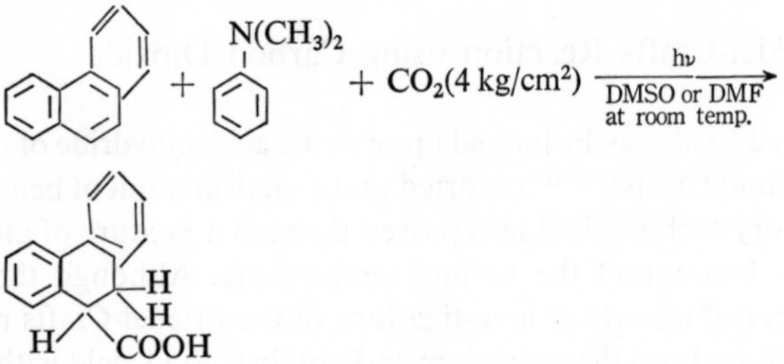

$$\text{(naphthyl-CH}_2\text{)} \; + \; \text{(C}_6\text{H}_4\text{N(CH}_3)_2) \; + \; CO_2(4 \text{ kg/cm}^2) \xrightarrow[\substack{\text{DMSO or DMF} \\ \text{at room temp.}}]{h\nu}$$

$$\text{(phenanthrene-CH)} \begin{array}{c} \text{—H} \\ \text{H} \\ \text{H}^{\diagup} \text{COOH} \end{array} \tag{2.101}$$

In the electrocarboxylation of aromatic hydrocarbons or activated olefins, it is thought that the electrically formed anions or anion radicals play the principal part in the carboxylation. In the photoreaction of weak electron donor-weak electron acceptor systems, it has been recently established that anion radical-cation radical pairs are formed in solution, but there have been few reports of synthetic applications of these active species, and this carboxylation is a nonbiological system.

The results of photocarboxylation of phenanthrene in the presence of various amines are shown in Table 2.33.

Table 2.33.   Photocarboxylation of phenanthrene. [Phenanthrene] = 2.8 × 10^{-2} mol· dm^{-3} in Me$_2$SO; p(CO$_2$) = 4 kg cm^{-2}; irradiation for 14h.

Amine (mol dm^{-3})	Reaction (%)[†1]	Yield of carboxylic acid (%)	Conversion into carboxylic acid (%)[†2]
PhNMe$_2$ (0.40)	93[†3]	41c[†3]	45[†3]
PhNMe$_2$ (0.40)	96	46	48
PhNEt$_2$ (0.44)	47	26	55
PhNHMe (0.40)	57	30	53
Et$_3$N (0.39)	86	26	33
PhNH$_2$ (0.55)	63	0	0
Pyridine (0.40)	0	0	0

[†1] Based on phenanthrene consumed.   [†2] % Yield of carboxylic acid/% reaction. [†3] In HCONMe.
Source: *J. Chem. Soc. Chem. Commun.*, 237 (1975). Reproduced by kind permission of the Chemical Society (London).

Anthracene, pyrene, naphthalene, and diphenyl are also carboxylated by this photocarboxylation. The reduction of carbon dioxide to carboxylic acids with the same radical anion–radical cation pairs in the presence of reducing agents has been reported by Tazuke and Kitamura.[147]

Other photochemical reactions with carbon dioxide are discussed in Chapter 6 in connection with the model of photosynthesis.

2.4.6. Carboxylation of Aliphatic Amines

An attractive direct synthesis of α-amino acid from aliphatic amines has been reported.[148] Amine derivatives dissolved in ether or hydrous solvents are carboxylated by bubbling carbon dioxide at 10°–50° C through the mixture in the presence of hydrogen peroxide and inorganic salt (Eq. 2.102).

$$R-CH_2NH_2 + CO_2 \xrightarrow[10-50°C,HCl]{H_2O_2(30\%)} D,L \underset{\underset{NH_2}{|}}{RCHCOOH} \qquad (2.102)$$

An illustrative procedure is the following. Six ml of 30% hydrogen peroxide solution is added dropwise to a solution of 60 ml of dioxane, 20 ml of 35% hydrochloric acid, and 2.8 g of isobutylamine with bubbling carbon dioxide at 30° C. The solution is filtered after completing the uptake of carbon dioxide (about 1.5 h), concentrated to about 20 ml, and then diluted with 100 ml of water. This solution is treated with ion exchange resin (IRA-400) to give 2.15 g of crude D,L-valine hydrochloride.

D,L-Alanine can be similarly prepared from ethylamine.

Ethylenediamine reacts with pyruvic acid at 40° C in an acetate buffer solution to give an enamine derivative of oxaloacetic acid (Eq. 2.103).[149]

$$H_2NCH_2CH_2NH_2 + H_3C\overset{O}{\overset{\|}{C}}CCOOH \longrightarrow H_2C=C\underset{\underset{COO^-}{|}}{NHCH_2CH_2\overset{+}{N}H_3}$$

$$\xrightarrow{CO_2} {}^-OOCCH_2C\underset{\underset{COO^-}{|}}{=}\overset{+}{N}HCH_2CH_2\overset{+}{N}H_3 \qquad (2.103)$$

Ethylenediamine also catalyzes the decarboxylation of oxaloacetic acid by the reverse reaction mechanism.

2.4.7. The Reaction of Carbon Dioxide with Ylide

The reaction of alkylidenphosphoranes with carbon dioxide is shown in Eq. (2.104).[150]

$$\underset{R'}{\overset{R}{\diagdown}}C{=}P(C_6H_5)_3 + CO_2 \longrightarrow \underset{R'}{\overset{R}{\diagdown}}\overset{\overset{\textstyle COOH}{|}}{C}{-}P(C_6H_5)_3 \xrightarrow[\text{ii) } H_3O^+]{\text{i) } OH^-/H_2O} \underset{R'}{\overset{R}{\diagdown}}CHCOOH \qquad (2.104)$$

2.4.8. Carboxylation of Organic Compounds by Irradiation with γ-Rays

An irradiated mixture of amine derivatives and carbon dioxide is known to have been one of the important elements of chemical evolution in the primitive global atmosphere (see also Chapter 6). Many other organic compounds are carboxylated by γ-ray irradiation (Eq. 2.105, 2.106),[151,152] but no reactions applicable to syntheses have been found yet.

$$CH_4 + CO_2 \xrightarrow{\gamma\text{-ray}} H_3CCOOH \qquad (2.105)$$

$$CH_3OH + CO_2 \xrightarrow{\gamma\text{-ray}} HOCH_2COOH + CH_2O + HOCH_2CH_2OH \quad (2.106)$$

2.4.9. Cracking of Paraffins in Carbon Dioxide

Vapor-phase cracking of high molecular weight paraffins in the presence of carbon dioxide at 500–700° C gives a mixture of saturated carboxylic acids (Eq. 2.107).[153]

$$C_nH_{2n+2} + CO_2 \xrightarrow{\text{cracking}} RCOOH \qquad (2.107)$$

REFERENCES

1) J. Mathieu and J. Weill-Raylnal, *Formation of C-C Bonds*, vol. 1, p. 281, Georg Thieme Publishers (1973).
2) S. Inoue, *Kagakuzokan* (Japanese), **63**, 29, Kagakudojin (1974).
3) T. Osa (ed.), *Tansangasu no Kagaku* (Japanese), Kyoritsu Shuppan (1976).
4) K. Kigoshi *et al.*, *Kagakukogyo* (Japanese), **26**, 1583 (1972).
5) T. Saegusa and T. Tsuda, *Kagakukogyo* (Japanese), **26**, 562 (1975).
6) Y. Hirakawa, *et al.*, *Yukigosei Kagaku Kyokaishi* (Japanese), **34**, 271 (1976).
7) H. Baubigny, *Z. Chem.*, **4**, 481 (1868); *Ann. Chim. Phys.*, [4] **19**, 221 (1870).
8) V. Prelog and U. Geyer, *Helv. Chim. Acta.*, **28**, 1677 (1945).
9) M. Stiles and H. L. Finkbeiner, *J. Am. Chem. Soc.*, **81**, 505 (1959).
10) M. Stiles, *J. Am. Chem. Soc.*, **81** 2598 (1959).
11) S. W. Pelletier, R. L. Chappell, P. C. Parthasarathy, and N. Lewin, *J. Org. Chem.*, **31**, 1747 (1966).
12) S. N. Balasubrahmanyam and M. Balasubramanian, *Organic Syntheses Collect.*, vol. 5, p. 439, Wiley (1973).
13) S. Julia and Ch. Huynh, *Compt. Rend.* [C], **270**, 1517 (1970).

14) J. Martin, P. C. Watts, and F. Johnson, *Chem. Commun.*, **27** (1970).
15) H. L. Finkbeiner and M. Stiles, *J. Am. Chem. Soc.*, **85**, 616 (1963).
16) H. L. Finkbeiner and G. W. Wagner, *J. Org. Chem.*, **28**, 215 (1963).
17) H. L. Finkbeiner, *J. Am. Chem. Soc.*, **86**, 961 (1964).
18) B. J. Whitlock and H. W. Whitlock, Jr., *J. Org. Chem.*, **39**, 3144 (1974).
19) E. C. Taylor and A. McKillop, *Tetrahedron*, **23**, 897 (1967).
20) T. Ito and Y. Takami, *Chem. Lett.*, 1035 (1974).
21) S. Sifniades, A. Tunick, and H. C. Wohlers, U. S. Patent 4,123,446; *Chem. Abstr.*, **90**, 54481j (1979).
22) G. Bottaccio and G. P. Chiusoli, *Chem. Commun.*, 618 (1966).
23) G. Bottaccio, G. P. Chiusoli, and M. G. Felicioli, *Gazz. Chim. Ital.*, **103**(1-2), 105 (1973).
24) H. Mori, H. Yamamoto, and T. Kwan, *Chem. Pharm. Bull.*, **20**, 2440 (1972).
25) G. Bottaccio, S. Campolmi, and M. G. Felicioli, Ger. Offen. 2,809,230; *Chem. Abstr.*, **89**, 196974a (1978).
26) M. Kawamata and H. Tanabe, Japan. Kokai 75 71,625; *Chem. Abstr.*, **83**, 178325m (1975).
27) M. Kawamata, T. Honda, S. Fujikake, and N. Koga, Japan. Kokai 75 64,213; *Chem. Abstr.*, **83**, 192583a (1975).
28) M. Kawamata and H. Tanabe, Japan. Kokai 75 88,004; *Chem. Abstr.*, **83**, 178338t (1975).
29) M. Kawamata and H. Tanabe, Japan. Kokai 74 56,911; *Chem. Abstr.*, **81**, 104781u (1974).
30) M. Kawamata and H. Tanabe, Japan. Kokai 74 36, 612; *Chem. Abstr.*, **81**, 104782v (1974).
31) M. Kawamata, H. Tanabe, and T. Takahashi, Japan. Kokai 74 102,611; *Chem. Abstr.*, **82**, 86090g (1975).
32) E. L. Patmore, W. R. Siegart, and H. Chafetz, U. S. Patent 3,696,146; *Chem. Abstr.*, **78**, 3965q (1973).
33) E. L. Patmore and W. R. Siegart, U. S. Patent 3,954,850; *Chem. Abstr.*, **85**, 62804d (1976).
34) E. L. Patmore, W. R. Siegart, and H. Chafetz, U. S. Patent 3,775,459; *Chem. Abstr.*, **80**, 47658n (1974).
35) E. L. Patmore, W. R. Siegart, and H. Chafetz, U. S. Patent 3,734,955; *Chem. Abstr.*, **79**, 66093v (1973).
36) E. L. Patmore, W. R. Siegart, and H. Chafetz, U. S. Patent 3,725,468; *Chem. Abstr.*, **79**, 78213x (1973).
37) T. Kwan, H. Yamamoto, H. Mori, and K. Mineda, Japan. Kokai 77 156,845; *Chem. Abstr.*, **88**, 190455q (1978).
38) E. L. Patmore, W. R. Siegart, and H. Chafetz, U. S. Patent 3,658,874; *Chem. Abstr.*, **77**, 100912g (1972).
39) E. L. Patmore, W. R. Siegart, and H. Chafetz, U. S. Patent 3,689,539; *Chem. Abstr.*, **78**, 57147t (1973).
40) E. L. Patmore, W. R. Siegart, and H. Chafetz, U. S. Patent 3,692,826; *Chem. Abstr.*, **78**, 4011n (1973).
41) G. Bottaccio, G. P. Chiusoli, and M. Marchi, Ger. Offen. 2,514,571; *Chem. Abstr.*, **84**, 89826d (1976).
42) E. Alneri, G. Bottaccio, V. Carletti, and G. Lana, Ger. Offen. 2,429,627; *Chem. Abstr.*, **82**, 155369u (1975).
43) G. Bottaccio, M. Marchi, and G. P. Chiusoli, *Gazz. Chim. Ital.*, **107**(9-10), 499 (1977).
44) E. J. Corey and R. H. K. Chen, *J. Org. Chem.*, **38**, 4086 (1973).
45) G. Bottaccio, G. P. Chiusoli, A. Coassolo, and V. Carletti, Ger. Offen. 2,245,892; *Chem. Abstr.*, **78**, 158946h (1973).
46) T. C. Bruce and S. J. Benkovic, *Bioorganic Mechanisms*, vol. 2, p. 380, W. A. Benjamin Inc. (1966).

47) T. C. Bruce and A. F. Hegarty, *Proc. Natl. Acad. Sci. USA*, **65**, 805 (1966).
48) Y. Otsuji, M. Arakawa, N. Matsumura, and E. Haruki, *Chem. Lett.*, 1193 (1973).
49) E. Haruki, M. Arakawa, and E. Imoto, 33rd Ann. Mtg. Chem. Soc. Japan (Japanese), 1975. Abstracts II, p. 903.
50) E. Haruki, M. Arakawa, and A. Shibata, unpublished data.
51) K. Chiba, T. Akama, K. Sakakibara, and K. Horie, *Chem. Lett.*, 1387 (1978).
52) H. Sakurai, A. Shirahata, and A. Hosomi, 3rd Ann. Mtg. App. CO₂ (Japanese), 1977. Abstracts, p. 23., *Tetrahedron Lett.*, **21**, 1967 (1980).
53) E. Haruki, M. Arakawa, N. Matsumura, Y. Otsuji, and E. Imoto, *Chem. Lett.*, 427 (1974).
54) E. Haruki, unpublished data.
55) E. L. Patmore, U. S. Patent 3,694,496; *Chem. Abstr.*, **78**, 28618s (1973).
56) E. L. Patmore, U. S. Patent 3,694,494; *Chem. Abstr.*, **78**, 4013q (1973).
57) E. Haruki and Y. Ueda, unpublished data.
58) E. Haruki, Japan. Kokai 77 203 (1977).
59) E. Haruki, A. Kimura, and H. Yoshikawa, unpublished data.
60) T. Tsuda and T. Saegusa, *Inorg. Chem.*, **11**, 2561 (1972).
    M. Hidai, T. Hikita, and Y. Uchida, *Chem. Lett.*, 521 (1972).
    S. Komiya and A. Yamamoto, *J. Organomet. Chem.*, **46**, C58 (1972).
61) T. Tsuda, Y. Chujo, and T. Saegusa, *J. Am. Chem. Soc.*, **100**, 630 (1978).
62) E. Haruki and A. Kimura, unpublished data.
63) E. Haruki, A. Kimura, H. Yoshikawa, and N. Nishimura, unpublished data.
64) E. Haruki and A. Kimura, unpublished data.
65) E. Haruki and H. Yoshikawa, unpublished data.
66) E. Haruki and A. Kimura, unpublished data.
67) E. Haruki and Y. Kajita, unpublished data.
68) N. Matsumura, Y. Yagyu, and E. Imoto, *Nippon Kagaku Kaishi* (Japanese) 1344 (1977).
69) E. Haruki, K. Sawada, and N. Nishimura, unpublished data.
70) E. Haruki, M. Hinenoya, and H. Yoshikawa, 3rd Ann. Mtg. App. CO₂ (Japanese), 1977. Abstracts, p. 27.
71) D. Seebach and D. Enders, *Angew. Chem., Int. Ed. Engli*, **14**, 15 (1975).
72) H. Stetter, *Angew. Chem., Int. Ed. Engl.*, **15**, 639 (1976).
73) E. Haruki, Y. Ueda, and E. Imoto, 1st Ann. Mtg. CO₂ (Japanese), 1975. Abstracts, p. 7.
74) B. Raecke, *Angew. Chem.*, **76**, 892 (1964).
75) K. Kudo and Y. Takezaki, *Kogyo Kagaku Zasshi* (Japanese), **70**, 2147 (1967).
76) E. Haruki, H. Shirono, and E. Imoto, 2nd Ann. Mtg. App. CO₂ (Japanese), 1976. Abstracts, p. 17.
77) E. Haruki and Y. Kajita, 4th Ann. Mtg. App. CO₂ (Japanese), 1978. Abstracts, p. 1.
78) H. Kolbe and E. Lautemann, *Ann.*, **113**, 125 (1860); **115**; 178 (1860).
79) R. Schmitt and E. Burkard, *Ber.*, **20**, 2699 (1877).
80) R. Schmitt, *J. Prakt. Chem.* [2] **31**, 397 (1885).
81) J. R. Johnson, *J. Am. Chem. Soc.* **55**, 3029 (1933).
82) J. L. Hales *et al.*, *J. Chem. Soc.* 3145 (1954).
83) E. A. Shilov *et al.*, *Chem. Abstr.*, **50**, 10692 (1956).
84) I. Hirao *et al.*, *Bull. Chem. Soc. Japan*, **46**, 3470 (1973).
85) I. Hirao *et al.*, *Kogyo Kagaku Zasshi*, **64**, 1213 (1961).
86) I. Hirao, K. Ota, S. Sueta, and Y. Hara, *Yukigosei Kagaku Kyokaishi* (Japanese), **24**, 1047 (1966).
87) I. Hirao, *Yukigosei Kagaku Kyokaishi* (Japanese), **24**, 1051 (1966).
88) K. Ota, I. Yuji, and I. Hirao, *Yukigosei Kagaku Kyokaishi* (Japanese), **26**, 992 (1968).
89) I. Hirao and T. Kito, *Asahigarasu Kogyo Gijutsu Shoreikai Kenkyuhokoku* (Japanese), **15**, 51 (1969).
90) R. Ueno and T. Miyazaki, Ger. Offen. 2,033,448; *Chem. Abstr.*, **77**, 20344h (1972).

91) R. Ueno and T. Miyazaki, Japan Patent 73 27,303; *Chem. Abstr.*, **79**, 136819n (1973).
92) W. Kashiwara, K. Muramoto, and I. Hirao, Japan Patent 77 12,185; *Chem. Abstr.*, **87**, 34650s (1977),.
93) R. Ueno and T. Miyazaki, Japan Patent 75 09,789; *Chem. Abstr.*, **83**, 113990e (1975).
94) K. Yoshida, K. Akunaga, O. Senda, Y. Maekawa, H. Yamamoto, K. Kodama, and H. Kato, Japan Patent 74 24,470; *Chem. Abstr.*, **82**, 111779y (1975).
95) S. Yura and T. Abe, Japan. Kokai 74 81,339; *Chem. Abstr.*, **81**, 169310j (1974).
96) R. Ueno and T. Miyazaki, Japan Patent 75 30,063; *Chem. Abstr.*, **84**, 121475b (1976).
97) I. Hirao, K. Ota, S. Sueta, and Y. Hara, *Yukigosei Kagaku Kyokaishi* (Japanese), **25**, 412 (1967).
98) I. Hirao, K. Ota, and S. Sueta, *Yukigosei Kagaku Kyokaishi* (Japanese), **25**, 1031 (1967).
99) J. I. Jones, *Chem. Ind.* (London), 228 (1958).
100) I. Hirao, T. Kondo, and T. Kito, *Kogyo Kagaku Zasshi* (Japanese), **72**, 692 (1969).
101) T. Kito, T. Kondo, H. Ago, S. Yamamoto, and I. Hirao, *Kogyo Kagaku Zasshi* (Japanese), **73**, 742 (1970).
102) T. Kito and I. Hirao, *Bull. Chem. Soc. Japan*, **44**, 3123 (1971).
103) T. Kito, Y. Hokamura, and I. Hirao, *Kogyo Kagaku Zasshi* (Japanese), **74**, 1651 (1971).
104) S. Uemura, N. Takamitsu, and T. Hashimoto, Japan. Kokai 77 95,627; *Chem. Abstr.*, **88**, 37438s (1978).
105) K. Sakata, T. Komoriya, and G. Yamashita, Japan. Kokai 73 96,553; *Chem. Abstr.*, **80**, 95519q (1974).
106) T. Hagihara, H. Tsuruta, and T. Yoshida, Japan. Kokai 77 122,336; *Chem. Abstr.*, **88**, 136324h (1978),.
107) H. Lind, J. Rody, and H. Brunnetti, Ger. Offen. 2,247,008; *Chem. Abstr.*, **78**, 159232j (1973).
108) W. I. Awad, M. K. El-Marsafy, M. F. Ismail, and Z. F. Hanna, *Egypt. J. Chem.*, **17**(1), 57 (1974); *Chem. Abstr.*, **83**, 9405g (1975).
109) Y. Takeda, A. Inuzuka, and Y. Chigira, Japan. Kokai 73 75,538; *Chem. Abstr.*, **80**, 59705p (1974).
110) W. Bachmann, C. Gnabs, K. Janecka, E. Mundlos, T. Papenfuhs, and G. Waese, Ger. Offen. 2,426,850; *Chem. Abstr.*, **85**, 20936t (1976).
111) Z. Weglinski and T. Talik, *Rocz. Chem.*, **51**(12), 2041 (1977); *Chem. Abstr.*, **89**, 43036w (1978).
112) F. Mutterer and C. D. Weis, *J. Heterocycl. Chem.*, **13**(5), 1103 (1976).
113) N. Yamazaki, F. Higashi, and T. Iguchi, *Tetrahedron Letters*, 1191 (1974).
114) N. Yamazaki, T. Iguchi, and F. Higashi, *Tetrahedron*, **31**, 3031 (1975).
115) N. Yamazaki, T. Tomioka, and F. Higashi, *Synthesis*, 384 (1975).
116) K. Shiina, 4th Ann. Mtg. App. CO$_2$ (Japanese), 1978, Abstracts, p. 9.
117) H. Ogura, K. Takeda, R. Tokue, and T. Kobayashi, *Synthesis*, 394 (1978).
118) Y. Akazaki, M. Hatano, and M. Fukuyama, 3rd Ann. Mtg. App. CO$_2$ (Japanese), (1977), Abstracts, p. 31.
119) T. Toda, *Chem. Lett.*, 957 (1977).
120) Y. Yoshida and S. Inoue, *Chem. Lett.*, 139 (1978).
121) S. Inoue and Y. Yoshida, *Asahigarasu Kogyo Gijutsu Shoreikai Kenkyuhokoku* (Japanese), **32**, 87 (1978).
122) Y. Yoshida and S. Inoue, *Bull. Chem. Soc. Japan*, **51**, 559 (1978).
123) M. H. Chisholm and M. W. Extine, *J. Am. Chem. Soci*, **99**, 782 (1977).
124) G. Chandra, A. D. Jenkins, M. F. Lappert, and R. C. Srivastava, *J. Chem. Soc.*, *A*, 2550 (1970).
125) Y. Yoshida and S. Inoue, *Chem. Lett.*, 1375 (1977).
126) J. Jordan and P. T. Smith, *Proc. Chem. Soc.*, 246 (1960).

127) S. Wawzonek and D. Wearring, *J. Am. Chem. Soc.*, **81**, 2067 (1959).
128) W. C. Neikam, U. S. Patent 3, 344, 045 (1967).
129) T. Osa and Y. Shinzaki, 2nd Ann. Mtg. App. CO_2 (Japanese), 1976, Abstracts, p. 23.
130) D. A. Tyssee, J. H. Wagenknecht, M. M. Baizer, and J. L. Chruma, *Tetrahedron Letters*, 4809 (1972).
131) D. A. Tyssee and M. M. Baizer, *J. Org. Chem.*, **39**, 2819 (1974).
132) S. Wawzonek and A. Gundersen, *J. Electrochem. Soc.*, **111**, 324 (1964).
133) V. J. Puglisi and A. J. Bard, *J. Electrochem. Soc.*, **120**, 748 (1973).
134) D. A. Tyssee and M. M. Baizer, *J. Org. Chem.*, **39**, 2823 (1974).
135) M. M. Baizer and J. L. Chruma, *J. Org. Chem.*, **37**, 1951 (1972).
136) P. Calas and A. Commeyras, *J. Electroanal. Chem.*, **89**, 363 (1978).
137) R. C. Hallcher and M. M. Baizer, *Justus Liebigs Ann. Chem.*, 737 (1977).
138) J. H. Wagenknecht, M. M. Baizer, and J. L. Chruma, *Synth. Commun.*, **2** (4), 215 (1972).
139) D. A. White, *J. Chem. Soc., Perkins Trans.* **1**, 1926 (1976).
140) C. Friedel and J. M. Crafts, *Compt. Rend.*, **86**, 1368 (1878).
141) C. Friedel and J. M. Crafts, *Ann. Chem. Phys.* (6), **14**, 433 (1883).
142) G. A. Olah and J. A. Olah, *Friedel-Crafts and Related Reactions*, vol. 3, part 2, p. 1269, Wiley (1964).
143) B. L. Lebedev, I. V. Pastukhova, and Y. T. Eidas, *Izu. Akad. Nauk SSSR. Ser. Khim.*, **4**, 967 (1972).
144) S. Fumasoni and M. Collepardi, *Ann. Chim.*, **54**, 1122 (1964); *Chem. Abstr.*, **62**, 7672c (1965).
145) T. Ito, N. Sugahara, Y. Kindaichi, and Y. Takami, *Nippon Kagaku Kaishi* (Japanese), 353 (1976).
146) S. Tazuke and H. Ozawa, *Chem. Commun.*, 237 (1975).
147) S. Tazuke and N. Kitamura, *Nature*, **275**, 301 (1978).
148) S. Enomoto, M. Inoue, and T. Ueyama, Japan. Kokai 73 28, 418; *Chem. Abstr.*, **79**, 79183z (1973).
149) T. S. Boiko and A. A. Yasnikov, *Dopov. Akad. Nauk Ukr. RSR, Ser. B: Geol. Khim. Biol. Nauki*, 1089 (1976); *Chem. Abstr.*, **86**, 120751t (1977).
150) H. J. Bestmann, T. Denzel, and H. Salbaum, *Tetrahedron Letters*, 1275 (1974).
151) M. A. Kurbanov, Z. B. Klapishevskaya, V. D. Timofeev, E. A. Borisov, B. G. Dzantier, S. P. Dobrovol'skii, and I. V. Zakharava, *Khim. Vys. Energ.*, **11** (1), 83 (1977); *Chem. Abstr.*, **86**, 139308a (1977).
152) I. A. Kapustin, V. I. Gergalov, I. S. Zhigunov, V. A. Kanash, and E. P. Petryaev, *Khim. Vys. Energ.*, 7(3), 252 (1973); *Chem. Abstr.*, **79**, 85582v (1973).
153) J. B. Wilkes, U. S. Patent 4, 016, 185; *Chem. Abstr.*, **87**, 5410q (1977).

3

Organometallic Reactions of Carbon Dioxide

Takashi Ito
Faculty of Engineering, Yokohama National University,
Yokohama-shi, Kanagawa 240, Japan

Akio Yamamoto
Research Laboratory of Resources Utilization, Tokyo Institute of
Technology, Yokohama-shi, Kanagawa 227, Japan

3.1. INTRODUCTION

The problems of oil shortage are becoming increasingly grave, and we must turn our attention to developing new methods for recycling the carbon resources which at present are simply released into the air in the form of carbon dioxide until the rest of the task is done by green plants, which fix carbon dioxide as carbohydrates. From the environmental point of view, the observed rise in ambient atmospheric carbon dioxide levels, due mainly to industrial combustion, could become increasingly serious.[1] There are at present only a few industrial processes utilizing carbon dioxide as a raw material; these include the synthesis of urea by reaction with ammonia, of salicylic acid by the Kolbe-Schmitt reaction, and of terephthalic acid by a Henkel process. Obviously, more effort should be concentrated on exploiting the available carbon dioxide.

Carbon dioxide is the ultimate oxidation product of organic compounds and is thus regarded as a highly stable compound. However, it does have unsaturated double bonds, and it should be theoretically possible to devise a chemical process to convert CO_2 into more useful organic compounds. First of all, let us examine the thermodynamic data to determine

Table 3.1. Enthalpy changes in some reactions involving carbon dioxide.[2]

Reactions	Enthalpy changes $\Delta H°$ (kcal/mol)
(1) CO_2 (g) $+$ H_2 (g) $\longrightarrow$ HCOOH (l)	-9.0
(2) CO_2 (g) $+$ $2H_2$ (g) $\longrightarrow$ HCHO (g) $+$ H_2O (l)	-2.0
(3) CO_2 (g) $+$ $3H_2$ (g) $\longrightarrow$ CH_3OH (l) $+$ H_2O (l)	-31.3
(4) CO_2 (g) $+$ $4H_2$ (g) $\longrightarrow$ CH_4 (g) $+$ $2H_2O$ (l)	-60.5
(5) $2CO_2$ (g) $+$ H_2 (g) $\longrightarrow$ $(COOH)_2$ (s)	-10.3
(6) $2CO_2$ (g) $+$ $6H_2$ (g) $\longrightarrow$ CH_3OCH_3 (g) $+$ $3H_2O$ (l)	-60.9
(7) CO_2 (g) $+$ H_2 (g) $+$ CH_3OH (l) $\longrightarrow$ $HCOOCH_3$ (l) $+$ H_2O (l)	-7.7
(8) CO_2 (g) $+$ H_2 (g) $+$ CH_3OH (l) $\longrightarrow$ CH_3COOH (l) $+$ H_2O (l)	-33.0
(9) CO_2 (g) $+$ $3H_2$ (g) $+$ CH_3OH (l) $\longrightarrow$ CH_3CH_2OH (l) $+$ $2H_2O$ (l)	-51.8
(10) CO_2 (g) $+$ H_2 (g) $+$ NH_3 (g) $\longrightarrow$ $HCONH_2$ (s) $+$ H_2O (l)	-24.9
(11) CO_2 (g) $+$ CH_4 (g) $\longrightarrow$ CH_3COOH (l)	-3.8
(12) CO_2 (g) $+$ CH_4 (g) $+$ H_2 (g) $\longrightarrow$ CH_3CHO (l) $+$ H_2O (l)	-2.3
(13) CO_2 (g) $+$ CH_4 (g) $+$ $2H_2$ (g) $\longrightarrow$ $(CH_3)_2CO$ (l) $+$ H_2O (l)	-15.7
(14) CO_2 (g) $+$ C_2H_2 (g) $+$ H_2 (g) $\longrightarrow$ CH_2=CHCOOH (l)	-52.0
(15) CO_2 (g) $+$ $CH_2 = CH_2$ (g) $\longrightarrow$ CH_2=CHCOOH (l)	-10.3
(16) CO_2 (g) $+$ CH_2=CH_2 (g) $\longrightarrow$ $\overline{CH_2CH_2O}$ (l) $+$ CO (g)	-3.1
(17) CO_2 (g) $+$ CH_2=CH_2 (g) $+$ H_2 (g) $\longrightarrow$ C_2H_5COOH (l)	-40.6
(18) CO_2 (g) $+$ CH_2=CH_2 (g) $+$ $2H_2$ (g) $\longrightarrow$ C_2H_5CHO (l) $+$ H_2O (l)	-39.8
(19) CO_2 (g) $+$ C_6H_6 (l) $\longrightarrow$ C_6H_5COOH (s)	-9.7
(20) CO_2 (g) $+$ C_6H_5OH (s) $\longrightarrow$ C_6H_4 (OH)COOH-m (s)	-6.4

the enthalpy changes accompanying some conceivable reactions using carbon dioxide, shown in Table 3.1. For comparison some selected reactions using carbon monoxide are summarized in Table 3.2. Table 3.1

Table 3.2 Enthalpy changes in some reactions involving carbon monoxide.[2]

Reactions	Enthalpy changes $\Delta H°$ (kcal/mol)
(21) CO (g) + 2H$_2$ (g) $\longrightarrow$ CH$_3$OH (l)	−30.7
(22) CO (g) + H$_2$O (l) $\longrightarrow$ HCOOH (l)	−8.3
(23) 2CO (g) + 2H$_2$O (l) $\longrightarrow$ (COOH)$_2$ (s) + H$_2$ (g)	−35.4
(24) CO (g) + CH$_3$OH (l) $\longrightarrow$ HCOOCH$_3$ (l)	−7.0
(25) CO (g) + CH$_3$OH (l) $\longrightarrow$ CH$_3$COOH (l)	−32.3
(26) CO (g) + NH$_3$ (g) $\longrightarrow$ HCONH$_2$ (s)	−24.2
(27) CO (g) + C$_2$H$_2$ (g) + H$_2$O (l) $\longrightarrow$ CH$_2$ = CHCOOH (l)	−51.3
(28) CO (g) + CH$_2$=CH$_2$ (g) + H$_2$ (g) $\longrightarrow$ C$_2$H$_5$CHO (l)	−39.1

clearly shows that some chemical reactions using carbon dioxide are thermodynamically favorable, despite the general impression that carbon dioxide is too stable to be utilized without the addition of energy. That some reactions beginning with carbon dioxide are thermodynamically possible does not mean that the reactions are kinetically feasible, however. In order to convert carbon dioxide into useful forms of organic compounds, three factors must be considered which will be briefly discussed in turn: activation of the carbon dioxide, activation of the organic compounds, and conversion of the carbon dioxide into other reactive forms.

3.1.1. Activation of the Carbon Dioxide

When transforming the stable carbon dioxide into an organic compound, activation of the carbon dioxide may often be necessary. A catalytic system having low kinetic barriers is particularly important in devising a feasible catalytic process utilizing carbon dioxide. In the search for a suitable system, information regarding the interaction of carbon dioxide with transition metal compounds is of vital importance. In this chapter we shall first discuss the reactions of carbon dioxide with non-transition metal compounds, then with transition metal compounds, and then turn to a discussion of catalytic systems.

Carbon dioxide has three possible sites at which it can interact with metallic compounds, as shown in Eq. (3.1). The modes of coordination will be discussed later, in Section 3.3.1, but the interactions of the B and C types are considered most important in reactions with transition metal

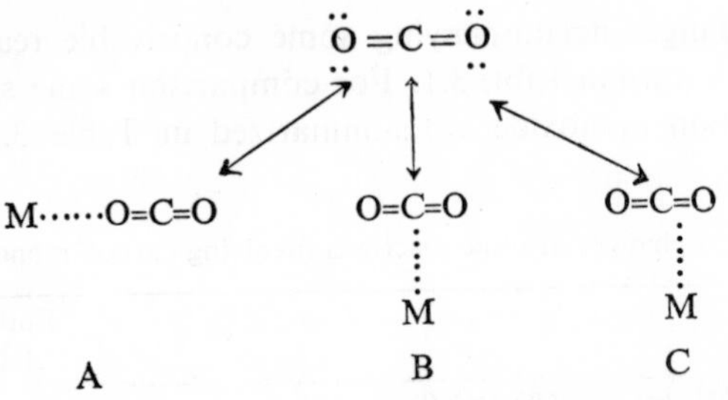

$$\ddot{O} = C = \ddot{O}$$

$$M\cdots\cdots O{=}C{=}O \qquad O{=}C{=}O \qquad O{=}C{=}O \qquad (3.1)$$

$$\underset{A}{} \qquad\qquad \underset{B}{M} \qquad \underset{C}{M}$$

complexes. The activated carbon dioxide may undergo further reactions such as insertion into the M–C and M–H bonds. Recent studies on organotransition metal complexes have accumulated basic information regarding these reactions, which may constitute elementary steps in catalytic processes.

3.1.2.   Activation of the Organic Compounds

If an organic compound could be selectively activated by interaction with a transition metal complex, resulting in the cleavage of a particular bond, for example, a C–H bond, a new reactive organometallic species having M–C and M–H bonds could be formed (Eq. 3.2).

$$-\overset{|}{\underset{|}{C}}{-}H \;+\; M \;\longrightarrow\; -\overset{|}{\underset{|}{C}}{-}M{-}H \qquad (3.2)$$

Interaction of carbon dioxide with the reactive species could then lead to the formation of new organic compounds carrying a CO_2 group. Reactions 11, 15, 19, and 20 in Table 3.1 represent some of these possibilities.

3.1.3.   Conversion of the Carbon Dioxide into Other Reactive Forms

Carbon dioxide molecules activated by coordination to transition metal complexes can be transformed to other forms of carbon oxides such as carbon monoxide. When carbon monoxide is formed, another oxygen atom may act as an oxidizing agent with other substrates. If one could devise a system in which carbon dioxide could be used as a source of CO and as a mild oxidizing agent at the same time, one would have a very attractive process for industrial use. Reaction 16 in Table 3.1 exemplifies one such possibility, although in this particular process the $\triangle G°$ value is positive because of the unfavorable entropy term.

When one looks at the processes shown in Table 3.1 more closely from an economic standpoint, it is apparent that some may not be practicable because of the cost of the raw materials. Other problems are illustrated by the system using CO_2 and H_2. By an equilibrium reaction, this is almost equivalent to a system using CO and H_2O, called the water-gas shift reaction, in which the forward and backward reactions are reversible with only a small enthalpy change (0.66 kcal mol^{-1}). In this case, a process requiring a lot of hydrogen to convert carbon dioxide into a useful organic material is less desirable than one employing carbon monoxide and water.

Since we are concerned in this chapter with organometallic reactions of carbon dioxide, we will not deal with other important aspects of the chemistry of carbon dioxide such as photochemical reactions and biomimetic systems.

It should be added here that, in doing laboratory experiments, one must be particularly careful about the purity of the carbon dioxide used to exclude the possibility of water and oxygen being introduced adventitiously into the system. Some published results, including our own, have later been found to have led to the wrong conclusions because this precaution was overlooked. In a survey of the literature it is difficult to differentiate the reliable papers from the unreliable ones in this respect, so we have had to cite most papers uncritically, although it seems certain that some results will have to be corrected in the future.

3.2. REACTIONS WITH NON-TRANSITION METAL COMPOUNDS

3.2.1. Alkali and Alkaline Earth Metal Compounds

The insertion of carbon dioxide into the metal-carbon bond of organic derivatives of magnesium or lithium is well known and constitutes one of the most important basic organic reactions for introducing a carboxylate group into an organic molecule and for characterizing organometallic species in a system.[3] Although Grignard reagents and organolithium compounds follow similar reaction pathways in which four centered intermediate species are involved (Eq. 3.3), features of the reactions differ somewhat.

$$R\text{-}M + CO_2 \longrightarrow \begin{matrix} \delta - \delta + \\ R\text{-}M \\ \vdots \quad \vdots \\ O\text{=}C\text{=}O \\ \delta + \delta - \end{matrix} \longrightarrow \overset{\displaystyle R}{\underset{\displaystyle |}{O\text{=}C\text{-}O\text{-}M}} \qquad (3.3)$$

Carbonation of Grignard reagents mainly affords carboxylic acids on hydrolysis, since the magnesium salt of the carboxylic acid thus formed is generally inert to a Grignard reagent (Eq. 3.4).

$$RMgX + CO_2 \longrightarrow RCOOMgX \xrightarrow{H^+} RCOOH \tag{3.4}$$

In contrast, organolithium compounds, on reaction with carbon dioxide, usually give ketones and carbinols as well as carboxylic acid. Because they react more readily than do Grignard reagents with carbonyl compounds, there is a reduction in the final yield of carboxylic acid (Eq. 3.5).

$$RLi + CO_2 \longrightarrow RCOOLi \xrightarrow{H^+} RCOOH$$
$$\downarrow RLi$$
$$\underset{R_2COLi}{\overset{OLi}{|}} \xrightarrow{-Li_2O} R_2C=O \tag{3.5}$$
$$\downarrow RLi$$
$$R_3COLi \xrightarrow{H^+} R_3COH$$

These reactions provide a convenient laboratory and industrial route to carboxylic acids, including aliphatic and aromatic carboxylic acids, and to olefinic and acetylenic acids. A recent systematic study of the distribution of products in the carbonation reactions of trifluoromethyl-substituted phenyllithiums found that the relative yields of acids, ketones, and carbinols are highly dependent on the position of the substituent on the phenyl ring, indicating the importance of the steric effect.[4] The steric course of the carbonation of these organometallic species has been investigated and found to proceed with a retention of the configuration.[5] From the point of view of synthetic organic chemistry, the utility of this carbonation reaction is significant when it is coupled with the *in situ* metalation of the organic molecules to which the carboxylic group is introduced. The organic molecules to be metallated are usually organic halides and, occasionally, hydrocarbons, as shown in Eqs. (3.6 and 3.7).[6,7]

$$PhNa + PhCH_3 \xrightarrow{reflux} PhCH_2Na \xrightarrow{CO_2} PhCH_2COONa \xrightarrow{H^+} PhCH_2COOH$$
$$\underset{\longrightarrow}{\overset{PhCH_2Na}{\vdash}} \xrightarrow{CO_2} \xrightarrow{H^+} PhCH(COOH)_2 \tag{3.6}$$

$$PhH + i\text{-}BuNa \longrightarrow \xrightarrow{CO_2} PhCOOH \tag{3.7}$$

In addition to such direct carbonation by the insertion of carbon dioxide into the metal-carbon bond, the carbonation reaction utilizing carbon dioxide activated with metal alkoxide is also very important. The production of hydroxybenzoic acids from carbon dioxide and alkali phenoxide, the Kolbe-Schmitt reaction, is one of the most important processes in which carbon dioxide is employed as a starting material in industrial organic chemistry (Eq. 3.8).

$$2 \underset{}{\text{OK}} + CO_2 \longrightarrow \underset{}{\text{OK}}-COOK + \underset{}{\text{OH}} \qquad (3.8)$$

In 1860 Kolbe devised a technique for the formation of salicylic acid by heating a mixture of phenol and sodium in the presence of carbon dioxide, and since then, many refinements, including that of Schmitt in 1884, have improved the method with respect to yield, selectivity in the position of carboxylation, and economy of the process.[8,9] The mechanism of this complicated carbonation reaction is still under debate, however.[9] It seems likely that carbon dioxide is first activated by complex formation with alkali phenolate, and is then inserted into the metal-oxygen bond as the carboxylic group is rearranged. The isomerization of the potassium salt of isophthalic acid into terephthalic acid and the disproportionation of potassium benzoate into terephthalic acid and benzene under carbon dioxide pressure with a cadmium or lead compound catalyst, which are industrially important Henkel processes, may proceed *via* a similar mechanism (Eqs. 3.9 and 3.10).

$$\underset{\text{COOK}}{\overset{\text{COOK}}{\bigcirc}} + CO_2 \xrightarrow{CdI_2} KOOC-\langle\_\rangle-COOK \qquad (3.9)$$

$$2 \langle\_\rangle-COOK + CO_2 \xrightarrow{CdI_2} KOOC-\langle\_\rangle-COOK + \langle\_\rangle \qquad (3.10)$$

The carbonation reaction using alkali metal alkylcarbonates, which can be prepared by the reaction of carbon dioxide and alkali metal alkoxide, produces hydroxybenzoic acids with a higher yield and selectivity than the ordinary Kolbe-Schmitt reaction (Eqs. 3.11 and 3.12).[10]

$$EtOK + CO_2 \longrightarrow EtOCOOK \qquad (3.11)$$

$$EtOCOOK + \underset{}{\overset{\text{OK}}{\bigcirc}} \longrightarrow \underset{}{\overset{\text{OK}}{\bigcirc}}^{\text{COOK}} + EtOH \qquad 3.12)$$

In principle, dipotassium malonate may be formed by an analogous reaction (Eq. 3.13).[11]

$$CH_3COOK + K_2CO_3 + CO_2 \longrightarrow CH_2(COOK)_2 + KHCO_3 \qquad (3.13)$$

In these reactions involving the carbonation of phenols, the activation of carbon dioxide through complexation with alkali metal phenoxide (or alkoxide) constitutes an important step that allows the reaction to proceed. Carbon dioxide activated in this way can react with organic compounds that possess active C–H bonds to give carboxylates (Eq. 3.14).[12,13]

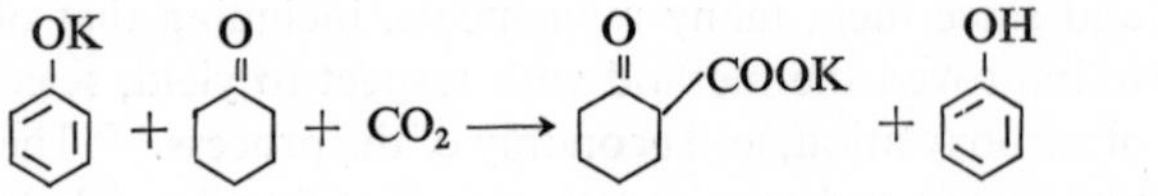

$$(3.14)$$

Since the potassium phenoxide in (Eq. 3.14) acts as a carbon dioxide carrier, it can be regarded as a kind of a model for biotin (Vitamin H), which fixes and transfers carbon dioxide to an organic molecule in a biological system. Other compounds in addition to the alkali metal phenoxides are known to activate carbon dioxide; these include alkali metal compounds such as sodium or potassium salts of alkyl-substituted hydroxythiazole,[12] potassium pyrolidone,[14,15] and lithium salts of some urea derivatives,[16] as well as non-metallic compounds such as dicyclohexylcarbodiimide-tetraalkylammonium hydroxide[16] and 1,8-diazabicyclo-[5.4.0]undec-7-ene.[17] Using these compounds as carbon dioxide carriers, it is possible to introduce a carboxyl group under very mild conditions to such compounds as cyclohexanone, acetophenone, indanone, indene, fluorene, nitroalkanes, acetone, methyl acetate, xanthate, and methyl pyruvate.

Apart from the alkali metal derivatives, magnesium alkoxide has been successfully used to form α-nitroacids by the carboxylation of α-nitroalkanes. The driving force of this reaction has been attributed to the ease with which magnesium ions are chelated with the nitroacetate dianion (Eq. 3.15).[18]

$$Mg(OCH_3)_2 + CO_2 \rightleftharpoons Mg(OCOOCH_3)_2 \xrightarrow{CH_3NO_2} \qquad (3.15)$$

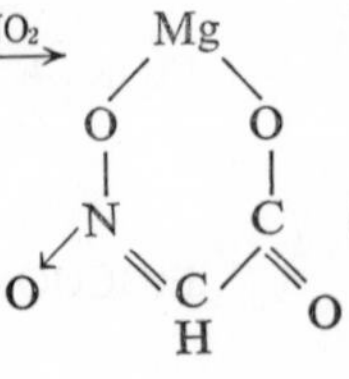

$$\xrightarrow{H^+} NO_2CH_2COOH$$

Magnesium methyl carbonate prepared by reaction (3.15) is a useful carboxylation agent important in organic synthesis.[19]

3.2.2. Aluminum and Boron Compounds

Organoaluminum compounds of the type R_3Al also react with carbon dioxide to give, after hydrolysis, carboxylic acid and, in some cases, carbinols (Eqs. 3.16 and 3.17).[20-22]

$$R_3Al + CO_2 \longrightarrow R_2AlOCOR \xrightarrow{H^+} RCOOH + 2RH \qquad (3.16)$$

$$R_2AlOCOR + R_3Al \longrightarrow R_2AlOCR_2OAlR_2 \xrightarrow{R_3Al} R_2AlOAlR_2$$

$$+ R_2AlOCR_3 \xrightarrow{H^+} R_3COH + 6RH \qquad (3.17)$$

Reaction (3.17) has been reported to be inhibited by the presence of amines such as ethylenediamine, bipyridine, and methylimidazole.[23] Since only one of three alkyl groups is carbonated in these reactions, and heating the system is sometimes necessary to complete the reaction, these reactions using organoaluminum compounds are not as important as those with organolithium and -magnesium compounds.

Several years ago it was shown that the insertion of carbon dioxide into the Al–C bond of tetraphenylporphinatoaluminum (TPPAl) ethyl in the presence of 1-methylimidazole in visible light produces a propionatoaluminum species, which gives methyl propionate on successive treatment with hydrogen chloride and diazomethane (Eq. 3.18).[24]

$$TPPAlEt + CO_2 \longrightarrow TPPAlOCOEt \xrightarrow{HCl} EtCOOH \xrightarrow{CH_2N_2} EtCOOCH_3 \quad (3.18)$$

Tetraphenylporphinatoaluminum methoxide, on the other hand, takes up carbon dioxide in the presence of 1-methylimidazole at room temperature to give the adduct. Though the mode of binding has not been clarified yet, the carbon dioxide in this adduct was found to be highly activated and to react with propylene oxide to give propylene carbonate under ambient conditions.[25] Insertion of carbon dioxide into the Al–N bond in $AlEt_2$-(NEt_2) was also accelerated by tertiary amines such as 1-methylimidazole.[23]

A Lewis acid such as $AlCl_3$ activates carbon dioxide to react with aromatic hydrocarbons at an elevated temperature, producing aromatic carboxylic acid (Eq. 3.19).[26] The Friedel-Crafts reaction of this type is thought to proceed *via* intermediary arylaluminum species and is applicable, although to a lesser extent, to the aliphatic hydrocarbons such as pentane.[27]

$$PhCH_3 + 2 AlCl_3 \longrightarrow [CH_3C_6H_4Al_2Cl_5] \xrightarrow{CO_2} \xrightarrow{H^+} CH_3C_6H_4COOH \quad (3.19)$$

Although trialkylborane does not react with carbon dioxide at room temperature, it does react at higher temperatures to give tributylboroxine (Eq. 3.20).[28]

$$(C_4H_9)_3B + CO_2 \xrightarrow[\text{(pressure)}]{270°-320°C} (C_4H_9BO)_3 + C_4H_{10} + C_4H_8 \quad (3.20)$$

In contrast, carbon dioxide is easily inserted into the B–N bond of amino-boranes, even under ambient conditions, to give carbamates (Eq. 3.21).[29]

$$PhB(NHR)_2 + CO_2 \xrightarrow{25°C} PhB(OCONHR)_2 \quad (3.21)$$

3.2.3.   Group IVa and Va Metal Compounds

Sanderson has divided representative elements (E) in the periodic table into two groups according to the reactivities of their methyl derivatives (EMe_n) toward carbon dioxide.[30] The methyl-element compounds with a more negative charge on methyl (more than -0.15), i.e., those with more positively charged metals, react with carbon dioxide, whereas those with a less negative methyl charge are inert to carbon dioxide. The elements included in the former group are (in decreasing order of the negative net charge on methyl) Cs, Rb, K, Na, Li, Ba, Sr, Ca, Mg, Be, and Al; while Cd, Zn, B, Si, In, Sn, Ga, P, Sb, Hg, Tl, Ge, Te, Pb, Bi, and As are included in the latter group.

As would be expected from this classification, there have been no reports of the insertion of carbon dioxide into the metal-carbon bonds of the organic derivatives of group IVa and Va metals. There have, however, been numerous examples of the insertion of carbon dioxide into the M–N bonds of the amides of Si, Ge, Sn, P, As, and Sb.

Insertions of carbon dioxide into Si–N bonds, yielding carbamato-silanes, have been described for Me_3SiNR_2 (R = Me[31] and Et[31,32]), $Me_2Si\{N(Et)R\}_2$ (R = H, Et),[33] and $Me_3SiNHNMe_2$[34] (Eq. 3.22). All these reactions proceed exothermally at room temperature. In the reactions

$$Si-NR_2 + CO_2 \longrightarrow SiOCONR_2 \quad (3.22)$$

with Me_3SiNR_2 and $Me_2Si(NEt_2)_2$, traces of free amine were found to be essential, hence an ionic chain mechanism involving the intermediary formation of carbamic acid, as shown in Eq. (3.23), has been proposed for these reactions.[31]

$$CO_2 + HNEt_2 \longrightarrow HOCONET_2$$

$$Me_3SiNEt_2 + HOCONEt_2 \longrightarrow Me_3SiOCONEt_2 + HNEt_2 \qquad (3.23)$$

Similar reactions for the formation of carbamato derivatives have been reported for germanium and tin (Eqs. 3.24–3.28).

$$R_3GeNR'_2 + CO_2 \longrightarrow R_3GeOCONR'_2 \ (R = Me^{[34]}, Et^{[35,36]}) \qquad (3.24)$$

$$(Et_3Ge)_2NH + CO_2 \longrightarrow (Et_3GeOCO)_2NH^{[36]} \qquad (3.25)$$

$$Sn(NMe_2)_4 + CO_2 \longrightarrow (Me_2N)_2Sn(OCONMe_2)_2^{[37]} \qquad (3.26)$$

$$R_2Sn(NR'_2)_2 + CO_2 \longrightarrow R_2Sn(OCONR'_2)_2 \ (R = Me, Ph)^{[37]} \qquad (3.27)$$

$$R_3SnNR'_2 + CO_2 \longrightarrow R_3SnOCONR'_2 \ (R = Me, Ph)^{[38]} \qquad (3.28)$$

Carbonato derivatives of tin have been also obtained by carbonation of alkoxytins (Eqs. 3.29–3.31).

$$Bu_3SnOMe + CO_2 \longrightarrow Bu_2SnOCOOMe^{[39]} \qquad (3.29)$$

$$Bu_3SnOSnBu_3 + CO_2 \longrightarrow Bu_3SnOCOOSnBu_3^{[39]} \qquad (3.30)$$

$$Bu_2Sn(OMe)_2 + CO_2 \longrightarrow Bu_2Sn(OCOOMe)_2^{[40]} \qquad (3.31)$$

The insertion of carbon dioxide into the metal-nitrogen bonds of tertiary amino-phosphine,[32] -arsine,[32] and -stibine[41] also occurs at room temperature to give the corresponding carbamates (Eq. 3.32).

$$M(NMe_2)_3 + CO_2 \longrightarrow M(OCONMe_2)_3 \ (M = P,^{[32]} As,^{[32]} Sb^{[41]}) \qquad (3.32)$$

In the case of diaminomethylphosphine and ethyl diaminophosphite, insertion of carbon dioxide occurs only at one of two P–N bonds, and the other P–N bond and the P–C and P–O bonds remain unattacked, yielding $RP(NMe_2) (OCONMe_2)$ (R = Me and OEt, respectively).[32]

The reaction of carbon dioxide with aminophosphorane, $Me(CF_3)_3$-$PNMe_2$, to produce carbamate, $Me(CF_3)_3P(OCONMe_2)$, has been reported recently. X-ray crystallograph of the carbamate revealed that the compound has a six-coordinate structure with the carbamato group attached to phosphorus in a bidentate fashion through two oxygen atoms.[42]

Yamazaki *et al.* showed that these reactions that result in the insertion of carbon dioxide into P–N bonds could be applied to organic synthesis. They were able to synthesize ureas from carbon dioxide and aniline in the presence of a stoichiometric amount of diphenyl phosphite and tertiary amine. They proposed a mechanism involving the insertion of carbon dioxide into the P–N bond of the carbamyl N-phosphonium salt of a tertiary amine (Eq. 3.33).[43,44]

$$
CO_2 + PhNH_2 + \underset{\underset{O}{\parallel}}{H-P(OPh)_2} + R_3N \longrightarrow Ph-NH-\underset{\underset{O}{\parallel}}{C}-O-\underset{\underset{H}{\overset{OPh}{\mid}}}{P}-\overset{+}{N}R_3(PhO^-)
$$

$$
\overset{PhNH_2}{\longrightarrow} PhNH-\underset{\underset{O}{\parallel}}{C}-NHPh + O=\underset{\underset{H}{\mid}}{\overset{OPh}{\overset{\mid}{P}}}-OH + PhOH \tag{3.33}
$$

The asymmetric urea $Me_2NCONHR$ was obtained by a stepwise reaction similar to (3.33) using hexaalkylphosphorus triamide (Eq. 3.34).[43,45]

$$
(Me_2N)_3P + CO_2 \xrightarrow{C_5H_5N} (Me_2N)_2POCONMe_2 \xrightarrow{RNH_2} Me_2NCONHR \tag{3.34}
$$
$$
(R = Ph, CH_2Ph, \text{ and cyclo-}C_6H_{11})
$$

These reactions are also applicable to the synthesis of polyurea with bifunctional amine, such as 4,4'-diaminodiphenylmethane in the place of aniline in Eq. (3.33), and with a pressure (20 atm) of carbon dioxide (Eq. 3.35).[46]

$$
CO_2 + H_2N-R-NH_2 + \underset{\underset{O}{\parallel}}{H-P(OPh)_2} + R'_3N \longrightarrow \underset{\underset{O}{\parallel}}{\{C}-NH-R-NH\}_n \tag{3.35}
$$

Polymerization involving carbon dioxide as a comonomer is described in more detail in Section 4.2.

3.2.4. Group IIb Metal Compounds

As described in Section 3.2.3, the methyl derivative of zinc has been known to be inert to the attack by carbon dioxide. However, it was found by Inoue and Yokoo[47] that diethylzinc can react with carbon dioxide under mild conditions provided that a complexing agent such as N,N,N',N'-tetramethylethylenediamine, 2,2′-bipyridine, or N-methylimidazole is present in the system (Eq. 3.36).

$$Et_2Zn + CO_2 \xrightarrow{\text{"N"}} EtZnOCOEt \qquad (3.36)$$

$$(\text{"N"} = Me_2NCH_2CH_2NMe_2, \quad \text{(structure)} \quad \text{or} \quad \text{(structure)} \quad)$$

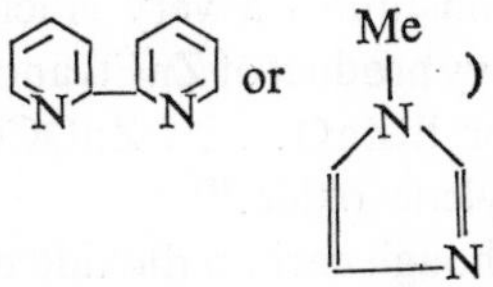

The fact that the complexation of N-methylimidazole to zinc considerably improves the reactivity of the zinc compound toward carbon dioxide suggests that the system can be regarded as a kind of model for the enzymatic activity of carbonic anhydrase, in which the complexation of histidine to zinc is thought to be important in the reversible hydration of carbon dioxide (Eq. 3.37).

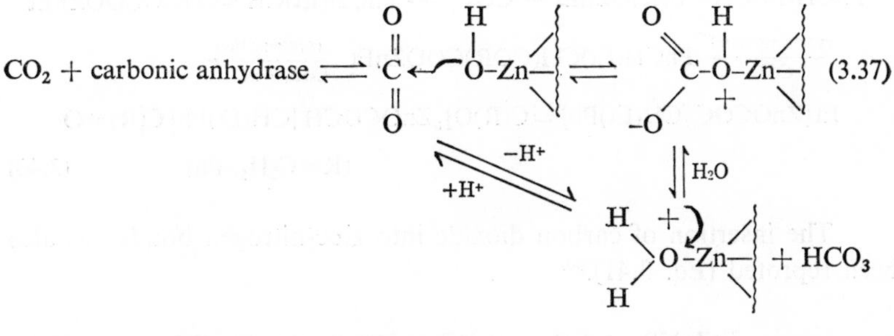

$$CO_2 + \text{carbonic anhydrase} \rightleftharpoons \cdots \rightleftharpoons \cdots \qquad (3.37)$$

The reactions of organozinc compounds with carbon dioxide have been studied extensively in an effort of elucidate the mechanism by which the alternate copolymerization of carbon dioxide and alkylene is initiated, as will be described in more detail in Section 4.2. These studies demonstrated the formation of (alkyl carbonato)-zinc alkoxide by a reaction between carbon dioxide and zinc dialkoxide, which was produced by the reaction of diethyl zinc with 2 molar equivalents of alcohol (Eqs. 3.38 and 3.39).[48]

$$ZnEt_2 + 2\,ROH \xrightarrow{C_6H_6} ROZnOR + 2\,C_2H_6 \qquad (3.38)$$

$$ROZnOR + CO_2 \longrightarrow ROZnOCOOR \qquad (3.39)$$

It is interesting to note that neither $ZnEt_2$ nor $ROZnEt$ reacts with carbon dioxide, and that only organic derivatives of zinc without a Zn–Et bond react with CO_2. The diethylzinc/alcohol system does not possess catalytic activity in the copolymerization of carbon dioxide and alkylene oxide because the alkylcarbonato-zinc alkoxide which is formed by reaction (3.39) reacts very slowly with alkylene oxide. However, the diethylzinc/water system constitutes a very efficient polymerization catalyst; $Et(ZnO)_nH$, the primary product of $ZnEt_2$ and water, gives carbonates such as $Et(ZnO)_n \cdot COOH$ or $EtZnO. \ldots ZnOCOOZnO. \ldots ZnOH$ which react readily with alkylene oxide.[48]

Although carbon dioxide can not be inserted into the Zn–O bond of ethylzinc alkoxide, alkylzinc enolates such as shown in Eq. (3.40) do react with carbon dioxide. Alkylzinc enolates, formed by the reaction between diethylzinc and ethyl(or phenyl) styryl ketone, react with carbon dioxide to give, first, carbonato derivatives, then polymeric zinc compounds through a keto-enol rearrangement, followed by intermolecular condensation reactions (Eq. 3.40).[49]

$$PhCH(Et)CH=C(R)OZnEt + CO_2 \xrightarrow{C_6H_6} PhCH(Et)CH=C(R)OCOOZnEt$$

$$\xrightarrow[C_6H_6]{rearrangement} PhCH(Et)CH(COR)COOZnEt \xrightarrow[C_6H_6, -nC_2H_6]{condensation}$$

$$Et\{ZnOCOC\{CH(Et)Ph\}=C(R)O\}_nZnOCOCH\{CH(Et)Ph\}C(R)=O$$

$$(R=C_2H_5, Ph) \qquad (3.40)$$

The insertion of carbon dioxide into zinc-nitrogen bonds has also been reproted (Eq. 3.41).[50]

$$EtZnNR_2 + CO_2 \longrightarrow RZnOCONR_2 \ (R=Ph, Et) \qquad (3.41)$$

The insertion of carbon dioxide into the Zn–carbon bond formed in situ from α-bromoesters and amalgamated zinc has been applied to the synthesis of malonic half-esters by the Reformatsky reaction (Eq. 3.42). A solvent such as dioxane or tetrahydrofuran is essential to promote this carboxylation reaction.[51]

$$BrCH_2COOR + Zn(Hg) + CO_2 \longrightarrow BrZnOCOCH_2COOR$$

$$\xrightarrow{H^+} HOCOCH_2COOR \qquad (3.42)$$

The insertion of carbon dioxide into the Hg–O bond of bis(alkyl-mercuric)oxide to give alkylmercuric carbonates is the only reported insertion reaction with group IIb metal compounds other than those of zinc (Eq. 3.43).[52]

$$(RHg)_2O + CO_2 \xrightarrow{\text{MeOH}} RHgOCOOHgR \qquad (3.43)$$

$$(R = Me, Et, n\text{-Pr}, i\text{-Pr, and } n\text{-Bu})$$

3.3. Reactions with Transition Metal Compounds

Most of the reactions of carbon dioxide with the non-transition metal compounds described in the preceding section proceed in a stoichiometric fashion. In order to extend these reactions to catalytic reactions, the employment of transition metal compounds is essential. Furthermore, reactions between the transition metal compounds and CO_2 usually occur stereospecifically under mild conditions. In contrast to the long history of studies on reactions of CO_2 with non-transition metal compounds, research on the interactions of CO_2 with transition metal complexes is relatively recent; the first report appeared in 1968, and the number of papers increased rapidly in the 1970s.

The role of transition metal compounds in the utilization of carbon dioxide for synthetic organic chemistry may be to activate both the organic molecule and the carbon dioxide by interacting with each molecule. Although the actual synthetic system may be composed of a combination of these two functions, information concerning the stoichiometric interactions of CO_2 with transition metal complexes is important for understanding the mechanisms of catalytic reactions involving carbon dioxide. These stoichiometric interactions that lead to the activation of carbon dioxide involve such reactions as coordination, reversible or irreversible insertion of carbon dioxide into metal-hydrogen, -carbon, -nitrogen, and -oxygen bonds, and formation of complexes with carbon monoxide, carbonato, or bicarbonato ligands. The activation of CO_2 by reversible insertion into transition metal complexes has attracted special attention as a possible mechanism for the biological carboxylation reaction performed in living organisms by enzymes containing biotin as a cofactor.[53] Similar studies with alkali and magensium salts of alcohol and carboxylic acid were described in the preceding section.[12-19]

This section summarizes the stoichiometric interactions of carbon dioxide with transition metal complexes. (For reviews covering this field, see refs. 54–65.)

3.3.1. Carbon Dioxide-Coordinated Complexes

Although a fairly large portion of the reactions between transition metal complexes and carbon dioxide are thought to proceed by the initial coordination of CO_2 to the metal, the number of CO_2-coordinated complexes that have been isolated and whose structures have been unequivocally determined is very limited. This contrasts with the situation of dinitrogen, also known to be a chemically inert molecule, with which a considerable number of stable N_2-coordinated transition metal complexes have been reported since the discovery of the complex of the type $[Ru(N_2)(NH_3)_5]^{2+}$ in 1965.[66] Carbon dioxide may be sufficiently activated during coordination to the metal that it can undergo further reactions such as insertion, oxidation, or reduction on the metal.

As CO_2 has canonical electronic structures as shown in Eq. (3.44), there are several possible ways of coordination to a transition metal. Some examples are shown in Fig. 3.1.

$$: \overset{..}{\underset{..}{O}} - C \equiv \overset{+}{O} : \quad \longleftrightarrow \quad : \overset{..}{O} = C = \overset{..}{O} : \quad \longleftrightarrow \quad : \overset{+}{O} \equiv C - \overset{..}{\underset{..}{O}} : \qquad (3.44)$$

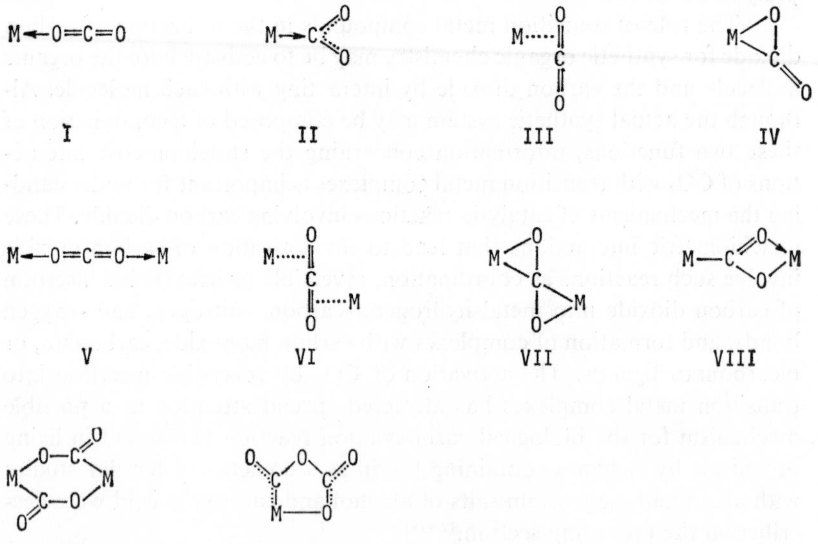

Fig. 3.1 Possible modes of coordination of carbon dioxide to a metal.

Carbon dioxide has a fairly high ionization potential (13.8 eV)[67] compared with isoelectronic compounds such as CS_2 (10.1 eV) and COS (11.3 eV), and it has a high electron affinity, as estimated by the level of the lowest unoccupied antibonding orbital (approx. 3.8 eV).[57] Therefore, modes of coordination such as I and V, in which a lone pair of electrons in oxygen donated to the metal, are considered less likely. A theoretical consideration of carbon dioxide coordination to a transition metal, based on the semiempirical LCAO-MO (Linear Combination of Atomic Orbitals-Molecular Orbitals) calculation within the CNDO (Complete Neglect of Differential Overlap) approximation, also predicted that coordination modes involving the carbon-oxygen double bond, such as types III, IV, VI, and VII, are energetically most favorable.[68]

Table 3.3 Transition metal complexes coordinated with carbon dioxide, and infrared spectral data characteristic of coordinated carbon dioxide.

Complex	$v(CO)_2$, cm^{-1}	Reference
Mononuclear complexes		
$[Mo(CO_2)_2(PMe_2Ph)_4]$	**(1)** 1760, 1510, 1335	69
$[Fe(CO_2)(PMe_3)_4]$	**(2)** 1620, 1108, 612 (Nujol)	70
$[Co(CO_2)(PPh_3)_3]$	**(3)** 1890	71
$[Rh(OH)(CO_2)(CO)(PPh_3)_2]$	**(4)** 1602s, 1351s, 821m (Nujol)	72
$[RhCl(CO_2)(P\text{-}n\text{-}Bu_3)_2]$	**(5)** 1668vs, 1630vs, 1165, 1120 (Nujol)	73
$[RhCl(CO_2)(PEtPh_2)_2]$	**(6)** 1665vs, 1625vs (Nujol)	73
$[RhCl(CO_2)(PEt_2Ph)_2]$	**(7)** 1658vs, 1620vs, 1238m, 827m (Nujol)	73
$[RhCl(CO_2)(PMe_2Ph)_2]$	**(8)** 1675vs, 1625vs, 1217m, 996m, 823m) (Nujol)	73
$[RhCl(CO_2)(PEtPh_2)_3]$	**(9)** 1675vs, 1625w, sh (Nujol)	73
$[RhCl(CO_2)(PEt_2Ph)_3]$	**(10)** 1670vs, 1635sh, 1255m, 963m, 780m (Nujol)	73
$[RhCl(CO_2)(PMePh_2)_3]$	**(11)** 1673vs, 1635sh, 1290m, 1000m, 822m, 760m (Nujol)	73
$[Ir(OH)(CO_2)(CO)(PPh_3)_2]$	**(12)** 1636m, 1310s, 815m (Nujol)	72
$[IrCl\{(CO_2)_2\}(PMe_3)_3]\cdot0.5C_6H_6$	**(13)** 1725s, 1680s, 1648sh, 1605m, 1290s, 1005m, 790m	74
$[IrCl(CO_2)(Me_2PCH_2CH_2PMe_2)_2]$	**(14)** 1550s, 1230s (Nujol)	75
$[Ir(CO_2)(AsMe_2C_6H_4)_2]Cl$	**(15)** 1550s, 1220s (Nujol)	75
$[Ni(CO_2)\{P(cyclo\text{-}C_6H_{11})_3\}_2]\cdot\cdot$ 0.75C$_6$H$_5$CH$_3$	**(16)** 1740vs, 1698vw, 1150s (Nujol)	76
$[Ni(CO_2)(PEt_3)_2]$	**(17)** 1660vs, 1635vs, 1203vs, 1009vs (Nujol)	77
$[Ni(CO_2)(P\text{-}n\text{-}Bu_3)_2]$	**(18)** 1660vs, 1632vs, 1200vs, 1008vs (Nujol)	77
$[Pt(CO_2)(PPh_3)_2]$	**(19)** 1640s, 1370s, 1320s	57
$[Cu(II)\text{-ephedrine-}CO_2]$	**(20)** 2350m, 2380s (C$_6$H$_6$)	78
$[Cu\{CH(NC)COOEt\}(CO_2)$- $(PPh_3)_2]$	**(21)** 1600vs, 1400m, 1365m, 1320s, 820w (KBr)	79
Binuclear complexes		
$[Rh_2(CO_2)(CO_2)(PPh_3)_3]$	**(22)** 1498s, 1368s, 813m (Nujol)	80
$[Rh_2Cl_2(CO_2)(PPh_3)_5]$	**(23)** 1630	57, 81
$[Rh_2(CO_2)_2(CO)_2(PPh_3)_3]\cdot C_6H_6$	**(24)** 1600s, 1355s, 825m	82
$[Ni_2(CO_2)\{P(cyclo\text{-}C_6H_{11})_3\}_4]$	**(25)** 1735 (KBr)	83
$(C_6H_5AgCO_2Ag)_n$	**(26)** 1496s, 1326s, 828s	84

Table 3.3 lists the carbon dioxide-coordinated transition metal complexes reported so far.

In the reaction of a transition metal complex with CO_2 there is a certain possibility that water or oxygen has not been completely excluded and may be involved in the reaction, leading to a misinterpretation of the results. Some of the structures assigned to CO_2-coordinated complexes in Table 3.3 are based only on infrared and chemical evidence and should be regarded as tentative until established by X-ray crystallography.

Most of the complexes listed in the Table 3.3 were prepared by direct treatment with carbon dioxide, except for **22**, which was synthesized through oxidation of one of the carbonyl ligands in $[Rh_2(CO)_4 (PPh_3)_4]$.[80] A typical method for the preparation of a carbon dioxide-coordinated complex is to replace the labile ligands, such as dinitrogen (**1, 3, 16,** and **25**), tertiary phosphines (**16–19, 23,** and **24**), and olefinic ligands (**5–11** and **13**), in the corresponding complexes with carbon dioxide under normal pressure. The remaining complexes in Table 3.3 were prepared by the simple addition of a carbon dioxide molecule to the original complex. Some examples of reaction schemes for the preparation of carbon dioxide-coordinated complexes are shown in Eqs. (3.45)–(3.52)

$$[Mo(N_2)_2(PMe_2Ph)_4] + CO_2 \xrightarrow[\text{room temp.}]{\text{toluene}} [Mo(CO_2)_2(PMe_2Ph)_4] + 2\,N_2 \quad (3.45)$$
$$(1)$$

$$[Ni_2\{P(cyclo\text{-}C_6H_{11})_3\}_4(N_2)] + CO_2 \xrightarrow[\text{room temp.}]{\text{Et}_2\text{O}} [Ni_2\{P(cyclo\text{-}C_6H_{11})_3\}_4(CO_2)]$$
$$(25)$$
$$+ N_2^{83)} \quad (3.46)$$

$$[Ni(PR_3)_n] + CO_2 \xrightarrow[-30°C]{\text{toluene}} [Ni(CO_2)(PR_3)_2] + (n-2)\,PR_3^{76,77)} \quad (3.47)$$
$$(\textbf{16}, R = cyclo\text{-}C_6H_{11}), (\textbf{17}, R = Et), (\textbf{18}, R = n\text{-Bu})$$

$$[Rh_2(CO)_2(PPh_3)_4(C_6H_6)] + CO_2 \xrightarrow[\text{room temp.}]{\text{benzene}} [Rh_2(CO_2)_2(CO)_2(PPh_3)_3(C_6H_6)]$$
$$+ PPh_3^{82)} \quad (\textbf{24}) \quad (3.48)$$

$$[Rh_2Cl_2(C_2H_4)_4] + nL + CO_2 \xrightarrow[\text{room temp.}]{\text{toluene}} 2\,[RhCl(CO_2)L_n] + 4\,C_2H_4^{73)} \quad (3.49)$$
$$(\textbf{5}, L = P\text{-}n\text{-Bu}_3, n = 2), (\textbf{6}, L = PEtPh_2, n = 2), (\textbf{7}, L = PEt_2Ph,$$
$$n = 2), (\textbf{8}, L = PMe_2Ph, n = 2), (\textbf{9}, L = PEtPh_2, n = 3), (\textbf{10}, L =$$
$$PEt_2Ph, n = 3), (\textbf{11}, L = PMePh_2, n = 3)$$

$$[IrCl\,(cyclo\text{-}C_8H_{14})(PMe_3)_3] + CO_2 \xrightarrow[\text{room temp.}]{\text{benzene}} [IrCl(C_2O_4)(PMe_3)_3]$$
$$+ cyclo\text{-}C_8H_{14}^{74)} \quad (\textbf{13}) \quad (3.50)$$

$$[Fe(PMe_3)_4] + CO_2 \xrightarrow[0°C]{pentane} [Fe(CO_2)(PMe_3)_4]^{70)} \qquad (3.51)$$

2

$$[M(OH)(CO)(PPh_3)_2] + CO_2 \xrightleftharpoons{solid, room temp.} [M(OH)(CO_2)(CO)(PPh_3)_2]^{72)}$$

(4, M = Rh), (12, M = Ir) (3.52)

Infrared absorption due to the coordinated CO_2 has been investigated extensively for the binuclear rhodium complexes **22** and **24** using $C^{18}O_2$- and $^{13}CO_2$-substitution techniques.[80,82) Both complexes [Rh$_2$-(CO_2) $(CO)_2$ $(PPh_3)_3$] **(22)** and [Rh$_2$ $(CO_2)_2$ $(CO)_2(PPh_3)_3$]. C_6H_6 **(24)** possess three medium to intense bands, as shown in Table 3.3. In view of the fact that gaseous carbon dioxide, which has a linear $D_{\infty h}$ symmetry, exhibits only two infrared absorptions, at 2349 (ν_3, asym ν_{CO_2}) and 667 cm^{-1} (ν_2, deg δ_{CO_2}, split by Fermi resonance with $2\nu_2$),[85) the appearance of three infrared absorptions in complexes **22** and **24** indicates that the carbon dioxide ligands in these complexes are in the bent forms. Thus, the three bands can be assigned to, in order of descending frequency, asym ν_{CO_2}, sym ν_{CO_2}, and δ_{CO_2}, respectively. Furthermore, the fact that the value of the difference between the two ν_{CO2} bands for complex **22** (asym ν_{CO_2} — sym ν_{CO_2} = 130 cm^{-1}) is much less than that for **24** (245 cm^{-1}) suggests that the carbon dioxide ligand in **22** may have a coordination mode like that of VIII of Fig. 3.1, whereas the mode for **24** may be IX. That both oxygen atoms of the CO_2 participate in bonding to the metal (type VIII) may considerably reduce the difference between the two ν_{CO_2} bands; this contrasts with coordination type IX in which only one oxygen is coordinated, leaving another oxygen as a free carbonyl group. These events resemble those with carboxylate ligands, which coordinate to the central metal either in a bidentate or a monodentate fashion.[86)

The complex with the type VIII coordination mode has also been considered by Vol'pin and Kolomnikov[57) for the complex [Rh$_2$Cl$_2$(CO$_2$)-(PPh$_3$)$_5$] **(23)** because it yielded methyl acetate on reaction with methyl iodide, although it has only one CO_2 band at 1630 cm^{-1}. However, the binuclear nickel complex [Ni$_2$ {P($cyclo$-C$_6$H$_{11}$)$_3$} $_4$(CO$_2$)] **(25)** which also has only one CO_2 band, is thought to have a symmetrical structure with a linear carbon dioxide ligand like that of type VI in Fig. 3.1.[83)

Studies of mononuclear CO_2 complexes have a much more solid basis than do studies of the binuclear complexes because the molecular structure of some of these complexes has been successfully determined. Aresta *et al.*[76) have prepared [Ni(CO$_2$) {P($cyclo$-C$_6$H$_{11}$)$_3$} $_2$]·0.75 toluene **(16)** by a reaction of carbon dioxide with either [Ni {P($cyclo$-C$_6$H$_{11}$)$_3$} $_3$] or [Ni$_2$ {P ($cyclo$-C$_6$H$_{11}$)$_3$} $_4$(N$_2$)] in toluene at room temperature, and they have succeeded for the first time in establishing its molecular structure by X-ray crystallography. As shown in Fig. 3.2, complex **16** has a distorted square planar structure, and the ligating carbon dioxide is markedly bent and linked to the metal through the carbon and one of the oxygen atoms. A similar molecular geometry has been established for the palladium[87] and platinum[88] complexes of carbon disulfiede, which is isoelectronic to carbon dioxide (Fig. 3.3). The bond length (1.22 Å) between the C and O of the coordinated CO moiety was found to be significantly different from that of the free carbon dioxide molecule (1.16 Å).

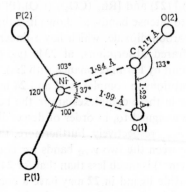

Fig. 3.2. Mode of bonding of CO_2 in [Ni(CO_2) {P(*cyclo*-C_6H_{11})$_3$}$_2$]·0.75toluene (**16**). (Source: ref. 76. Reproduced by kind permission of the Chemical Society (London).)

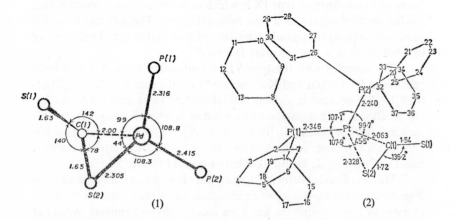

Fig. 3.3 Molecular geometry of the complexes (1) [Pd(CS_2)(PPh_3)$_2$] and (2) [Pt(CS_2)-(PPh_3)$_2$]. (Source: refs. 87 and 88. Reproduced by kind permission of the Chemical Society of Japan (1) and the Chemical Society (London). (2))

On the basis of infrared bands characteristic of the ligating carbon dioxide, and a consideration of their coordination numbers and chemical properties, complexes **2**, **5–8**, **17**, and **18** are assumed to have structures of type IV (Fig. 3. 1), as does complex **16**, whereas complexes **9–11**, **14**, and **15** are thought to be type II complexes in which carbon dioxide is attached to the metal through the carbon atom. Furthermore, infrared study of the reaction system consisting of $[NiL_4]$ (L = PEt_3 and P-*n*-Bu_3) and CO_2 in toluene that produces $[Ni(CO_2)L_2]$ (**17** and **18**) suggested the presence in solution of the intermediate species $Ni(CO_2)L_3$, to which the structure II has been tentatively assigned.

The molecular structure of complex $[IrCl\{(CO_2)_2\}\ (PMe_3)_3]\cdot0.5C_6H_6$ (**13**) has also been established by means of X-ray crystallography. It was found to have the novel type of coordination involving 2 molecules of carbon dioxide shown in Eq. (3.53) (type X of Fig. 3.1).[74]

$$[IrCl(cyclo\text{-}C_8H_{14})\ (PMe_3)_3] + CO_2 \longrightarrow IrCl(CO_2)\ (PMe_3)_3 \xrightarrow{CO_2}$$

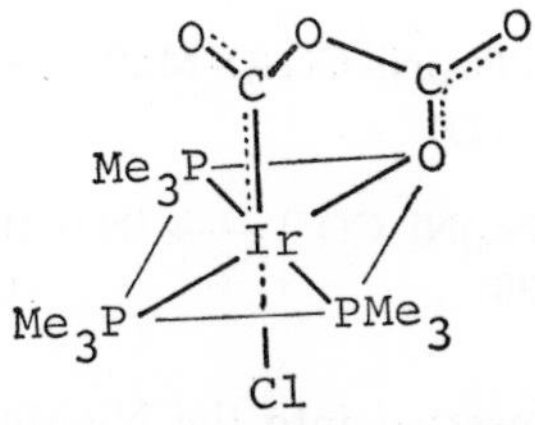

$$(3.53)$$

(**13**)

The assignment of the seven infrared bands to the coordinated $(CO_2)_2$ group of **13** in Table 3.3 was confirmed by experiments using $^{13}CO_2$ and $C^{18}O_2$ in place of CO_2 in Eq. (3.53). Presumably, complex **13** is formed stepwise *via* the intermediary formation of a CO_2 adduct with 1 mole of CO_2 followed by the addition of another mole of CO_2. Because complex **13** decomposes on heating at 150° C to yield a mixture of products consisting of Ir-CO and Ir-CO_2 species, as can be determined from its infrared spectrum, **13** is assumed to be a compound that represents an intermediate in the conversion of ligating carbon dioxide to the carbonyl or carbonato ligand, as will be described in more detail in the following section.

Most complexes with a carbon dioxide ligand release an almost quantitative amount of CO_2 either on heating or on treatment with an acid such as H_2SO_4 and HCl or with an oxidizing agent such as I_2. The coordination of CO_2 in complexes **4**, **9**, **12**, and **20** was found to be reversible; their ligated CO_2 was easily released by simply pumping the solid complex[72,73] or by sweeping the solution containing the complex with argon gas.[78]

3.3.2. Carbonyl, Carbonato, and Peroxocarbonato Complexes Derived from Carbon Dioxide

Some transition metal complexes yield carbonyl complexes on reaction with CO_2 rather than the carbon dioxide complexes mentioned in the preceding section. The formation of carbonyl complexes is considered to proceed *via* the reduction of CO_2 which is activated by coordination to a transition metal. It is interesting to note that most carbonyl-formation reactions occur with complexes that have transition metal-non-transition metal sigma bonds. A typical example of such a system is the reaction of CO_2 with the so-called "transition metal Grignard reagent," as shown in Eq. (3.54).[89]

$$[Ni(\eta\text{-}C_5H_5)Br(PPh_3)] + Mg \longrightarrow [(\eta\text{-}C_5H_5)(PPh_3)Ni\text{-}MgBr]$$
$$\qquad\qquad (27) \qquad\qquad\qquad\qquad\qquad\qquad (28)$$

$$\xrightarrow[\text{MgBr}_2]{\text{CO}_2} \{(\eta\text{-}C_5H_5)(PPh_3)Ni\text{-}CO\text{-}O\text{-}MgBr\} \longrightarrow$$
$$\qquad\qquad\qquad (29)$$

$$\longrightarrow \{(\eta\text{-}C_5H_5)(PPh_3)Ni\text{-}CO^+\} \longrightarrow [Ni(CO)_2(PPh_3)_2] \qquad (3.54)$$
$$\qquad\qquad (30) \qquad\qquad\qquad\qquad (31)$$

Carbon dioxide is first inserted into the Ni–Mg bond of **28**, formed *in situ* (the insertion of CO_2 will be described in subsequent sections in more detail), to give the metal carboxylate intermediate **29** which, on further reaction, is converted into the stable, known dicarbonyl complex **31**.[89] Equations (3.55) and (3.56) are similar reactions which may proceed analogously.

$$[Fe(\eta\text{-}C_5H_5)Br(Ph_2PCH_2CH_2PPh_2)] + Mg \longrightarrow [(\eta\text{-}C_5H_5)(PPh_3)Fe\text{-}MgBr,$$

$$3THF] \xrightarrow[\text{MgBr}_2]{\text{CO}_2} \xrightarrow{\text{AgBF}_4} [Fe(\eta\text{-}C_5H_5)(Ph_2PCH_2CH_2PPh_2)(CO)]BF_4{}^{[90]} \qquad (3.55)$$
$$(32) \qquad\qquad\qquad\qquad\qquad (33)$$

(THF = tetrahydrofuran)

$$[Mo(\eta\text{-}C_5H_5)_2H_2] + i\text{-PrMgBr} \xrightarrow{\text{Et}_2\text{O}} \xrightarrow{\text{hot THF}} [(\eta\text{-}C_5H_5)_2HMo\text{-}MgBr\text{-}(THF)_2]$$
$$\qquad\qquad\qquad\qquad\qquad\qquad (34)$$

$$\xrightarrow{\text{CO}_2} [Mo(\eta\text{-}C_5H_5)_2(CO)]^{[91]} \qquad\qquad\qquad\qquad\qquad\qquad (3.56)$$
$$\qquad (35)$$

Non-transition metals other than magnesium also combine with transition metals to give transition metal-non-transition metal bonds into which the insertion of carbon dioxide occurs and produces, ultimately, a carbonyl complex (Eqs. 3.57–3.61).

$$[Mo(\eta\text{-}C_5H_5)_2H_2] + AlMe_3 \longrightarrow [Me_2Al(C_5H_4)_2Al(Me)\text{-}Mo_2\text{-}Al(Me)\text{-}$$

$$(C_5H_4)_2AlMe_2] \xrightarrow{CO_2} \xrightarrow{H_2O} \xrightarrow{HBr} \xrightarrow{PF_6^-} [Mo(\eta\text{-}C_5H_5)_2Br(CO)]^+PF_6^{-92)} \quad (3.57)$$
$$(36) \qquad\qquad\qquad\qquad\qquad\qquad\qquad (37)$$

$$[Ti(\eta\text{-}C_5H_5)_2Cl] + 2\ CO_2 + 5/3\ Al \longrightarrow [Ti(\eta\text{-}C_5H_5)_2(CO)_2] + 1/3\ AlCl_3$$
$$(38)$$
$$+\ 2/3\ Al_2O_3{}^{93)} \qquad (3.58)$$

$$Na[Fe(\eta\text{-}C_5H_5)(CO)_2] + CO_2 \longrightarrow \{Fe(\eta\text{-}C_5H_5)(CO)_2(CO_2)\}^-Na^+$$
$$(39)$$

$$\xrightarrow{CO_2} \{Fe(\eta\text{-}C_5H_5)(CO)_2(C_2O_4)\}^-Na^+ \xrightarrow[-Na_2CO_3]{}$$
$$(40)$$

$$\{Fe(\eta\text{-}C_5H_5)(CO)_3\}^+ \xrightarrow[-CO]{39} [Fe(\eta\text{-}C_5H_5)(CO)_2]^{94)} \qquad (3.59)$$
$$(41) \qquad\qquad\qquad\qquad (42)$$

$$[RhCl(PPh_3)_3] + HSi(OEt)_3 + CO_2 \longrightarrow [RhCl(CO)(PPh_3)_2]^{95)} \quad (3.60)$$

$$[RuCl_2(PPh_3)_3] + HSi(OEt)_3 + CO_2 \longrightarrow [RuCl_2(CO)(PPh_3)_3]^{95)} \quad (3.61)$$

In Eq. (3.57), the multinuclear intermediate 36 was isolated and its molecular structure was established through single-crystal X-ray study. Again, carbon dioxide is probably inserted into the Mo–Al bond of 36, followed by a splitting of the C–O bond on hydrolysis to give a CO-coordinated complex. Reaction (3.58) may proceed *via* an intermediary formation of a Ti–Al bond, although it was not detected in the system. The accumulation of ^{13}CO ligand in 42 during the reaction of 39 with $^{13}CO_2$ suggests that the disproportionation of $C_2O_4^{2-}$ ligand in 40 to CO and $CO_3{}^{2-}$, as shown in Eq. (3.59), is in fact taking place. In this reaction, too, the initial incorporation of CO_2 into 39 may be the insertion of CO_2 into the Fe–Na bond. Equations (3.60) and (3.61) are both considered to represeat the insertion of CO_2 into the Si–Rh or Si–Ru bond of the intermediate RhH{Si(OEt)$_3$} Cl(PPh$_3$)$_2$ or RuH{Si(OEt)$_3$} Cl$_2$(PPh$_3$)$_2$ species.

The transition metal centers of the starting complexes in Eqs. (3.54) –(3.61) are all highly electron-rich due to the electron flow from the non-transition metal elements directly attached to the transition metal centers.

The basic transition metal center may attack the carbon atom of carbon dioxide to form an intermediate of the type M_T–C(O)–O–M_N, where M_T is the transition metal center and M_N is the non-transition metal; the subsequent splitting at the C–O bond of M_T–C(O)–O–M_N would give M_T(CO) species. [Ni(CO)$_2$ {P(cyclo-C$_6$H$_{11}$)$_3$}$_2$] can be synthesized with a longer reaction period than is used for [Ni(CO$_2$) {P(cyclo-C$_6$H$_{11}$)$_3$}$_2$]·0.75toluene (16) from [NiBr$_2$ {P (cyclo-C$_6$H$_{11}$)$_3$}$_2$] and Na-disperison;[76] the mechanism for this synthesis is probably similar.

There may be two other pathways for the formation of carbon monoxide from the carbon dioxide on transition metal complexes. One is to reduce carbon dioxide, which is activated by interaction with the transition metal center, with a mild reducing agent. In the following examples (Eqs. 3.62–3.64), tertiary phosphines, which are present in the system as a result of the dissociation or decomposition of the starting complex, are assumed to play the role of the reducing agent.

$$[\text{Mo}(C_2H_4)(Ph_2PCH_2CH_2PPh_2)_2] + CO_2 \xrightarrow[110°C]{\text{toluene}} \textit{cis-}[\text{Mo}(CO)_2-$$

$$(Ph_2PCH_2CH_2PPh_2)_2] + Ph_2P(O)CH_2CH_2P(O)Ph_2 + C_2H_4{}^{[96]} \qquad (3.62)$$

$$[\text{CoMe}(PMe_3)_4] + CO_2 \longrightarrow [\text{CoMe}(CO)(PMe_3)_3] + O=PMe_3{}^{[97]} \qquad (3.63)$$

$$[\text{RhCl}(CO_2)(P\text{-}n\text{-}Bu_3)_2] \text{ (5)} \xrightarrow{\text{CH}_2\text{Cl}_2} [\text{RhCl}(CO)(OP\text{-}n\text{-}Bu_3)(P\text{-}n\text{-}Bu_3)]^{[73]} \qquad (3.64)$$

Another possible pathway for the formation of carbon monoxide is the reductive disproportionation of carbon dioxide into carbon monoxide and carbonate (Eq. 3.65).

$$2 CO_2 + 2 e \longrightarrow CO_3{}^{2-} + CO \qquad (3.65)$$

Although this type of reaction does not occur readily with free carbon dioxide, it may be facilitated by the presence of the transition metal center. In fact, the following examples demonstrate the conversion of carbon dioxide into carbon monoxide and carbonate on the metal.

The bis (carbon dioxide) complex of molybdenum, [Mo(CO$_2$)$_2$(PMe$_2$-Ph)$_4$] (1), is so unstable chemically that, when recrystallization from tetrahydrofuran is attempted, it rearranges to the binuclear carbonyl complex with two bridging carbonato groups, [(PMe$_2$Ph)$_3$(CO) Mo(CO$_3$)$_2$-Mo(CO) (PMe$_2$Ph)$_3$] (43).[69] The molecular structure of 43 has been established by means of X-ray crystallography, as shown in Fig. 3.4.
The authors assumed that the reductive disproportionation shown in Eq. (3.65) took place on the molybdenum atom.[69]

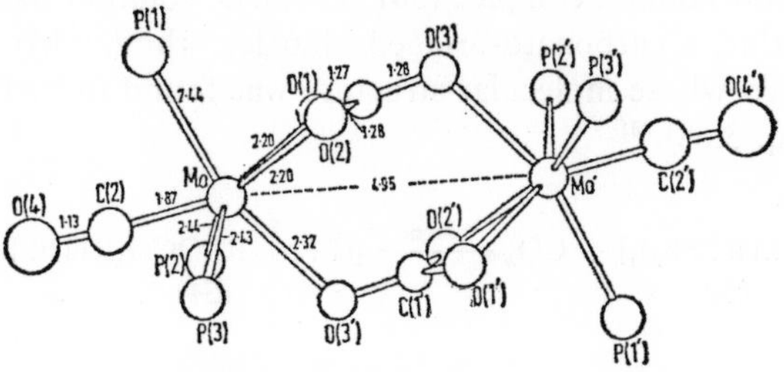

Fig. 3.4 The molecular structure of $[(PMe_2Ph)_3(CO)Mo(CO_3)_2Mo(CO)(PMe_2Ph)_3]$. (Source: ref. 69. Reproduced by kind permission of the Chemical Society (London).)

Another example of the rearrangement of coordinated carbon dioxide to a carbonyl-carbonato complex can be seen in the iron complex **2**, as shown in Eq. (3.66).

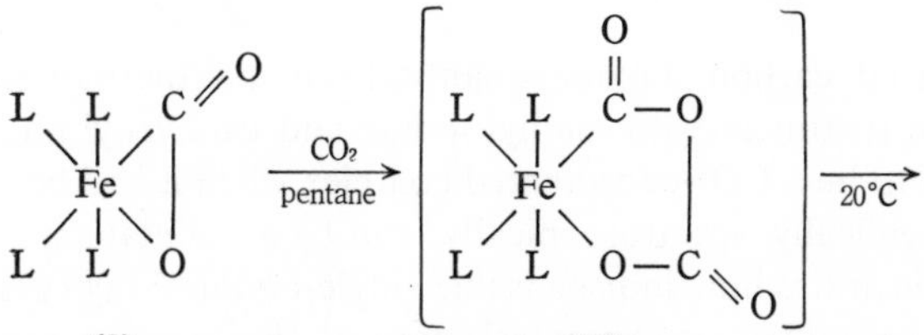

In this reaction Karsch presumes the presence of the C_2O_4 intermediate **44**, which is analogous to the isolated Ir complex **13** (Table 3.3 and Eq. 3.53). If **2** is thermally decomposed in a solid state, ligating carbon dioxide is reduced with PMe_3 to give iron carbonyl species and trimethylphosphine oxide.[70]

There are several other examples of the formation of carbonato complexes by direct interaction between CO_2 and transition metal complexes. Treatment of an equimolar mixture of dicyclopentadienylchlorotitanium (**46**) and zinc with carbon dioxide gave binuclear carbonato complexes of titanium(III) (**47** and **48**) together with a small amount of titanium(II) carbonyl species (**49**) (Eq. 3.67).[93]

$$[Ti(\eta\text{-}C_5H_5)_2Cl] + Zn + CO_2 \longrightarrow [(\eta\text{-}C_5H_5)_2TiOCO_2Ti(\eta\text{-}C_5H_5)_2]-$$
$$(46) \hspace{4.5cm} (47)$$

$$(ZnCl_2)_2 \cdot 2\,Et_2O + [\{Ti(\eta\text{-}C_5H_5)_2CO_3\}_2Zn] + [Ti(\eta\text{-}C_5H_5)_2(CO)_2] \quad (3.67)$$
$$(48) \hspace{3cm} (49)$$

When hydridorhodium complex (50) in toluene solution was treated with carbon dioxide, a carbonato-bridged binuclear rhodium(I) complex (51) was produced whose molecular structure was found to be that shown in Fig. 3.5 (Eq. 3.68).[98]

$$[RhH(PPh_3)_4] + CO_2 \xrightarrow{\text{toluene}} [(PPh_3)_3RhOCO_2Rh(PPh_3)_2] \qquad (3.68)$$
$$(50) \qquad\qquad\qquad\qquad (51)$$

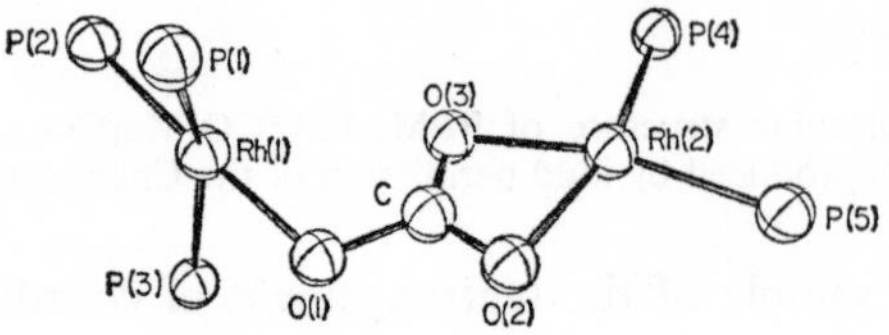

Fig. 3.5. The molecular structure of $[Rh_2(CO_3)(PPh_3)_5]\cdot C_6H_6$. (Source: ref. 98. Reproduced by kind permission of the American Chemical Society.)

As complex 51 released carbon dioxide quantitatively on thermolysis and acidolysis and on treatment with methyl iodide and triphyenyl phosphite, it was thought to be a CO_2-coordinated complex when it had been characterized only chemically, spectroscopically, and by elemental analysis.[99] This example illustrates how indispensable single-crytal X-ray crystallography is for an unambiguous characterization of the reaction products of transition metal complexes and CO_2.

Some examples of the formation of a bridging carbonato complex by the rearrangement of a carbamido or hydrogencarbonato ligand, produced by reactions with CO_2, have been reported in recent years (Eqs. 3.69–3.74).

$$[PdMe_2(PPh_3)_2] + NH_3 + CO_2 \xrightarrow[\text{room temp.}]{\text{THF}} [PdMe(OCONH_2)(PPh_3)_2]$$
$$(52) \qquad\qquad\qquad\qquad\qquad (53)$$
$$+ CH_4^{100)} \qquad\qquad\qquad\qquad (3.69)$$

$$2 \quad 53 \xrightarrow[\text{80°C}]{\text{toluene}} [(PPh_3)_2MePdOCO_2PdMe(PPh_3)]^{100)} \qquad (3.70)$$
$$(54)$$

$$[RhH(P–i-Pr_3)_3] + H_2O + CO_2 \xrightarrow{\text{THF}} [RhH_2(O_2COH)(P–i-Pr_3)_2]^{101)} \quad (3.71)$$
$$(55) \qquad\qquad\qquad\qquad\qquad (56)$$

$$2 \quad 56 + CO_2 \longrightarrow [(P–i-Pr_3)_2(CO)RhOCO_2Rh(CO)(P–i-Pr_3)_2]^{101)} \quad (3.72)$$
$$(57)$$

$$[\text{Cu(O–}t\text{-Bu)}] + CO_2 + 3 \; t\text{-BuNC} \longrightarrow [\text{Cu(O}_2\text{CO–}t\text{-Bu)}(t\text{-BuNC})_3] \xrightarrow{H_2O}$$
$$\text{(58)} \hspace{5.5cm} \text{(59)}$$

$$\text{Cu(O}_2\text{COH)}(t\text{-BuNC})_n{}^{102)} \hspace{3cm} \text{(3.73)}$$
$$\text{(60)}$$

$$\textbf{60} + \textbf{58} \longrightarrow \text{Cu}_2\text{CO}_3(t\text{-BuNC})_n{}^{102)} \hspace{2.5cm} \text{(3.74)}$$
$$\textbf{(61)}$$

The formation of carbamido and hydrogen- or alkyl-carbonato complexes will be discussed in more detail in the following section. The NMR spectroscopic findings from **54** suggest that the compound has the asymmetrical structure shown overleaf, which is very similar to **51** (Fig. 3.5). A close similarity in the infrared bands due to the μ-CO_3 groups (Table 3.4)

Table 3.4   The carbonato and the peroxocarbonato complexes of transition metals derived directly or indirectly from carbon dioxide, and relevant infrared bands.

Complex	IR bands, cm^{-1}	Reference
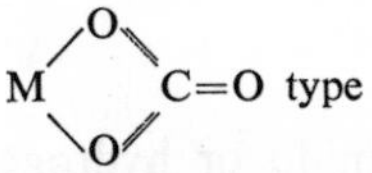 $M{\overset{O}{\underset{O}{\Big\langle}}}C{=}O$ type		
[Fe(CO$_3$)(CO)(PMe$_3$)$_3$]	**(45)** 1604, 1239, 1013, 832 (Nujol)	70
[Pd(CO$_3$)(PPh$_3$)$_2$]	**(62)** 1670s (Nujol)	103
[Pt(CO$_3$)(PPh$_3$)$_2$]	**(63)** 1680vs, 1180m, 815w, 760m (Nujol)	103
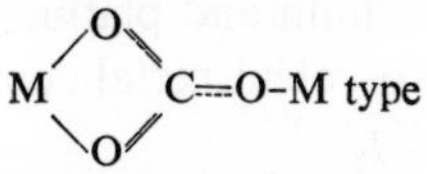 $M{\overset{O}{\underset{O}{\Big\langle}}}C{\cdots}O{-}M$ type		
[Ti$_2$(η-C$_5$H$_5$)$_4$(μ-CO$_3$)]- (ZnCl$_2$)$_2$·2Et$_2$O	**(47)** †	93
[Ti$_2$(η-C$_5$H$_5$)$_4$(μ-CO$_3$)$_2$]Zn	**(48)** †	93
[Mo$_2$(CO)$_2$(PMe$_2$Ph)$_6$(μ-CO$_3$)$_2$]	**(43)** 1835 (Nujol)	69
[Rh$_2$(PPh$_3$)$_5$(μ-CO$_3$)]·C$_6$H$_6$	**(51)** 1485s, 1460vs, 1360m, 1350m, 820w 500m (KBr)	98
[Rh$_2$(CO)$_2$(P-i-Pr$_3$)$_4$(μ-CO$_3$)]	**(57)** †	101
[Pd$_2$(Me)$_2$(PPh$_3$)$_3$(μ-CO$_3$)]	**(54)** 1475vs, 1360m, 820w (KBr)	100
[Cu$_2$(CN–t-Bu)$_n$(μ-CO$_3$)]	**(61)** †	102
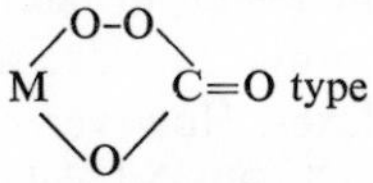 $M{\overset{O-O}{\underset{O}{\Big\langle}}}C{=}O$ type		
[Pt(CO$_4$)(PR$_3$)$_2$] (PR$_3$ = P($cyclo$-C$_6$H$_{11}$)$_3$, P-t-Bu$_2$-n-Bu, P-t-Bu$_2$Me, P-i-Pr$_3$)	**(65)** ~1680	105
[Ni(CO$_4$){P($cyclo$-C$_6$H$_{11}$)$_3$}$_2$]	**(66)** 1625s, br, 1162s, 858m, 750m (Nujol)	77

† No mention of the infrared data.

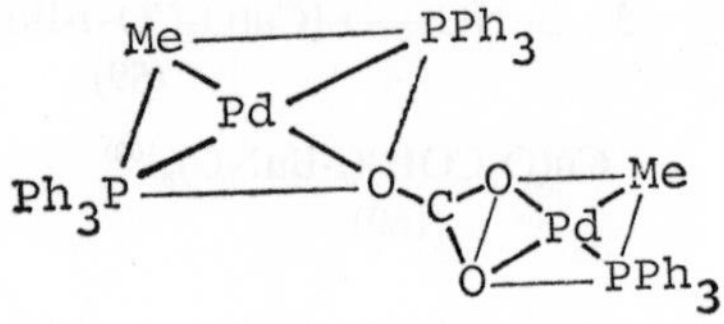

(54)

in **51** and **54** provides further evidence for this structure. On the other hand, complex **57** has been shown to have a symmetrical η^2-coordination of the bridging carbonato ligand in solution which is in equilibrium with the asymmetrical η^3-type coordination, as shown in Eq. (3.75).

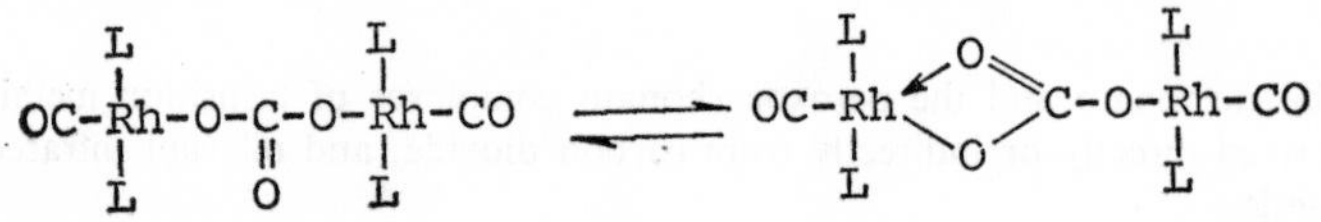

$$(3.75)$$

(**57**, L = P-i-Pr$_3$)

The mechanism for the rearrangement of the carbamido or hydrogen carbonato complex to the bridging carbonato complex has not been clarified.

The mononuclear carbonato complexes of palladium and platinum have been formed by reactions between CO_2 and zero-valent metal complexes in the presence of oxygen (Eqs. 3.76 and 3.77).[103]

$$[Pd(PPh_3)_4] + O_2 + CO_2 \longrightarrow [Pd(CO_3)(PPh_3)_2] + O{=}PPh_3 \qquad (3.76)$$
$$\mathbf{62}$$

$$[Pt(PPh_3)_3] + O_2 + CO_2 \longrightarrow [Pt(CO_3)(PPh_3)_2] + O{=}PPh_3 \qquad (3.77)$$
$$\mathbf{63}$$

Neither the spectroscopic evidence (Table 3.4) nor the chemical finding that the acid hydrolysis of **63** evolved carbon dioxide provided a decisive basis for assigning **62** and **63** to the carbonato complexes. However, the observation that the reaction of Ag$_2$CO$_3$ with cis-[PtCl$_2$(PPh$_3$)$_2$] yielded the same compound as **63** strongly supported the validity of this choice.[103] Wilkinson et al. suggested that the mechanism might involve the initial formation of PtO$_2$ (PPh$_3$)$_2$, followed by the insertion of CO_2 into a Pt–O bond, giving the peroxocarbonato intermediate **64**.[104]

$$\text{Ph}_3\text{P} \diagdown \underset{\text{Pt}}{\diagup} \overset{\text{O–O}}{\diagdown} \underset{\diagdown}{\diagup} \text{C=O}$$
$$\text{Ph}_3\text{P} \diagup \qquad \diagdown \underset{\text{O}}{\diagup}$$

(64)

Peroxocarbonatoplatinum complexes with alkylphosphines have been successfully isolated by the reaction of dioxygen complexes with carbon dioxide (Eq. 3.78).[105]

$$trans\text{-}[\text{PtCl}_2(\text{PR}_3)_2] \xrightarrow[-2\,\text{NaCl}]{2\,\text{Na–naphthalene}} [\text{Pt}(\text{PR}_3)_2] \xrightarrow{\text{O}_2} [\text{PtO}_2(\text{PR}_3)_2]$$

$$\xrightarrow{\text{CO}_2} [\text{Pt}(\text{CO}_4)(\text{PR}_3)_2] \qquad\qquad (3.78)$$
$$\textbf{(65)}$$
$$(\text{PR}_3 = \text{P}(cyclo\text{-C}_6\text{H}_{11})_3, \text{ P-}t\text{-Bu}_2\text{-}n\text{-Bu}, \text{ P-}t\text{-Bu}_2\text{Me}, \text{ P-}i\text{-Pr}_3)$$

Alternatively, peroxocarbonato complex can also be obtained by treating CO_2-coordinated complex **16** with dioxygen (Eq. 3.79)[77]

$$[\text{Ni}(\text{CO}_2)\{\text{P}(cyclo\text{-C}_6\text{H}_{11})_3\}_2] + \text{O}_2 \longrightarrow [\text{Ni}(\text{CO}_4)\{\text{P}(cyclo\text{-C}_6\text{H}_{11})_3\}_2] \quad (3.79)$$
$$\textbf{(66)}$$

3.3.3. Reversible Insertion of Carbon Dioxide into Transition Metal Complexes

Although the study of interactions between CO_2 and transition metal began relatively recently, there were many reports during the 1970s of the insertion of CO_2 into transition metal-E bonds, where E is hydrogen, carbon, nitrogen, and oxygen. Most of these studies are concerned with irreversible insertion reactions, and they provide important information on catalytic and non-catalytic synthetic organic processes involving carbon dioxide. Studies of the reversible insertion of CO_2 into M–E bonds are limited, but special attention has been paid to these reactions in recent years in an effort to mimic the biotin-dependent carboxylations in biological systems. For this reason, the reversible insertion of CO_2 into transition metal complexes is treated in this section separately from the general insertion reactions, most of which are irreversible and which will be described in the next section.

The enzymes named biotin carboxylases contain biotin (or Vitamin H, **67**) as a cofactor and are known to catalyze the carboxylation of substrates such as α-keto acids or acyl derivatives of coenzyme A (Eq. 3.80 and 3.81).[53,106]

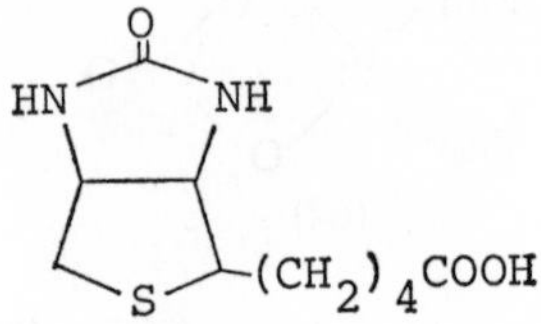

$$\text{(67)}$$

structure (67): biotin ring with HN-C(=O)-NH, fused S-containing ring bearing $-(CH_2)_4COOH$

$$\text{E-biotin} + \text{ATP} + \text{HCO}_3^- \xrightarrow{M^{2+}} \text{E-biotin} \sim \text{CO}_2 + \text{ADP} + \text{PO}_4^{3-} \quad (3.80)$$

$$\text{E-biotin} \sim \text{CO}_2 + \text{substrate} \rightleftharpoons \text{E-biotin} + \text{substrate} \sim \text{CO}_2 \quad (3.81)$$

In these reactions biotin-containing enzymes act as carbon dioxide carriers, and the metal (M^{2+}) ions such as manganese and/or magnesium are known to play an important role in combining and activating carbon dioxide on the enzyme.

The following examples of reversible CO_2 insertion into transition metal complexes have been reported. Ruthenium hydride complexes (68-70) absorb carbon dioxide in benzene solution at room temperature to give hydrido (formate) complex (71). As shown in Eq. (3.82), the reactions are reversible and each starting complex is recovered on treatment of 71 with triphenylphosphine, nitrogen, or hydrogen, respectively.[107,108]

$$[RuH_2(PPh_3)_4] + CO_2 \underset{+PPh_3}{\overset{-PPh_3}{\rightleftarrows}}$$
$$\text{(68)}$$

$$[RuH_2(N_2)(PPh_3)_3] + CO_2 \underset{+N_2}{\overset{-N_2}{\rightleftarrows}} \quad [RuH(OCOH)(PPh_3)_3] \quad (3.82)$$
$$\text{(69)} \qquad\qquad\qquad\qquad\qquad \text{(71)}$$

$$[RuH_4(PPh_3)_3] + CO_2 \underset{+H_2}{\overset{-H_2}{\rightleftarrows}}$$
$$\text{(70)}$$

These results suggest that complex 71, whose molecular structure has been shown by X-ray crystallography study to possess a formato ligand (Fig. 3.6),[108] is in equilibrium in the solution (Eq. 3.83).

$$[(PPh_3)_3Ru\underset{O}{\overset{H \quad O}{\diagdown}}CH] \rightleftharpoons (PPh_3)_3Ru(CO_2)H_2 \rightleftharpoons RuH_2(PPh_3)_3 + CO_2$$
$$\text{(71)} \qquad\qquad\qquad\qquad\qquad\qquad\qquad\qquad (3.83)$$

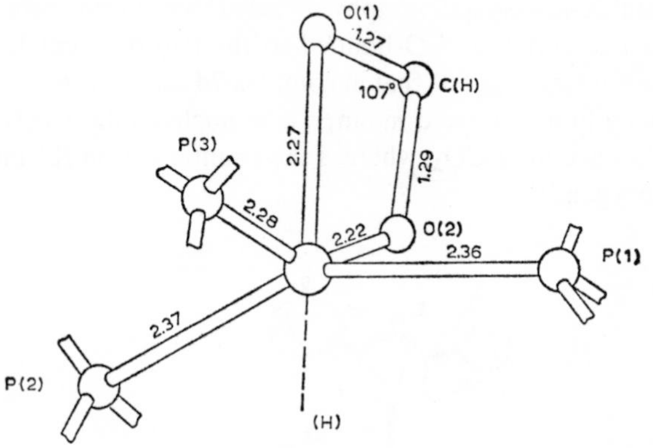

Fig. 3.6. The molecular structure of [RuH(OCOH)(PPh$_3$)$_3$]. (Source: ref. 108. Reproduced by kind permission of Elsevier Scientific Publishing Co. (The Netherlands)).

A similar reversible CO_2 insertion in the metal hydride has been reported for a dihydridoplatinum complex, as shown in Eq. (3.84).[109]

$$\textit{trans-}[PtH_2\{P(\textit{cyclo-}C_6H_{11})_3\}_2] + CO_2 \xrightleftharpoons[N_2]{C_6H_6,\ 80°C} \textit{trans-}[PtH(O_2CH)-$$

$$\{P(\textit{cyclo-}C_6H_{11})_3\}_2] \qquad (3.84)$$

The product formate complex has been structurally established by means of X-ray crystallography.[109]

The alkali metal salts of N,N'-ethylenebis(salicylidene iminato) cobaltate anion [NaCo (salen)] (72) were also found to reversibly absorb CO_2 to give the alkali salts of cobalt carboxylic acid, probably as a result of the insertion of CO_2 into the ionic bond between cobalt and the alkali metal (Eq. 3.85).[110]

$$[NaCo(salen)] \xrightleftharpoons[-CO_2]{+CO_2} [Co(salen)CO_2Na] \qquad (3.85)$$
$$\quad (72) \qquad\qquad\qquad (73)$$

The pr-salen derivative of 73, [Co(pr-salen)CO$_2$K(THF)] (74), has been similarly obtained beginning with 75.

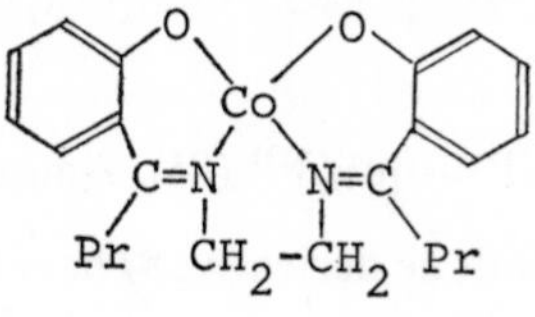

[Co(pr-salen)] (75)

Single crystal X-ray crystallography revealed that, in this compound, CO_2 is C-bonded to cobalt and O-bonded to the two different K^+ ions (Fig. 3.7).[111] This finding suggests that in complex 74 carbon dioxide is activated bifunctionally in a concerted manner. The nucleophilic Co(I) attacks the electrophilic carbon of CO_2, whereas the counter cation K^+ interacts with the basic oxygen.[111]

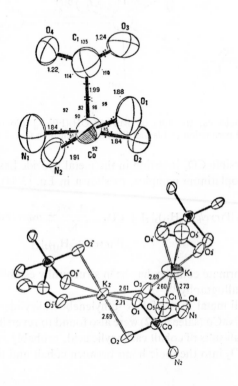

Fig. 3.7.   The molecular geometry of [Co(pr-salen)CO_2K(THF)]. (Source: ref. 111. Reproduced by kind permission of the American Chemical Society.)

Several copper(I) complexes which reversibly combine with carbon dioxide have been reported: [Cu(O_2CO–t-Bu) (CN–t-Bu)$_3$][112]   (76); [Cu(O_2CC≡CPh)(P–n-Bu$_3$)$_n$][113]   (77);   [Cu(O_2CCH$_2$CN)(P–n-Bu$_3$)][114] (78); [Cu(O_2COH)(PPh$_3$)$_n$][102]   (79); [Cu(O_2COH)(PEt$_3$)$_n$][102]   (80); and

$[(n\text{-Bu}_3P)_n\text{CuO}_2\text{CN} \overset{\overset{\displaystyle O}{\parallel}}{\underset{\underline{\qquad}}{}} \text{NCO}_2\text{Cu(P–}n\text{-Bu}_3)_n]^{53)}$ (81). Among these, cyano-acetato complex 78 and 2-imidozolidone derivative 81 have been successfully utilized in introducing a carboxylic group into organic molecules with an active methylene group.

$$[Cu(CH_2CN)(P\text{-}n\text{-}Bu_3)_n] + CO_2 \rightleftharpoons [Cu(O_2CCH_2CN)(P\text{-}n\text{-}Bu_3)_n]^{114)} \quad (3.86)$$
$$(78)$$

$$[LCuN\overset{\overset{\displaystyle O}{\|}}{\diagup}\diagdown NCuL] + CO_2 \rightleftharpoons [LCuO_2CN\overset{\overset{\displaystyle O}{\|}}{\diagup}\diagdown NCO_2CuL]^{53)} \quad (3.87)$$
$$(82 \qquad (L = (P\text{-}n\text{-}Bu_3)_n) \qquad (81)$$

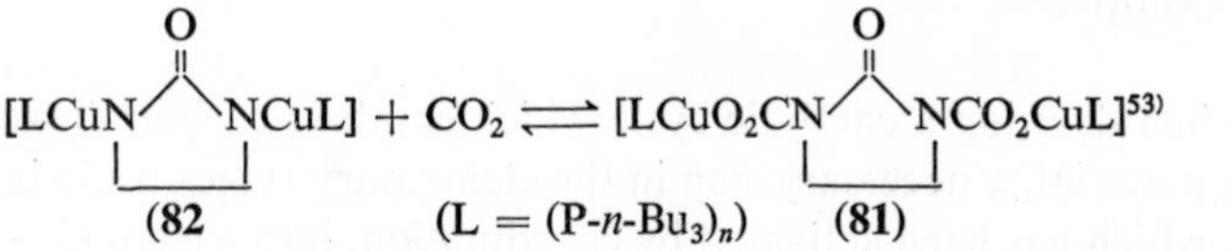

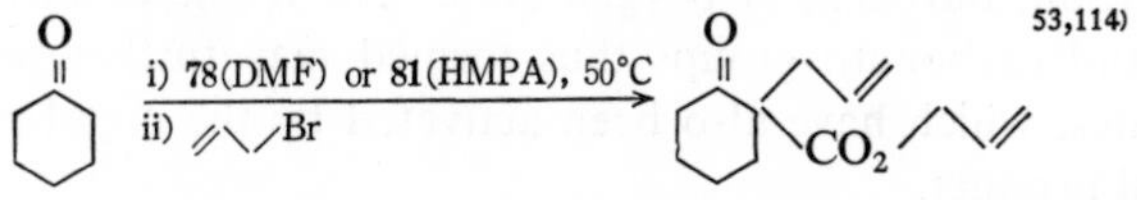

$$(3.88)$$

The reaction path for the transcarboxylation reaction (3.88) has been suggested to be that shown in Eq. (3.89).[114]

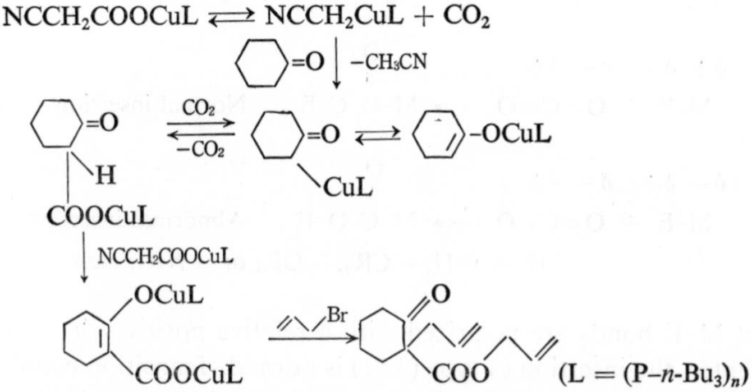

$$(3.89)$$

Related to the transcarboxylation reaction with complex **78** is the carbonation of propylene oxide with copper cyanoacetate to give propylene carbonate (Eq. 3.90). In this reaction, copper cyanoacetate is regarded as an activated carbon dioxide carrier.[115]

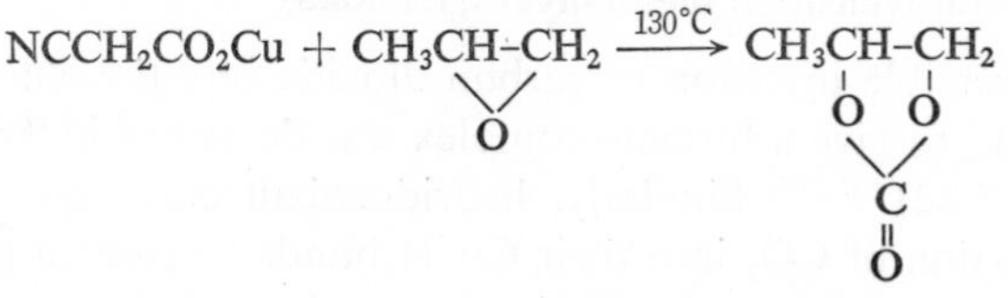

$$(3.90)$$

In addition to the reversible insertion reactions described here, there are several examples of the reversible coordination of CO_2 to transition metal complexes such as complexes **4, 9, 12,** and **20** in Table 3.3 or those in Eq. (3.52) in Section 3.3.1.

3.3.4. Irreversible Insertion of Carbon Dioxide into Transition Metal Complexes

When devising a catalytic cycle which employs carbon dioxide as a starting material, a necessary step in the elementary stages is the insertion of CO_2, which has been activated by coordination, into a transition metal-hydrogen, carbon, nitrogen, or oxygen bond. The formato, carboxylato, carbamido, and carbonato groups thus formed may further react with other substrates, which have also been activated by the metal center, to yield the final product.

An insertion reaction can be regarded as a nucleophilic attack of a transition metal complex on the electron-deficient carbon atom of carbon dioxide. This implies that the insertion of carbon dioxide may follow two directions, depending on the polarity of the complex.

$$\overset{\delta+}{M}\!-\!\overset{\delta-}{E} + \overset{\delta-}{O}\!=\!C\!=\!\overset{\delta+}{O} \longrightarrow M\!-\!O\!-\!\overset{\overset{\textstyle O}{\|}}{C}\!-\!E \qquad \text{Normal insertion} \qquad (3.91)$$

$$\overset{\delta-}{M}\!-\!\overset{\delta+}{E} + \overset{\delta-}{O}\!=\!C\!=\!\overset{\delta+}{O} \longrightarrow M\!-\!\overset{\overset{\textstyle O}{\|}}{C}\!-\!O\!-\!E \qquad \text{Abnormal insertion} \quad (3.92)$$

$$(E = -H, -CR_3, -OR, \text{ or } -NR_2, \text{ etc.})$$

As most M–E bonds are polarized with a relative positive charge at the metal center, the insertion of type (3.91) is normal. Transition metal complexes with highly π-accepting ligands, and transition metal complexes with a positively polarized hydride ligand, are possible candidates for abnormal insertion, as in Eq. (3.92). Reported examples of abnormal insertions are so far very limited despite their apparent importance in catalytic systems.

A. Insertion into transition metal-hydrogen bonds

The reversible insertion of carbon dioxide into the Ru–H bond of $[RuH_2(PPh_3)_4]$ to give a formate complex was described in the preceding section (Eq. 3.82).[107,108] Similarly, hydridocobalt complexes **83** and **84** undergo insertion of CO_2 into their Co–H bonds to give formatobis (triphenylphosphine) cobalt (**85**). The insertion of CO_2 is irreversible in this case (Eq. 3.93).[116-118]

$$[CoH(N_2)(PPh_3)_3] + CO_2 \xrightarrow{-N_2}$$
$$(83)$$

$$\xrightarrow{PPh_3} [Co(OOCH)(PPh_3)_3] \qquad (3.93)$$
$$(85) \quad \xrightarrow{\text{MeI}} HCOOMe$$

$$[CoH_3(PPh_3)_3] + CO_2 \xrightarrow{-H_2}$$
$$(84)$$

In these reactions, it is necessary to add triphenylphosphine to the system in order to prevent the dismutation of complex **85** to Co(II) diformate $Co(OOCH)_2(PPh_3)_x$ and carbonyl complex $[Co(CO)(PPh_3)_3]_2$.[116] The formato complex **(85)**, which showed strong infrared bands at 1620 and 1300 cm^{-1}, was found to be spectroscopically identical with the complex derived from **83** by reaction with formic acid. On treatment with methyl iodide and hydrogen chloride, **85** afforded methyl formate and formic acid, respectively. In addition to the normal insertion in Eq. (3.93), there has been a brief report of an abnormal insertion of carbon dioxide into complex **83**; on treatment of the reaction system with methyl iodide, acetic acid and its methyl ester were detected (Eq. 3.94).[57]

$$[CoH(N_2)(PPh_3)_3] + CO_2 \longrightarrow (PPh_3)_nCo-COOH \xrightarrow{\text{MeI}}$$
$$(83) \qquad\qquad\qquad (86)$$

$$MeCOOH + MeCOOMe \qquad\qquad (3.94)$$

Hydridoiron complexes **87** and **88** behaved similarly to the cobalt complexes **83** and **84** and yield bis(formato) complex **89** on reaction of CO_2 under irradiation of light[119]. The isolated bis(formato) complex **(89)** had infrared bands characteristic of formato groups at 1590 (ν_{as} HCOO) and 1370 cm^{-1} (ν_s HCOO) and yielded ethyl formate on reaction with ethyl bromide (Eq. 3.95).[119]

$$[FeH_2(N_2)(PEtPh_2)_3] + CO_2$$
$$(87)$$

$$\xrightarrow{h\nu} [Fe(OOCH)_2(PEtPh_2)_2] \qquad (3.95)$$
$$(89)$$

$$[FeH_4(PEtPh_2)_3] + CO_2$$
$$(88)$$

$$\xrightarrow{\text{EtBr}} FeBr_2 + 2 HCOOEt + PEt_2Ph_2Br$$

Tetrakis(trimethylphosphine)iron(0) is reversibly converted into the internally metallated Fe (II) species in a polar solvent (Eq. 3. 96).[70]

$$[Fe(PMe_3)_4] \underset{\text{non-polar solvent}}{\overset{\text{polar solvent}}{\rightleftharpoons}} [FeH(PMe_2\overline{CH_2})(PMe_3)_3] \qquad (3.96)$$
$$(90a) \qquad\qquad\qquad (90b)$$

When complex **90a** was treated with CO_2 in pentane at $0° C$, a CO_2-coordinated complex, [Fe (PMe$_3$)$_4$ (CO$_2$)] (2), was isolated (Eq. 3. 51). In contrast, when the reaction with CO$_2$ was carried out in a polar solvent such as THF, it proceeded through the stepwise insertion of carbon dioxide first into the Fe–C bond, then into the Fe–H bond of 90b (Eq. 3.97).[70]

(3.97)

(90b)	(91)	(92)
L = PMe$_3$	1610, 1592, and 1328 cm^{-1}	1600, 1345, and 1312 cm^{-1}

The formation of a formato rhenium complex (94) as a result of the interaction of hydridorhenium complex 93 with CO$_2$ has been also reported (Eq. 3. 98).[120]

$$[\text{ReH(CO)}_2(\text{PPh}_3)_3] + \text{CO}_2 \longrightarrow [\text{Re(OOCH)(CO)}_2(\text{PPh}_3)_2] + \text{PPh}_3 \quad (3.98)$$
$$\text{(93)} \qquad\qquad\qquad\qquad\qquad \text{(94)}$$

All the reactions between transition metal hydride complexes and carbon dioxide described above resulted in the formation of metal formates, except for one which gave a metallo-carboxylic acid (Eq. 3.94). Although the resulting metal formates can yield alkyl formates on treatment with alkyl halide, no process utilizing CO$_2$ insertion into the M–H bond has been successfully devised.

The reduction of CO$_2$ to formaldehyde and even to methoxide with zirconium hydride complex 95 was achieved in the past few years.[121] Although the reactions (Eqs. 3.99 and 3.100) are stoichiometric, these findings are important in terms of the effective utilization of carbon dioxide as an industrial material because the system relates to a reduction of carbon monoxide known as the Fischer-Tropsch process.

$$2\,[\text{ZrH}(\eta\text{-C}_5\text{H}_5)_2\text{Cl}] + \text{CO}_2 \xrightarrow{\text{THF}} [\{\text{Zr}(\eta\text{-C}_5\text{H}_5)_2\text{Cl}\}_2\text{O}] + \text{HCHO} \quad (3.99)$$
$$\text{(95)} \qquad\qquad\qquad\qquad\qquad \text{(96)}$$

$$3\,[\text{ZrH}(\eta\text{-C}_5\text{H}_5)_2\text{Cl}] + \text{CO}_2 \xrightarrow{\text{THF}} [\{\text{Zr}(\eta\text{-C}_5\text{H}_5)_2\text{Cl}\}_2\text{O}] + [\text{Zr(OMe)}-$$
$$\text{(95)} \qquad\qquad\qquad\qquad\qquad \text{(96)}$$
$$(\eta\text{-C}_5\text{H}_5)_2\text{Cl}] \qquad\qquad\qquad\qquad (3.100)$$
$$\text{(97)}$$

The formation of formaldehyde is assumed to proceed *via* a formato intermediate formed as a result of the insertion of carbon dioxide into the Zr–H bond. Formatozirconium then reacts with **95** to give formaldehyde and **96** (Eq. 3.101).

$$\text{Zr--H} + CO_2 \longrightarrow \text{ZrOCOH} \xrightarrow{\ 95\ } \text{ZrOZr} + \text{HCHO} \qquad (3.101)$$
$$\textbf{(95)} \hspace{5.5cm} \textbf{(96)}$$

The formation of zirconium methoxide **(97)** was ascribed to the insertion of the carbonyl group of formaldehyde, a product of Eq. (3. 101), into the Zr–H bond of **95** just as CO is inserted into the Zr–H bond to give formylzirconium species, as shown in Eq. (3. 102).[121] This was confirmed by a separate experiment between **95** and CH_2O which also yielded **97**.

$$[\text{ZrH}(\eta\text{-}C_5H_5)_2Cl] + CO \xrightarrow{\text{THF}} \text{Zr--CO--H} \xrightarrow{\ 95\ } [\{\text{Zr}(\eta\text{-}C_5H_5)_2Cl\}_2CH_2O]$$
$$\textbf{(95)} \hspace{4.5cm} \textbf{(98)} \hspace{3.5cm} (3.102)$$

The unusual behavior of **95** toward carbon dioxide, or carbonyl compounds in general, may be ascribed partly to the highly polarized nature of the Zr–H bond and partly to the high stability of the Zr–O bond. The inertness of the Zr–O bond in **96** may hinder the use of this type of reaction for catalytic applications.

B. Insertion into transition metal-carbon bonds

At present there are many examples of the insertion of CO_2 into a transition metal-carbon bond, although less than ten years has been passed since the first finding of such a reaction. A fairly large number of carboxylato type complexes produced by reactions between organotransition metal complexes and CO_2 have been isolated and characterized chemically and spectroscopically. The molecular structures of some of them have been established by means of the X-ray crystallography. The mode of insertion is normal (Eq. 3. 91) in all cases, as with insertion into metal hydrides, except for one case in which abnormal insertion is claimed to have been observed.[57,122] The reaction of ethylcobalt complex $[\text{CoEt}(CO)(PPh_3)_2]$ **(99)** with carbon dioxide was reported to yield, in addition to propionato complex $[\text{Co}(OCOEt)_2(PPh_3)_2]$ **(100)**, a complex with an ethoxycarbonyl ligand $[\text{Co}(CO_3)_n(COOEt)(PPh_3)]$ **(101)**, which afforded ethyl acetate on treatment with methyl iodide.[122]

The normal insertion of CO_2 into a transition metal-carbon bond was reported for the first time for benzyl complexes $M(CH_2Ph)_4$ (M = Ti, **102**, and Zr, **103**). Although the CO_2-inserted complexes were not isolated, additional evidence for the insertion of carbon dioxide was the detection of equimolar amounts of phenylacetic acid and tribenzylcarbinol after hydrolysis of the system (Eq. 3.103).[123]

$$[M(CH_2Ph)_4] + CO_2 \longrightarrow (PhCH_2)_3MOCOCH_2Ph \xrightarrow{H_2O} PhCH_2COOH$$
$$\text{(104)}$$
$$+ (PhCH_2)_3COH \qquad (3.103)$$
$$(M = Ti, \textbf{102}; Zr, \textbf{103})$$

Formation of the tertiary alcohol may be due to the high reactivity of M–CH$_2$Ph species toward carbonyl compounds. The reaction may proceed in a manner similar to that shown in Eq. (3.5).

Carboxylato complex **107** was obtained by the reaction of diphenyl-titanocene (**105**) with CO$_2$ in xylene at 80°C (Eq. 3.104).[124,125]

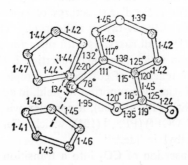

$$[Ni(\eta\text{-}C_5H_5)_2Ph_2] \xrightarrow[-C_6H_6]{80°C} (\eta\text{-}C_5H_5)_2Ti \xrightarrow{CO_2} (\eta\text{-}C_5H_5)_2Ti$$

(105)   (106)   (107)   O–C=O

(3.104)

Complex **107** showed infrared bands characteristic of carboxylato groups at 1660vs, 1620s, 1280vs, 1130s, and 880m cm^{-1}, and yielded benzoic acid and methyl benzoate on treatment with dilute H$_2$SO$_4$ and BF$_3$/MeOH, respectively, at an elevated temperature. A reaction path involving the insertion of CO$_2$ into the Ti–C bond of titaniumdehydrobenzene intermediate **106**, produced by the thermal decomposition of diphenyltitanocene, has been proposed.[124] The molecular structure of **107** has been determined, as shown in Fig. 3.8.

Fig. 3.8.   The molecular geometry of the complex, [Ti(η-C$_5$H$_5$)$_2$(OCOC$_6$H$_4$)] (**107**). (Source: ref. 124. Reproduced by kind permission of the Chemical Society (London).)

Phthalate complex **108** was also obtained by the additional insertion of CO_2 into the Ti–C bond of **107** (Eq. 3.105).[124]

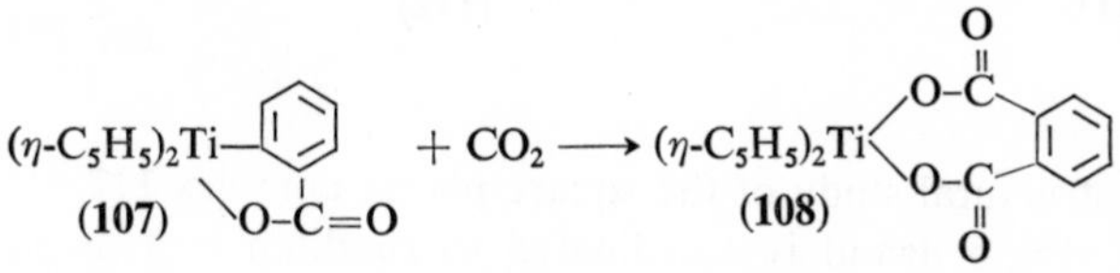

$$(\eta\text{-}C_5H_5)_2Ti\underset{\underset{(107)}{\diagdown}O\text{-}C=O}{\overset{\diagup}{\diagup}} + CO_2 \longrightarrow (\eta\text{-}C_5H_5)_2Ti\underset{O\text{-}C}{\overset{O\text{-}C}{\diagdown}} \quad (3.105)$$
$$(108)$$

Analogous reactions using bis (*m*- and *p*-tolyl) titanocene yielded, after treatment with BF_3/MeOH, methyl esters of *m*- and *p*-benzoic acid, indicating that the intermediary dehydrotoluene complex of titanocene is of type **109** and not type **110** for both isomers (Eq. 3.106).[126]

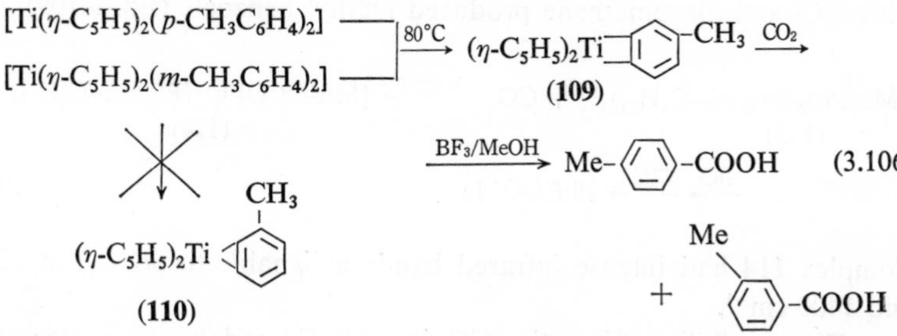

$$[Ti(\eta\text{-}C_5H_5)_2(p\text{-}CH_3C_6H_4)_2]$$
$$[Ti(\eta\text{-}C_5H_5)_2(m\text{-}CH_3C_6H_4)_2]$$
$$\xrightarrow{80°C} (\eta\text{-}C_5H_5)_2Ti\underset{(109)}{\boxed{}}\text{-}CH_3 \xrightarrow{CO_2}$$

$$(\eta\text{-}C_5H_5)_2Ti\underset{(110)}{\overset{CH_3}{\boxed{}}}$$

$$\xrightarrow{BF_3/MeOH} Me\text{-}\langle\text{-}\rangle\text{-}COOH \quad (3.106)$$

$$+ \quad \overset{Me}{\underset{}{\langle\text{-}\rangle}}\text{-}COOH$$

In contrast to ethylcobalt complex **99**, into which CO_2 was claimed to have been abnormally inserted,[122] the apparently normal insertion of CO_2 into the cobalt-methyl bond has been observed for methylcobaloxime. When the reaction of methylcobaloxime with CO_2 was carried out in the presence of dithiol, a small amount of acetic acid, in addition to methane, was detected in the reaction product (Eq. 3.107).[127]

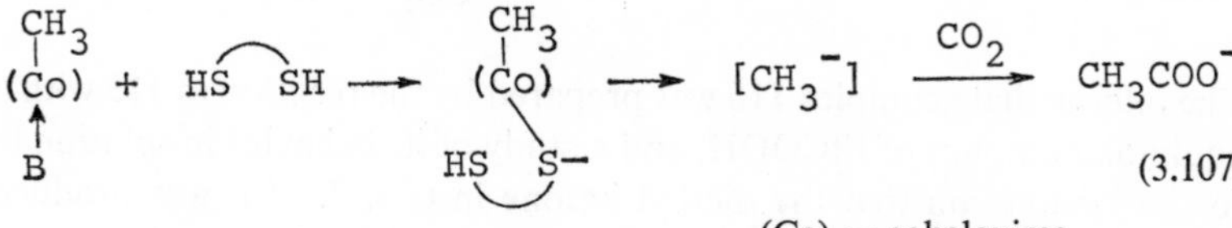

$$\underset{B}{\overset{CH_3}{\underset{\uparrow}{(Co)}}} + HS\frown SH \longrightarrow \underset{HS\diagdown S-}{\overset{CH_3}{(Co)}} \longrightarrow [CH_3^-] \xrightarrow{CO_2} CH_3COO^-$$
$$(3.107)$$

(Co) = cobaloxime,
B = pyridine

Benzoatorhodium complex **112** was derived from phenylrhodium complex **111** by the insertion of CO_2 under pressure into its Rh-carbon bond (Eq. 3.108).[128]

$$[Rh(Ph)(PPh_3)_3] + CO_2 \ (20 \ atm) \longrightarrow [Rh(OCOPh)(PPh_3)_3] \begin{array}{l} \xrightarrow{H^+} PhCOOH \\[2mm] \xrightarrow[BF_3]{MeOH} PhCOOMe \end{array}$$
$$\text{(111)} \qquad\qquad\qquad\quad \text{(112)}$$

$$(3.108)$$

Molecular structural study of the square planar complex **112** has shown that the benzoato ligand is coordinated to rhodium in a monodentate fashion. The complex has infrared bands due to the carboxylato group at 1608s, 1570s, and 1360vs cm^{-1}, and afforded benzoic acid and methyl benzoate on treatment with acid and MeOH/BF$_3$, respectively.[128] Similarly, CO$_2$ was found to be inserted into the phenyl-manganese σ-bond of [Mn(Ph)$_2$ {P (cyclo-C$_6$H$_{11}$)$_3$}] **(113)** under ambient conditions to give bis(benzoato)manganese complex **(114)**, which on successive treatment with HCl and diazomethane produced methyl benzoate (Eq. 3.109).[129]

$$[Mn(Ph)_2\{P(cyclo\text{-}C_6H_{11})_3\}] + CO_2 \xrightarrow{toluene} [Mn(OCOPh)_2\{P(cyclo\text{-}C_6H_{11})\}_3]$$
$$\text{(113)} \qquad\qquad\qquad\qquad\qquad\qquad\qquad \text{(114)}$$

$$\xrightarrow{HCl} \xrightarrow{CH_2N_2} PhCOOMe \qquad\qquad\qquad (3.109)$$

Complex **114** had intense infrared bands assignable to $\nu(CO_2)$ at 1550 and 1400 cm^{-1}.

The possibility of inserting CO$_2$ into Ni–C bonds has been suggested by the detection of diethyl ketone and nickel propionate in the reaction system between diethylnickel complex **115** and CO$_2$ at 40–50° C (Eq. 3.110).[130]

$$[NiEt_2(bpy)] + CO_2 \longrightarrow (bpy)Ni \begin{array}{l} \diagup Et \\ \diagdown OCOEt \end{array} \begin{array}{l} \xrightarrow{CO_2} Ni(OCOEt)_2(bpy) \xrightarrow{H^+} EtCOOH \\ \diagdown Ni(CO_3)(bpy) + EtCOEt \end{array}$$
$$\text{115} \qquad\qquad\quad \text{116} \qquad\qquad\qquad \Big\downarrow H^+$$
$$\text{(bpy} = 2{,}2'\text{-bipyridine)} \qquad\qquad\qquad\qquad CO_2 \qquad (3.110)$$

The intermediate complex **116** was prepared by the reaction of **115** with an equimolar amount of EtCOOH, and a study of its behavior in solution led to the assumption that the diethyl ketone in (Eq. 3. 110) was produced through an intramolecular rearrangement of **116**. A similar reaction with [NiMe$_2$(bpy)] gave nickel acetate.[130]

Bis (η-allyl) nickel **(117)** reacted with CO$_2$ in dimethylformamide at 140° C to give γ-butyrolacetone. The reaction was considered to proceed via the normal insertion of CO$_2$ into the Ni–C bond of the σ-allyl intermediate, followed by intramolecular cyclization (Eq. 3.111).[131]

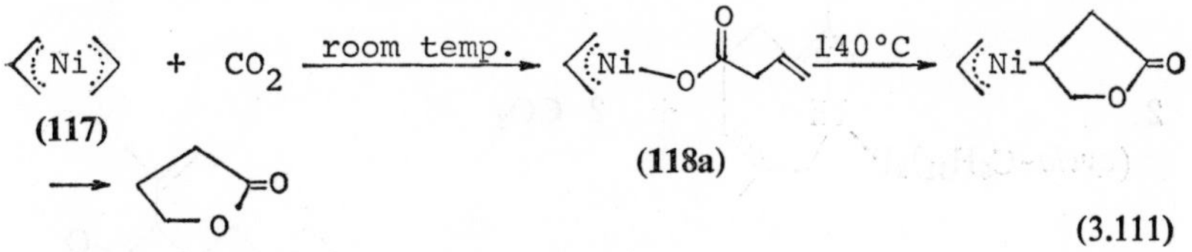

$$(3.111)$$

η-Methallyl(meth allylcarboxylato) complex **118b**, which corresponds to intermediate **118a** in (Eq. 3.111), was isolated when the reaction was carried out at a low temperature in the presence of tertiary phosphine (Eq. 3.112).[132]

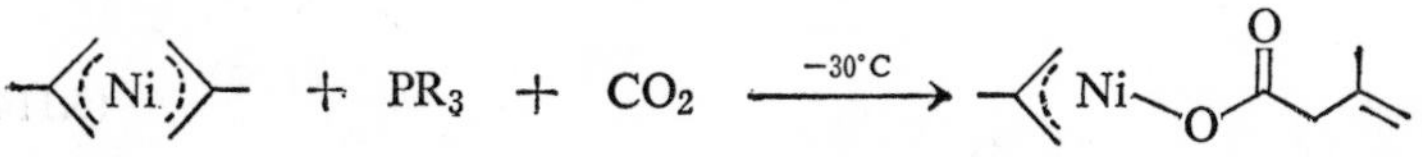

(R = CH₃, *cyclo*-C₆H₁₁)   ν(C=O) 1610 cm⁻¹   (3.112)

The allyl (carboxylato) type complexes of nickel, **119, 120,** and **121,** have been prepared in a similar way, and the molecular structures of **120** and **121** have been confirmed by X-ray structure analyses (Eqs. 3.113–3.115).[132]

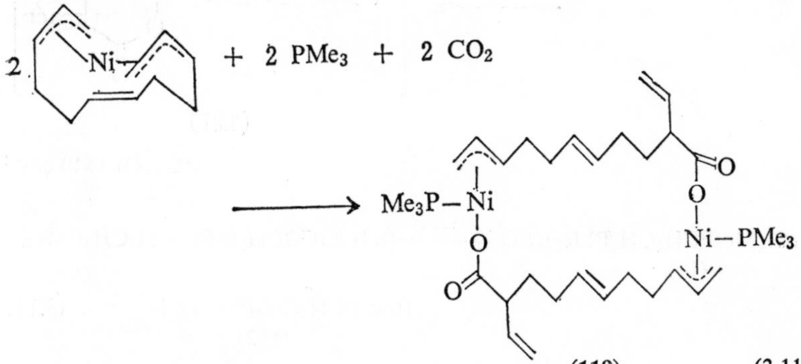

(119)   (3.113)

Carbon dioxide insertion into the iron-carbon bond of the metallated phosphine ligand has been reported for tetrakis (trimethylphosphine)-iron (0) (**90a**), as shown in Eqs. (3. 96) and (3. 97).[70] Similarly, when the CO₂-coordinated complex [IrCl(Me₂PCH₂CH₂PMe₂) (CO₂)] (**14**) was heated in solution at 120° C, metallation of the phosphine ligand took place accompanied by carboxylation of the Ir–C bond.[75]

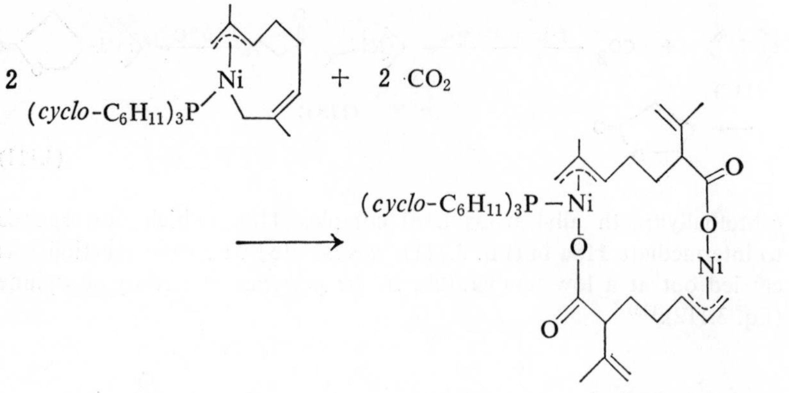

$$2 \quad (cyclo\text{-}C_6H_{11})_3P \diagup \overset{\diagup}{Ni} \diagdown + \ 2 \ CO_2 \longrightarrow \tag{3.114}$$

(120)

$\nu(C{=}O)$ 1615 cm^{-1}

$$4 \quad (cyclo\text{-}C_6H_{11})_3P \diagup \overset{\diagup}{Ni} \diagdown + \ 4 \ CO_2$$

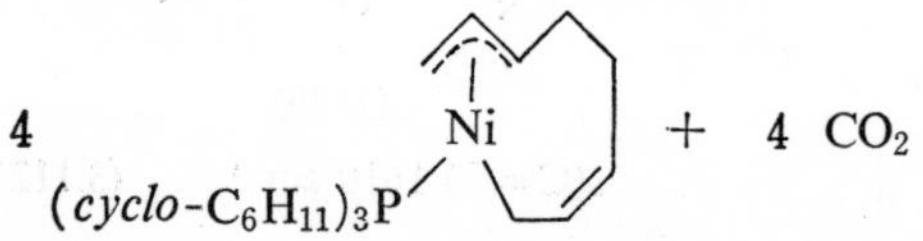

$$\longrightarrow \ 2 \ [(cyclo\text{-}C_6H_{11})_3PNiC_8H_{12}CO_2]_2 \tag{3.115}$$

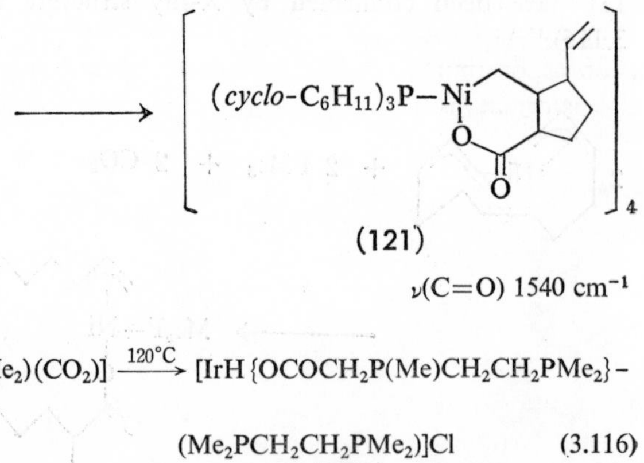

(121)

$\nu(C{=}O)$ 1540 cm^{-1}

$$[IrCl(Me_2PCH_2CH_2PMe_2)(CO_2)] \xrightarrow{120^{\circ}C} [IrH\{OCOCH_2P(Me)CH_2CH_2PMe_2\}-$$
(14)

$$(Me_2PCH_2CH_2PMe_2)]Cl \tag{3.116}$$
(122)

$\nu(Ir{-}H)$ 2180 cm^{-1}, $\nu(C{=}O)$ 1640 cm^{-1}

It is noteworthy that metallation does not occur in the absence of CO_2. The complex $[Ir(depe)_2]_2Cl$ (depe $= Et_2PCH_2CH_2PEt_2$) (123) was found to reversibly combine with CO_2 in acetonitrile solution. Formation of cyanoacetate complex 124 was indicated by the detection of methyl cyano-acetate after treatment of the system with HCl and $BF_3/MeOH$ (Eq. 3.117).[133)]

$$[Ir(depe)_2]Cl + CH_3CN + CO_2 \rightleftharpoons [Ir(H)(O_2CCH_2CN)(depe)_2]Cl$$
$$\textbf{(123)} \qquad\qquad\qquad\qquad \textbf{(124)}$$

$$\xrightarrow{HCl} \xrightarrow{BF_3/MeOH} NCCH_2COOMe \qquad (3.117)$$

The following are analogous reactions to (3.117).[133]

$$[Rh(dmpe)_2]Cl + CH_3NO_2 + CO_2 \rightleftharpoons [Rh(H)(O_2CCH_2NO_2)(dmpe)_2]Cl \quad (3.118)$$
$$\textbf{(125)} \qquad (dmpe = Me_2PCH_2CH_2PMe_2) \qquad \textbf{(126)}$$

$$[Ir(PMe_3)_4]Cl + CH_3CN + CO_2 \rightleftharpoons [Ir(H)(O_2CCH_2CN)(PMe_3)_4]Cl \quad (3.119)$$
$$\textbf{(127)} \qquad\qquad\qquad\qquad \textbf{(128)}$$

$$[FeH(naphthyl)(dmpe)_2] + CH_3CN \xrightarrow{-C_{10}H_8} [FeH(CH_2CN)(dmpe)_2] \xrightarrow{CO_2}$$
$$\textbf{(129)}$$

$$Fe(O_2CCH_2CN) \xrightarrow{BF_3/MeOH} NCCH_2COOMe^{[134]} \qquad (3.120)$$

In reaction (3.119), the failure to form **128** by the reaction of [Ir(H)(CH$_2$-CN) (PMe$_3$)$_4$] Cl with carbon dioxide suggests that the carboxylation did not take place by CO$_2$ insertion into the metal-carbon bond, at least in this case.[133] The attack of an adduct between CO$_2$ and CH$_3$CN on the metal complex, as happens with reactions of CO$_2$ in the presence of water, alcohols, or amines, should be considered here (see subsequent sections).

Normal insertions of CO$_2$ were observed for some alkylcopper complex with tertiary phosphine ligands **(130)** (Eq. 3.121).[135,136]

$$[CuR(PR'_3)_2] + CO_2 \xrightarrow[THF]{-20° \text{ to } 0°C} [Cu(OCOR) (PR'_3)_2] \qquad (3.121)$$
$$\textbf{(130)} \qquad\qquad\qquad \textbf{(131)}$$
$$(PR'_3 = PPh_3: R = Me, Et, n\text{-Pr}, i\text{-Bu};$$
$$PR'_3 = P(cyclo\text{-}C_6H_{11})_3, PPh_2Me: R = Me)$$

The product carboxylato complex **131** was identified by the synthesis of a similar complex from **130** and the corresponding carboxylic acid. Prolonged reaction of CO$_2$ with **129** at room temperature gave another type of product which was thought to be [Cu(OCOR)(CO$_2$)(PR$_3$)$_2$]. Later study revealed, however, that water had been involved in the reaction, and the product was formulated as a hydrogen carbonato complex, [(HOCOO)-Cu(PR$_3$)$_2$].

In addition to reversible CO$_2$ insertion into organocopper complexes such as [PhC≡CCu (P-n-Bu$_3$)$_3$][113] and [NCCH$_2$Cu (P-n-Bu$_3$)$_n$],[114] which can be utilized as carbon dioxide carriers (Section 3.3.3), carboxylic groups may also be introduced into organic molecules in the same manner in which carbon dioxide is inserted into copper- or silver-carbon bonds (Eqs. 3.122–3.125).

$$PhC \equiv CH \xrightarrow{t\text{-BuOM}} PhC \equiv CM \xrightarrow{P\text{-}n\text{-Bu}_3} \xrightarrow{CO_2} \xrightarrow{MeI} PhC \equiv CCOOMe^{137)} \quad (3.122)$$
$$(M = Cu \text{ and } Ag)$$

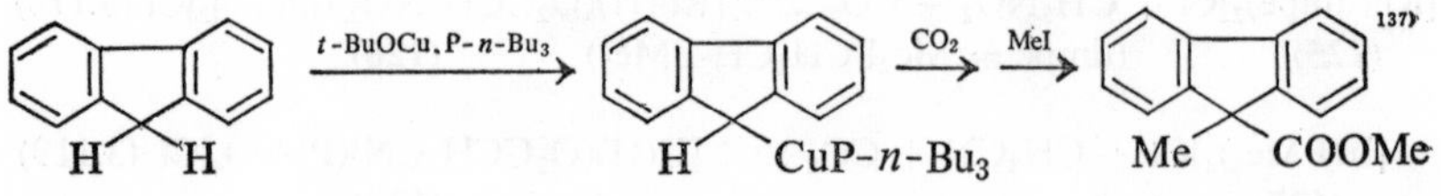

(3.123)

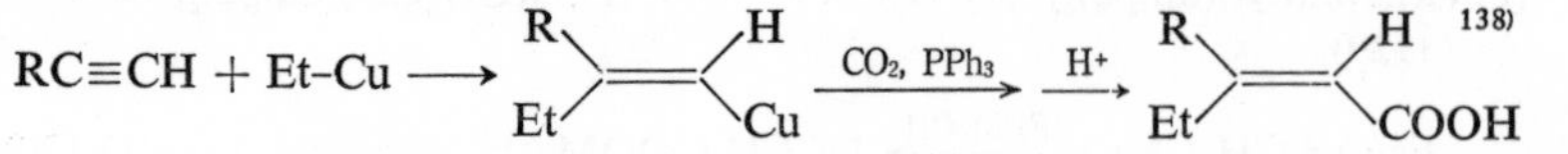

$$(R = n\text{-Bu, Me}) \tag{3.124}$$

$$(CF_3)_2CFAg \xrightarrow[]{CH_3CN, \ CO_2} \xrightarrow{PhCH_2Br} (CF_3)_2CFCOOCH_2Ph^{139)} \quad (3.125)$$

C. Insertion into Transition Metal-Nitrogen Bonds

Carbon dioxide insertion into a transition metal-nitrogen bond was first observed with titanium (IV) amide (132) (Eq. 3.126).[140)]

$$[Ti(NMe_2)_4] + CO_2 \longrightarrow [Ti(OCONMe_2)_4] \quad (3.126)$$
$$(132) \qquad\qquad (133)$$

The resulting titanium tetracarbamate (133) showed infrared bands at 1590 and 1690 cm^{-1}, assignable to $\nu_{sym}(CO_2)$ and $\nu_{asym}(CO_2)$, respectively; hence the monodentate mode of coordination of the carbamato group in 133 was postulated. A similar reaction of CO_2 with triisopropoxotitanium (IV) diethylamide (134) yielded monocarbamato complex 135, which showed strong infrared absorptions characteristic of bidentate carbamato ligands at 1560–1475 cm^{-1}, indicating the five coordination of the complex (Eq. 3.127).[141)]

$$[Ti(O\text{-}i\text{-Pr})_3(NEt_2)] + CO_2 \xrightarrow[0°C]{pentane} [Ti(O\text{-}i\text{-Pr})_3(OCONEt_2)] \quad (3.127)$$
$$(134) \qquad\qquad\qquad (135)$$

These types of CO_2 insertions into transition metal amides have been investigated extensively by Chisholm and his coworkers for a variety of early transition elements (Eq. 3.128).[142–147)]

$$[M(NMe_2)_m] + nCO_2 \longrightarrow [M(NMe_2)_{m-n}(OCONMe_2)_n] \quad (3.128)$$
$$\text{(136)} \qquad\qquad\qquad \text{(137)}$$

(M = Ti, Zr, V: $m = n = 4$; M = Nb, Ta: $m = n = 5$; M = Mo, W: $m = 6, n = 3$)

In the case of tungsten, a partially carbonated complex, [W(NMe_2)_3-(OCONMe_2)_3], was obtained even in the presence of excess CO_2, and its structure was confimred by X-ray crystallography.[142,145] A carbamate structure was also established for [Nb (OCONMe_2)_5].[145] A NMR study of the system containing 137 and $^{13}CO_2$ revealed that the CO_2 group of the carbamato ligand in 137 is readily exchanged with the added carbon dioxide gas, shown in Eq. (3.129).

$$ML^*_n + x^{12}CO_2 \rightleftharpoons ML_mL^*_{n-m} + m^{13}CO_2 + (x - m)^{12}CO_2 \quad (3.129)$$
$$(L = O_2{}^{12}CNMe_2; L^* = O_2{}^{13}CNMe_2)$$

The activation energy required for the exchange was estimated by a dynamic NMR measurement to be 13 ± 2 kcal mol^{-1} for 137 when M = Zr, $m=n=4$, and approximately 23 kcal mol^{-1} for 137 when M=W, $m=6$, $n=3$.[143] In addition to the CO_2 exchange reaction shown in Eq. (3.129), complexes of Ti and Zr underwent exchange reactions of the type seen in Eqs. (3.130) and (3.131).[145]

$$MO_2{}^{13}CNMe_2 + MO_2{}^{12}CNR_2 \rightleftharpoons MO_2{}^{12}CNMe_2 + MO_2{}^{13}CNR_2 \quad (3.130)$$

$$MO_2CNMe_2 + M'O_2CN(CD_3)_2 \rightleftharpoons MO_2CN(CD_3)_2 + M'O_2CNMe_2 \quad (3.131)$$

Investigation of these exchange reactions using isotopes, together with studies of isolated intermediate species such as [Ti(NMe_2)_n(OCO-NMe_2)_{4-n}] ($n= 1$ and 2), [V(NMe_2) (OCONMe_2)_3], and [Ta(NMe_2)_2-(OCONMe_2)_3] showed that exchange reactions (3.129–3.131) are catalyzed by the amine HNMe_2 present in the system in a trace amount.[146,147] Thus reactions (3.129) – (3. 131) may proceed *via* an attack by the carbamic acid formed by the interaction between CO_2 and an amine fortuitously introduced, as in Eqs. (3.132) and (3.133).

$$^{12}CO_2 + HNMe \rightleftharpoons HO_2{}^{12}CNMe_2 \quad (3.132)$$

$$M-O_2{}^{13}CNMe_2 + HO_2{}^{12}CNMe_2 \rightleftharpoons M-O_2{}^{12}CNMe_2 + HO_2{}^{13}CNMe_2 \quad (3.133)$$

Althouth the insertion of CO_2 into the metal amide (Eq. 3.128) was first thought to proceed in a manner similar to that of olefin insertion into metal-carbon bonds,[142] it is now apparent, in the case of the tungsten complex at least, that the insertion reaction itself occurs via an amine-catalyzed mechanism, as shown in Eqs. (3.134) and (3.135).[146]

$$CO_2 + HNR_2 \rightleftharpoons HO_2CNR_2 \qquad (3.134)$$

$$MNR_2 + HO_2CNR_2 \rightleftharpoons MO_2CNR_2 + HNR_2 \qquad (3.135)$$

A reaction between chromium (IV) amide and CO_2 afforded two different types of binuclear carbamato complexes (139 and 140). The resulting complex depended on the relative concentration of CO_2. The structures of complexes 139 and 140 have been established by X-ray analysis. In order to account for the reduction of Cr (IV) complex 138 to the Cr (III), it was proposed that the intermediary $Cr^{II}(NEt_2)(O_2CNEt_2)$ is formed, which on reaction with 138 gives Cr(III) species (139) and Cr (V) species (140). The formation of the Cr(II) species was explained by sequential β-hydrogen elimination and reductive elimination of Et_2NH (Eq. 3.136).[148]

$$[Cr(NEt_2)_4] + CO_2 \longrightarrow Cr(NEt_2)_3(O_2CNEt_2) \xrightarrow{-EtN=CHMe}$$

$$Cr(H)(NEt_2)_2(O_2CNEt_2) \xrightarrow{-NHEt_2} Cr^{II}(NEt_2)(O_2CNEt_2) \xrightarrow{138}$$

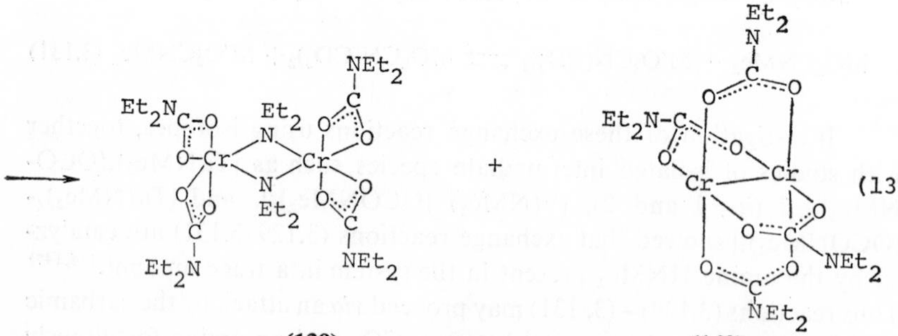

(139)   (140)   (3.136)

The binuclear complexes of tungsten[149,150] and molybdenum[151] with μ-carbamato ligands have been similarly obtained, and their molecular structures have been established.

The titanium carbamate formation reaction (3.126) has been utilized in combination with 1,2-epoxycyclohexane for the synthesis of carbamic ester (Eq. 3.137).[152]

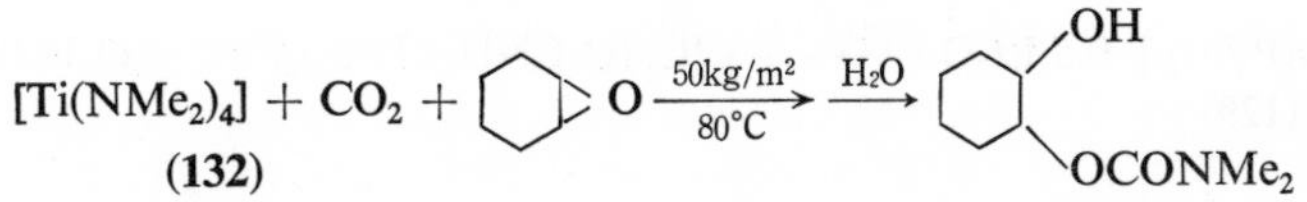

$$[Ti(NMe_2)_4] + CO_2 + \underset{(132)}{\text{[epoxide]}} \xrightarrow[80°C]{50kg/m^2} \xrightarrow{H_2O} \text{[product]} \qquad (3.137)$$

Since the epoxide did not react with **132** in the absence of CO_2, a mechanism involving the insertion of epoxide into the Ti–O bond of carbamate complex **133**, yielding $Ti(O\text{—}[\text{cyclohexyl}]\text{—}O\overset{O}{\overset{\|}{C}}NMe_2)_4$ **(141)**, has been proposed.[152]

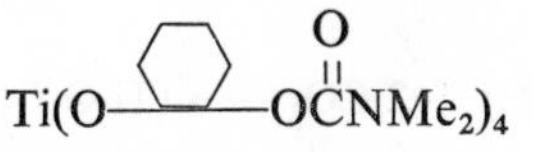

Dimethylbis (tertiary phosphine) palladium (II) **(142)** in THF reacted with amines in an atmosphere of CO_2 at room temperature to afford carbamatopalladium complexes **(143)** (Eq. 3.138).[100]

$$\underset{(142)}{[PdMe_2L_2]} + HNRR' + CO_2 \longrightarrow \underset{(143)}{[PdMe(O_2CNRR')L_2]} + CH_4 \quad (3.138)$$
$$(L = PEt_3, PMePh_2, PPh_3; R = H, Et; R' = H, Et, n\text{-Bu}, CH_2Ph, Ph)$$

In a manner somewhat analogous to the mechanism of the reaction between early transition metal amides and CO_2 (Eqs. 3.134 and 3.135), this type of reaction is thought to proceed *via* the attack on the dimethylpalladium complex by carbamic acid formed from amine and CO_2 because **142** did not react with amines in the absence of CO_2, whereas an acetato complex was formed on reaction with acetic acid. When diphenylamine was employed in reaction (3.138)(R=R'=Ph), only diphenylamido complex $[PdMe(NPh_2)L_2]$ **(144)** was isolated, presumably because the primary product, $[PdMe(O_2CNPh_2)_2L_2]$, is so unstable that it is decarboxylated easily to give **144**.[100]

The following reactions (Eqs. 3.139–3.143) may follow similar mechanisms to those shown in Eqs. (3.134) and (3.135).

$$\underset{(93)}{[ReH(CO)_2(PPh_3)_3]} + CO_2 + \underset{(R'=Et, n\text{-Pr})}{NR'_2H} \longrightarrow \underset{(145)}{[Re(O_2CNR'_2)(CO)_2(PPh_3)_2]}^{120)} \quad (3.139)$$

$$\underset{(146)}{[RuH(PMe_2Ph)_5]PF_6} + CO_2 + NMe_2H \longrightarrow \underset{(147)}{[Ru(O_2CNMe_2)(PMe_2Ph)_4]PF_6}^{153)} \quad (3.140)$$

$$\underset{(68)}{[RuH_2(PPh_3)_4]} + CO_2 + NMe_2H \longrightarrow \underset{(148)}{[RuH(O_2CNMe_2)(PPh_3)_3]}^{153)} \quad (3.141)$$

$$[CuO\text{-}t\text{-Bu}] + CO_2 + RR'NH \longrightarrow [CuO_2CNRR']$$
$$\xrightarrow{t\text{-BuNC}} [Cu(O_2CNRR')(CN\text{-}t\text{-Bu})_n] \xrightarrow{MeI} RR'NCO_2Me^{154)} \quad (3.142)$$
$$(R = R' = H, Et, n\text{-Bu}; R = H, R' = n\text{-Bu}, t\text{-Bu})$$

$$[MeCu(PPh_3)_2], 0.5 \ Et_2O \xrightarrow{NH_3} \xrightarrow{CO_2} [Cu(O_2CNH_2)(PPh_3)_2]^{79)} \quad (3.143)$$
(129)

The reaction of CO_2 with imidohydridobis (dicyclopentadienyltitanium) gave a carbamato type complex, which, on treatment with HCl, CH_3I, or O_2, evolved CO_2 gas (Eq. 3.144).[155]

$$[\{Ti(\eta\text{-}C_5H_5)_2NH\}_2H] + 2 \ CO_2 \longrightarrow (Ti\text{-}O_2CNH_2) \text{ type complex} \quad (3.144)$$

An alternative mechanism to Eqs. (3.134) and (3.135) has been proposed recently for the formation of uranium carbamate by the reaction of UCl_4 with CO_2 in the presence of amine. In this reaction, the fast preliminary formation of dialkylammonium carbamate has been postulated on the basis of the amount of CO_2 gas consumed per amine.[156]

$$2 \ R_2NH + CO_2 \longrightarrow NH_2R_2{}^+NR_2CO_2{}^-$$

$$4 \ NH_2R_2{}^+NR_2CO_2{}^- + UCl_4 \longrightarrow 4 \ NH_2R_2{}^+Cl^- + U(O_2CNR_2)_4$$

The formation of carbamato complexes by the reaction of transition metal complexes with CO_2 in the presence of amines provides important information about the elementary steps of catalytic formamide formation reactions in a homogeneous system, which will be described in the next section.

A reaction between CO_2 and a transition metal amide, accompanied by CO_2 deoxygenation, has been reported (Eq. 3.145).[157]

$$[CuN(SiMe_3)_2] + CO_2 + L \xrightarrow{0°C} [Cu(NCO)L_{2-3}] + (Me_3Si)_2O \quad (3.145)$$
(149)   (150)

(L = t-BuNC, P-n-Bu$_3$, and PEt$_3$)

The reaction of the resulting isocyanato complex (150) with alkyl halide afforded alkyl isocyanate, which, on further treatment with alcohol or amine, yielded carbamic ester and urea, respectively (Eq. 3.146).[157]

$$[Cu(NCO)L_{2-3}] + RI \longrightarrow RNCO \underset{RNH_2}{\overset{R'OH}{=\!=\!=\!=}} \begin{array}{l} \nearrow RNHCO_2R' \\ \searrow RNHCONHR \end{array} \quad (3.146)$$
(150)

D. Insertion into Transition Metal-Oxygen Bonds

The study of CO_2 insertion into transition metal-oxygen bonds is important for understanding the mechanism of CO_2 uptake by carbonic anhydrase, shown in Eq. (3.37).

Kinetic study of pH-dependent CO_2 uptake by hydroxocobalt(III) ion revealed that the reaction could be expressed by the following schemes, in which the intermediate formation of cobalt carbonate species occurs (Eqs. 3.147–3.150).[158,159]

$$Co(tetren)(OH_2)^{3+} \rightleftharpoons Co(tetren)(OH)^{2+} + H^+ \qquad (3.147)$$
(tetren = tetraethylenepentamine)

$$Co(tetren)(OH)^{2+} + CO_2 \underset{k_1}{\overset{k_2}{\rightleftharpoons}} Co(tetren)(CO_3H)^{2+} \qquad (3.148)$$

$$Co(tetren)(CO_3H)^{2+} \rightleftharpoons Co(tetren)(CO_3)^+ + H^+ \qquad (3.149)$$

$$CO_2 + H_2O \rightleftharpoons H^+ + HCO_3^- \rightleftharpoons 2\,H^+ + CO_3^{2-} \qquad (3.150)$$

The isolation of hydrogencarbonato complexes prepared by an interaction between hydroxy complex and carbon dioxide has been reported for the first time for Ir and Rh complexes (Eq. 3.151).[160]

$$[M(OH)(CO)(PPh_3)_2] + CO_2 \xrightarrow{EtOH} [M(OCO_2H)(CO)(PPh_3)_2] \quad (3.151)$$
$$(151,\ M = Rh,\ Ir) \qquad\qquad\qquad (152)$$

When the same reaction was carried out in the solid state, CO_2-coordinated complexes **4** and **12** were obtained (Eq. 3.52),[72] whereas the reaction in benzene yielded the binuclear complex **22** (Table 3.3). The fact that the dissolution of **4** and **12** in EtOH gives the hydrogencarbonato complexes **152** indicates that the insertion of CO_2 into the M–O bond proceeds stepwise, passing through the coordination of CO_2 to the metal. Treatment of

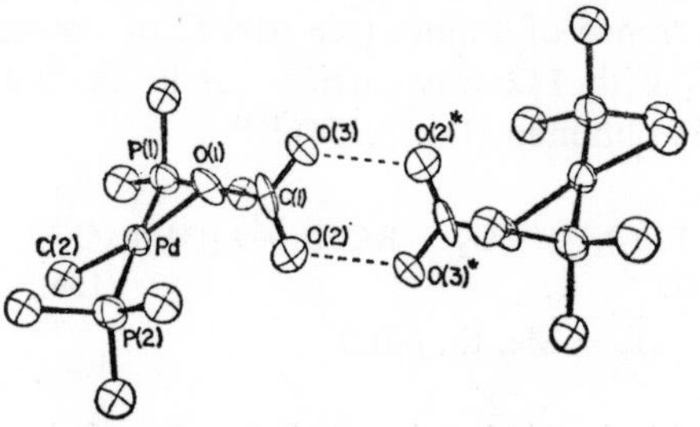

Fig. 3.9. The molecular structure of $[PdMe(O_2COH)(PEt_3)_2]$, (**155**). (Source: ref. 162. Reproduced by kind permission of Elsevier Scientific Publishing Co. (The Netherlands).)

152 with NaOEt yielded the ionic carbonato complex $Na[M(O_2CO)(PPh_3)_3]$ (**153**).[160]

Dialkylpalladium complex, $[PdR_2L_2]$ (R = Me, **142**; R = Et, **154**; L = PEt_3, $PMePh_2$), in hydrocarbon solvent reacted with CO_2 at room temperature (when R = Me) or at $-20°C$ (when R = Et), evolving 1 mole of methane and ethane, respectively, to give complexes which were formulated as CO_2-adducts.[161] The resulting complex later turned out to be hydrogencarbonato complex **155**, as was established during the molecular structural study of **142** (L = PEt_3) (Fig. 3.9).[162]

$$[PdR_2L_2] + CO_2 \xrightarrow{H_2O} [PdR(O_2COH)L_2] + RH \qquad (3.152)$$
$$(\textbf{142}, R = ME) \qquad\qquad (\textbf{155})$$
$$(\textbf{154}, R = Et) \qquad\qquad (L = PEt_3, PMePh_2)$$

As the yield of **155** increased on addition of 1 mole of water per mole of complex to the system, the formation of alkyl hydrogencarbonato complex **155** from dialkyl complex was thought to proceed by the initial attack on **142** or **154** by a small amount of water that remained in the CO_2, yielding an alkyl hydroxo intermediate.[162] The insertion of CO_2 into the Pd–O bond may yield **155** (path A of Eq. 3. 153). The alternative path involves the initial formation of carbonic acid, produced by the interaction between water and CO_2, which attacks the dialkylpalladium complex to give **155** (path B).

$$PdR_2L_2 \xrightarrow[-RH]{H_2O} PdR(OH)L_2 \xrightarrow{CO_2} [PdR(O_2COH)L_2] \qquad (A)$$

$$CO_2 + H_2O \longrightarrow HCO_3H \xrightarrow{PdR_2L_2} [PdR(O_2COH)L_2] + RH \quad (B) \quad (3.153)$$

Path B is analogous to the mechanism proposed for the formation of carbamato complexes by the interaction of CO_2 with transition metal complexes in the presence of amines (see part C of this section). When the reaction of CO_2 with **142** was carried out in alcohol, alkyl carbonato complex **156** was obtained (Eq. 3.154).[162]

$$[PdMe_2(PEt_3)_2] + CO_2 + ROH \longrightarrow [PdMe(O_2COR)(PEt_3)_2] \quad (3.154)$$
$$(\textbf{142}) \qquad\qquad\qquad (\textbf{156})$$
$$(R = Me, Et, \textit{n}\text{-Bu})$$

Similarly, a hydrido (methyl carbonato) complex of platinum was obtained by reaction (Eq. 3.155).[109]

$$trans\text{-}[PtH_2\{P(cyclo\text{-}C_6H_{11})_3\}_2] + CO_2 \xrightarrow{\text{MeOH}} trans\text{-}[PtH(O_2COCH_3)\text{-}$$
$$\mathbf{(157)}$$

$$\{P(cyclo\text{-}C_6H_{11})_3\}_2] \qquad (3.155)$$
$$\mathbf{(158)}$$

The molecular structure of **158** was established by means of single-crystal X-ray crystallography.[109] Reaction (3. 156) may be another example of this type of reaction.[163]

$$[RuH(PMe_2Ph)_5]PF_6 + CO_2 \xrightarrow{\text{ROH}} [Ru(O_2COR)(PMe_2Ph)_4]PF_6 \quad (3.156)$$
$$\mathbf{(146)} \qquad\qquad\qquad \mathbf{(159)}$$

A hydration of carbon dioxide analogous to Eq. (3.152) was reported recently for rhodium(I) hydrido complex (Eq. 3.157).[101]

$$[RhH(P\text{-}i\text{Pr}_3)_3] \text{ or } [Rh_2H_2(\mu\text{-}N_2)\{P(cyclo\text{-}C_6H_{11})_3)_4\}] + CO_2$$
$$+ H_2O \longrightarrow [RhH_2(O_2COH)L_2] \qquad (3.157)$$
$$(L = P\text{-}i\text{-}Pr_3, \mathbf{160}; L = P(cyclo\text{-}C_6H_{11})_3, \mathbf{161})$$

The hydrogencarbonato complex (**161**) thus produced further reacts with CO_2, resulting in the dehydration from hydridic hydrogen and carbon dioxide and giving carbonyl complex **162,** which undergoes reversible decarboxylation (Eq. 3.158).[101]

$$[RhH_2(O_2COH)\{P(cyclo\text{-}C_6H_{11})_3\}_2] \xrightarrow{CO_2, -H_2O} [Rh(CO)(O_2COH)\text{-}$$
$$\mathbf{(161)} \qquad\qquad\qquad \mathbf{(162)}$$

$$\{P(cyclo\text{-}C_6H_{11})_3\}_2] \underset{+CO_2}{\overset{-CO_2}{\rightleftarrows}} [Rh(OH)(CO)\{P(cyclo\text{-}C_6H_{11})_3\}_2] \quad (3.158)$$

A hydrogencarbonato complex of copper analogous to **60** has been prepared by the reaction of alkoxocopper complex with CO_2 in the presence of H_2O (Eq. 3.159).[79]

$$2[Cu(OR)(PPh_3)_2] + 2 CO_2 + 2 H_2O \xrightarrow{\text{THF}} [Cu(O_2COH)(PPh_3)_2]_2 + 2 ROH$$
$$(R = Et, Ph) \qquad\qquad\qquad \mathbf{(163)} \qquad\qquad (3.159)$$

The resulting complex (**163**) was reversibly converted into μ-carbonato complex, $[Cu_2(PPh_3)_4(\mu\text{-}CO_3)]$, on heating at $100°C$.[79] When reaction (3. 159) was carried out in a mixture of Et_2O and EtOH in the absence of H_2O, ethylcarbonato complex, $[Cu(O_2COEt)(PPh_3)_2]$, was obtained.[79]

Tertiary butoxycopper reversibly added CO_2 to give butylcarbonato complex **76** (Eq. 3.160).[112]

$$[Cu(O\text{-}t\text{-}Bu)(CN\text{-}t\text{-}Bu)] + 2\ t\text{-}BuNC + CO_2$$
$$(58)$$

$$\underset{C_6H_6,\ refl.}{\overset{C_6H_6,\ room\ temp.}{\rightleftharpoons}} [Cu(O_2CO\text{-}t\text{-}Bu)(CN\text{-}t\text{-}Bu)_3] \quad (3.160)$$
$$(76)$$

Heating either complex **76** or a mixture of $CuO\text{-}t\text{-}Bu/t\text{-}BuNC/CO_2$ in a hydrocarbon solvent at 100–150° C yielded CO and t-BuNCO. This means that $CuO\text{-}t\text{-}Bu$ converts CO_2 into CO in the presence of t-BuNC, where **76** acts as a carrier of activated carbon dioxide.[164] More recently, hydration of CO_2 by $CuO\text{-}t\text{-}Bu$ to give $[Cu(O_2COH)(t\text{-}BuNC)_n]$ **(60)** has been reported (Eq. 3.73).[102]

The binuclear molybdenum alkoxides **(164)** reacted reversibly with CO_2 both in solution and in the solid state to give insertion products **(165)** (Eq. 3.161).[165,166]

$$[Mo_2(OR)_6] + 2\ CO_2 \underset{high\ temp.}{\overset{room\ temp.}{\rightleftharpoons}} [Mo_2(OR)_4(O_2COR)_2] \quad (3.161)$$
$$(164) \qquad\qquad\qquad (165)$$
$$(R = Me_3Si,\ Me_3C,\ Me_2CH,\ Me_3CCH_2)$$

Single crystal X-ray crystallography of **165** ($R = CMe_3$) revealed that it has the structure shown below with Mo-to-Mo triple bonds. The kinetic studies suggested that CO_2 insertion in the solid state proceeded *via* a direct attack on the CO_2 molecule, as shown in Eq. (3.162).[166]

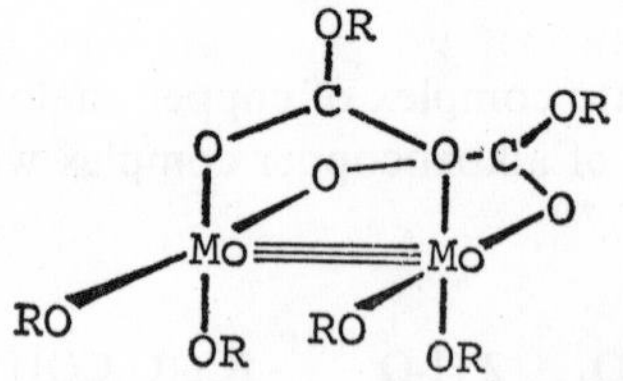

$$(165)$$

$$M\text{-}OR + CO_2 \rightleftharpoons (CO_2)M\text{-}OR \rightleftharpoons M\text{-}OR \rightleftharpoons MoO_2COR \quad (3.162)$$

In constrast, CO_2 insertion in solution is thought to proceed *via* a kinetically more favorable pathway, one involving an alcohol-catalyzed chain mechanism (Eqs. 3.163 and 3.164) analogous to that postulated for the reactions between metal amides and CO_2 (Eqs. 3.134 and 3.135) and for those between palladium complexes and CO_2 in the presence of water (path B in Eq. 3.153).[166]

$$ROH + CO_2 \rightleftharpoons ROCOOH \qquad (3.163)$$

$$MoOR + ROCOOH \longrightarrow Mo\text{-}O_2COR + ROH \qquad (3.164)$$

When metal alkoxides in solution were exposed to one atmosphere of carbon dioxide at 30°C, a certain amount of CO_2 uptake was observed. Because diethyl carbonate was formed on treatment of the system with EtI, the insertion of CO_2 into the M–O bond seemed likely.[167]

$$M(OEt)_n + CO_2 \rightleftharpoons M(OEt)_{n-1}(O_2COEt) \xrightarrow{\text{EtI}} EtO_2COEt \qquad (3.165)$$
$$(M = Fe\ (n = 3),\ Nb\ (n = 5),\ Zr\ (n = 4))$$

Infrared spectroscopic study has been done on the insertion of CO_2 into the Mn–O bond formed by the co-condensation of manganese vapor and ROH (R = Me and Et). Analysis of the infrared bands due to the organic carbonate group observed at 1700–800 cm^{-1} suggested that the CO_2 insertion proceeded according to the following scheme (Eq. 3. 166).[168]

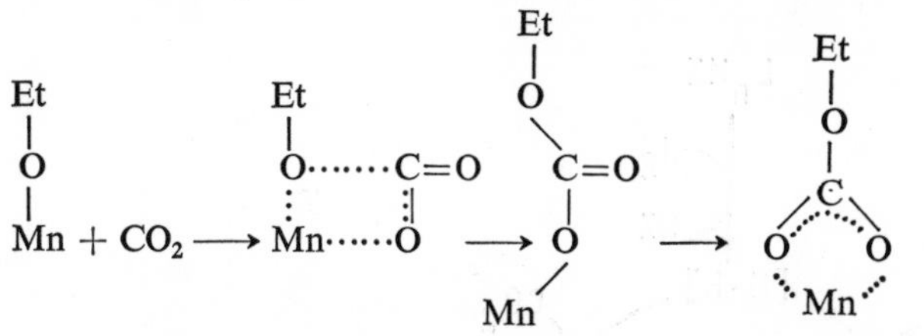

3.4. CATALYTIC REACTIONS USING CO_2 PROMOTED BY TRANSITION METAL COMPLEXES

As Table 3. 1 shows, there should be a variety of catalytic processes that are possible from a thermodynamic point of view. In reality, however, only a limited number of catalytic processes using carbon dioxide have been devised. Let us first examine the reported examples of catalytic CO_2 utilization before discussing the feasibility of various conceivable catalytic processes.

3.4.1. Catalytic Processes Promoted by Transition Metal Complexes using Hydrogen and Carbon Dioxide

The first catalytic process converting amine to formamide using CO_2 and H_2 in the presence of transition metal complexes was reported in 1970.[169)]

$$R_2NH + CO_2 + H_2 \xrightarrow{C_6H_6, \ cat.} HCONR_2 + H_2O \qquad (3.167)$$

On heating the reaction mixture containing amine and the catalyst in benzene under pressure of CO_2 and H_2 (28 atm each at room temperature) at 125°C for 17h, 1,000 or more moles of dimethylformamide (DMF, when R = Me) per mole of catalyst was produced. Among the transition metal complexes investigated, [IrCl (CO) $(PPh_3)_2$], [CoH $(Ph_2PCH_2CH_2PPh_2)_2$], and [CuCl $(PPh_3)_3$] were found to be especially effective as catalysts for the reaction, followed by [RhCl$(PPh_3)_3$], [Pd$(CO_3)(PPh_3)_2$], [Pt(CO_3)-$(PPh_3)_2$], [Pt$(PPh_3)_3$], [RuCl$_2(PPh_3)_3$], and [H$_2$PtCl$_4$·6H$_2$O]. The mechanism shown in Fig. 3. 10 (BH = R_2NH) has been proposed for the reaction. Additional evidence for this mechanism was the observation that, when D_2 was used in the place of H_2, equal amounts of DMF-d_1 and DMF-d_0 were produced together with a scrambled mixture of D_2 and H_2.[169)] Secondary amines other than dimethylamine, such as dipropylamine,

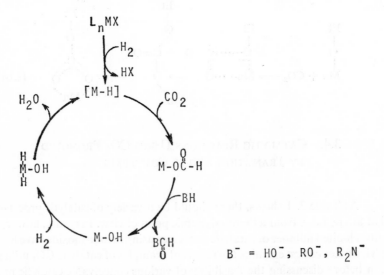

Fig. 3.10.   A mechanism for the formation of formic acid, formates, and formamides.

pyrrolidine, and piperidine, are also converted to the corresponding formamides according to Eq. (3.167).

A significant improvement in the yield and the rate of the reaction was subsequently achieved by carrying it out in methyl celloslove in the presence of a base such as $KHCO_3$, K_2CO_3, and cyclic tertiary amine.[170] $PdCl_2$ was found to be an effective catalyst in this system. Because potassium formate was detected in the reaction products after treatment with water, and because some DMF was obtained even in the absence of CO_2, it was thought that the reaction might proceed according to the pathway shown in Fig. 3.11 rather than that in Fig. 3.10.

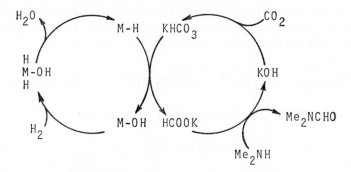

Fig. 3.11. A proposed mechanism for the formation of Me_2NCHO from Me_2NH and CO_2 in the presence of $KHCO_3$.[170]

When reaction (Eq. 3.167) was carried out in the presence of alcohol instead of secondary amines, formic esters were catalytically formed.[171–173]

$$ROH + CO_2 + H_2 \xrightarrow{\text{cat.+cocat.}} HCOOR + H_2O \qquad (3.168)$$

BF_3 is required as a cocatalyst when the reaction is carried out at 100°C in benzene, along with $[IrH_3(PPh_3)_3]$, $[RuCl_2(PPh_3)_3]$, $[RuHCl(PPh_3)_3]$, $[RuH_2(PPh_3)_4]$, or $[RuCl_2(PPh_3)_4]$ as a catalyst.[171] Another group of investigators reported a similar type of reaction using tertiary amines such as trialkylamines, N-methylpyrrolidine, N-methylpiperidine, and 1,4-diazabicyclo[2.2.2]octane as cocatalysts.[172] In addition to the complexes mentioned above, $[Pd(dpe)_2]$ ($dpe = Ph_2PCH_2CH_2PPh_2$) was found to be effective as a catalyst for this solvent-free system. Under a pressure of 50 atm (at 25°C) at 140°C for 21h, 20–40 moles of HCOOR (R = Me, Et, and n-Pr) per mole of $[Pd(dpe)_2]$ was obtained with this reaction.[172]

A cocatalyst, either BF_3, or R_3N, is required for catalytic reaction

(3.168) to proceed with considerable efficiency. A system which is less efficient, but which does not need a cocatalyst, was found that employs anionic iron complexes such as $[Fe(H)(CO)_{11}]^-$ $[N(PPh)_3]_2^+$, $[Fe(H)-(CO)_{11}]^-$ $[HNEt_3]^+$, and $[Fe(H)(CO)_4]^-$ $[N(PPh_3)_2]^+$.[173] These results suggest that the interaction of CO_2 with the metal center of the catalyst is an important step in these systems, because the metal complexes with a high electron density at the metal center, like the anionic complexes, were more effective than the others.

The reaction of CO_2 and H_2 (at 40 atm at room temperature) at 100°C in hexamethylphosphoramide in the presence of $[RhCl (PPh_3)_3]$ yielded the known monocarbonyl complex of rhodium (I).[174] Both CO_2 and H_2 are necessary for this reaction to proceed, and it is accelerated when water is added, so a pathway involving the intermediate formation of formic acid has been proposed (Eqs. 3.169–3.171).[174]

$$[RhClL_3] \underset{L}{\overset{S}{\rightleftharpoons}} RhClL_2 \text{ (S)} \xrightarrow{H_2} H_2RhClL_2 \text{ (S)} \xrightarrow{CO_2} HRh(O_2CH)ClL_2 \text{ (S)}$$

$$\xrightarrow{H_2O} HRh(OH)ClL_2 \text{ (or } Rh(H_2O)ClL_2) + HCOOH \qquad (3.169)$$

$$HCOOH \rightleftharpoons H_2O + CO \qquad (3.170)$$

$$RhClL_2 \text{ (S)} + CO \longrightarrow [RhCl(CO)L_2] \qquad (3.171)$$
$$(L = PPh_3; S = \text{solvent})$$

A similar reaction is known which yields rhodium carbonyl complex by the interaction of carbon dioxide and $[RhCl(PPh_3)_3]$ in the presence of a reducing agent (Eq. 3.60).[95]

One important factor which should be mentioned here in connection with reactions between CO_2 and H_2 is their relationship with the water-gas shift reaction. In recent years there has been considerable interest in water-gas shift reactions utilizing transition metal complexes, particularly for application to the energy problem, and some catalytic processes, although low in efficiency, have been developed to convert carbon monoxide and water to carbon dioxide and hydrogen.[175] Since the water-gas shift reaction is a reversible process with a small ΔH value, the reverse process should be also catalyzed by similar transition metal complexes. In considering the mechanisms of the catalytic processes discussed in this section, the reader should bear in mind that the transition metal complexes may be acting as promotors for the reverse water-gas shift reaction, and the mechanisms proposed by authros of individual papers may be discussed in the light of the catalytic process using carbon monoxide as a component.

When a reaction similar to (3.169) was carried out in the presence of a base such as tertiary amines, sodium hydroxide, and sodium hydrogen carbonate, together with a very small amount of water in benzene, formic acid was produced catalytically (up to about 90 moles of HCOOH per mole of catalyst at room temperature) (Eq. 3.172).[176] $[Pd(dpe)_2]$, $[Ni(dpe)_2]$, $[Pd(PPh_3)_4]$, $[RhCl(PPh_3)_3]$, $[IrH_3(PPh_3)_3]$, and $[RuH_2(PPh_3)_4]$ could be used as catalysts in this system, among which the last complex gave the highest yeild.[76]

$$CO_2 + H_2 \xrightarrow[\text{base, H}_2\text{O}]{\text{cat.}} HCOOH \qquad (3.172)$$

A possible mechanism of this reaction is that shown in Fig. 3.10 (B = OH), but an alternative mechaism (Eq. 3.173) cannot be ruled out.[176]

$$CO_2 + H_2O + NR_3 \rightleftharpoons HOCO_2NHR_3 \xrightarrow[\text{cat.}]{H_2} HCOOH + HONHR_3 \quad (3.173)$$
$$(\text{or } H_2O + HCOONHR_3)$$

The catalytic formation of potassium formate from CO_2 (40 atm), H_2(106 atm), and KOH in an aqueous medium at 240° C was achieved by using $PdCl_2$ as a catalyst.[177] The observation that the reaction proceeded without CO_2 when $KHCO_3$ was added to the system suggests that a mechanism involving the initial hydration of CO_2 with KOH to give $KHCO_3$, like that shown in Fig. 3.11, is more reasonable than the one shown in Fig. 3.10, where the initial insertion of CO_2 to M–H is involved. A kinetic study of this system provided evidence for the mechanism shown in Eqs. (3.174)–(3.176).[177]

$$KOH + CO_2 \underset{}{\overset{k_1}{\rightleftharpoons}} KHCO_3 \underset{}{\overset{k_2}{\rightleftharpoons}} HCO_3^- + K^+ \qquad (3.174)$$

$$M + H_2(1) \underset{}{\overset{k_3}{\rightleftharpoons}} M\cdot H_2^* \qquad (3.175)$$
$${\scriptstyle K_H} \Big\downarrow$$
$$H_2(g)$$

$$HCO_3^- + M\cdot H_2^* \xrightarrow{k_4} [HCO_3^- \cdot M\cdot H_2^*] \xrightarrow[+K^+]{\text{fast}} HCOOK + M + H_2O \quad (3.176)$$

When reactions (3.169)–(3.171) were carried out in the presence of methyloxirane (propylene oxide), CO_2 was catalytically fixed to form 1.2-propanediol formates together with 4-methyl-1,3-dioxolane-2-one (propylene carbonate) and 1,2-propanediol (Eq. 3.177).[178]

$$CO_2 + H_2 + \underset{\underset{\displaystyle O}{\diagdown\diagup}}{H_2C-CH}\overset{CH_3}{\vert} \xrightarrow[100-170°C]{RhCl(PPh_3)_3} \underset{\underset{\displaystyle O}{\Vert}}{\underset{HCO\ \ OH}{H_2C-CH}}\overset{CH_3}{\vert} + \underset{\underset{\displaystyle O}{\Vert}}{\underset{HO\ \ OCH}{H_2C-CH}}\overset{CH_3}{\vert} + \underset{\underset{\displaystyle O\ \ \ O}{\Vert\ \ \ \Vert}}{\underset{HCO\ \ OCH}{H_2C-CH}}\overset{CH_3}{\vert}$$

$$+\ \underset{\underset{\underset{\displaystyle O}{\Vert}}{\underset{C}{\diagup}}}{\underset{O\ \ \ O}{H_2C-CH}}\overset{CH_3}{\vert} +\ \underset{HO\ \ OH}{H_2C-CH}\overset{CH_3}{\vert} \qquad (3.177)$$

A mechanism that assumes the presence of a minor amount of water was proposed for this system (Fig. 3.12).

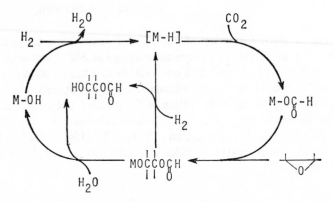

Fig. 3.12.   A proposed mechanism for the formation of 1,2–propanediol formate from H_2, CO_2, and an oxirane, with [RhCl(PPh₃)₃] as catalyst.[178]

The catalytic formation of alkylene carbonate has been described in a similar system without the addition of hydrogen. Zero-valent nickel complexes such as NiL_n (L = PPh₃, $n = 2$; L = P($cyclo$-C₆H₁₁)₃, $n = 2,3$; L = $cyclo$-C₈H₁₂, $n = 2$) were found to be the active catalysts for this reaction (Eq. 3.178).[179]

$$R^1R^2C-CH_2 + CO_2 \xrightarrow[\text{cat.}]{C_6H_6,\ 100°C} \underset{\underset{\underset{\displaystyle O}{\Vert}}{\underset{C}{\diagup}}}{\underset{O\ \ \ O}{R^1R^2C-CH_2}} \qquad (3.178)$$

The catalyst system $TiCl_4 \cdot 2THF/Mg/THF$ has been reported to fix carbon dioxide and hydrogen under mild conditions to give magnesium formate (Eqs. 3.179–3.181).[180]

$$n\ [THFCl_2Mg_2Ti]_2 + n\ H_2 \longrightarrow n\ [THFCl_2Mg_2H_2Ti]_2$$

$$\xrightarrow{n Mg} n\ [THFCl_2Mg_2Ti]_2 + n\ MgH_2 \quad (3.179)$$

$$[THFMg_2Cl_2H_2Ti_2] + 4\ CO_2 + 2\ Mg \longrightarrow [THFCl_2Mg_2Ti]_2 \quad (3.180)$$
$$+ 2\ Mg(OOCH)_2$$

$$MgH_2 + 2\ CO_2 \longrightarrow Mg(OOCH)_2 \quad (3.181)$$

Since the presence of MgH_2 is essential for the system to be catalytic, it was necessary to alternately feed hydrogen and carbon dioxide. Introduction of carbon dioxide without prior treatment with hydrogen afforded titanium formate $[(THF)Ti(OOCH)_2MgCl_{1.5}]$.[181]

3.4.2. Effect of Carbon Dioxide on the Catalytic Telomerization of Dienes and the Formation of Lactones and Related Compounds

$[Pt(PPh_3)_3]$ or $[Pd(PPh_3)_4]$ is known to catalyze the dimerization of butadiene to give primarily 4-vinylcyclohexene (**166**) (90–97%) together with a minor amount of 1,3,7-octatriene (**167**) (3–10%). The introduction of CO_2 into the system significantly enhances the catalytic activity, and a linear tetraene (**167**) is produced (80%) rather than the cyclic isomer (**166**) (eq. 3.182).[182]

$$2\ \text{\/\/}\ \xrightarrow[CO_2]{[Pt(PPh_3)_3]}\ \text{(166, 20\%)} + \text{(167, 80\%)} \quad (3.182)$$

166 (20%)   **167** (80%)

$[Pd(PPh_3)_4]$ showed a similar effect, whereas the $[Ni(PPh_3)_4]$-catalyzed reaction gave 1,5-cyclooctadiene in 66% yield in the absence of CO_2; 4-vinylcyclohexene was produced in 93% yield in the presence of CO_2. There are several other instances in which presence of CO_2 modified the reaction path in a catalytic system involving diene. It was reported by Musco and Silvani that CO_2 enhanced the catalytic effect of PdL_n [where $L = PEt_3$ ($n = 3$), $P(cyclo\text{-}C_6H_{11})_3$ ($n = 2$), and PPh_3 ($n = 3$)] in the dimerization of butadiene to 1,3,7-octatriene (**167**) and, more notably, in the subsequent isomerization of **167** to 2,4,6-octatriene.[183] In this system, they assumed that, in the presence of butadiene, CO_2 assisted the formation of a Pd(II) species which was active both for the dimerization and the isomerization.[183] The PdL_n/CO_2 system was also found to be an effective catalyst for the tail-to-tail dimerization of isoprene. In the absence of CO_2 the PdL_n complexes catalyze the dimerization only in very low conversion. As is the case for the butadiene dimerization, the catalyst system is especially effective for the isomerization of terminal olefins to internal olefins (Eq. 3.183)[184]

$$2 \text{ (structure)} \xrightarrow[\text{tail-to-tail} \atop \text{dimerization}]{PdL_n/CO_2} \text{ (structure)} \xrightarrow[\text{isomerization}]{PdL_n/CO_2} \text{ (structure)}$$

$$+ \text{ (structure)}$$

(1.183)

If the dimerization of butadiene by a palladium catalyst, such as [Pd-(acac)$_2$]/PPh$_3$ (acac = acetylacetonato), [Pd(PPh$_3$)$_4$], or [Pd(CO$_3$) (PPh$_3$)$_2$], is carried out in such solvents as t-butyl-alcohol, acetone, or acetonitrile in the presence of water, a small amount of octadienols is formed in addition to 1, 3, 7-octatriene as a major product (Eq. 3.184). Addition of CO_2 to this catalytic system changes the selectivity of the catalyst greatly, giving octa-2,7-diene-1-ol as the major product (Eq. 3.185),[185]

$$2 \text{ (structure)} + 4 H_2O \xrightarrow[\text{solvent}]{Pd} \text{ (structure)} + \begin{array}{c} \text{octadienes} \\ \text{(small amount)} \end{array}$$

(major)

(3.184)

$$2 \text{ (structure)} + 4 H_2O$$

$$\xrightarrow[\substack{\text{solvent} \\ 80\sim90°C}]{Pd \text{ cat.}/CO_2} \underset{65\%}{\text{(structure)}-OH} + \underset{19\%}{\text{(structure with OH)}} + \underset{4\%}{\text{(structure)}}$$

(when solvent=t-BuOH,   + octadienyl ethers
Pd cat.=Pd (acac)$_2$/3 PPh$_3$)   4%

(3.185)

The fact that low levels of CO_2 can enhance the reaction rate suggests that carbon dioxide interacts directly with the palladium catalyst, as in the similar system without water.[185]

The system $PdCl_2/PPh_3$, in the presence of CO_2 and/or methanol, showed significant catalytic activity for the formation of 2:1 adducts of isoprene with amines such as 168, 169, and 170, where 169 is the main product. A similar system without CO_2 yielded only 1:1 adducts such as **171** and **172** in a low yield (Eq. 3.186).[186]

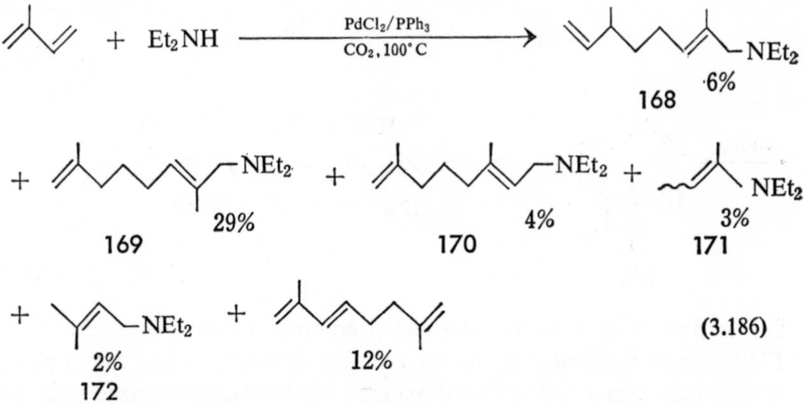

$$\text{(3.186)}$$

Reports in recent years have described the incorporation of CO_2 groups into organic molecules in the course of catalytic telomerization of dienes or acetylenes to give cyclic or linear esters. The lactone, 2-ethylidene–5-hepten-4-olide (**173**) was detected in the products of a catalytic system consisting of butadiene, carbon dioxide, and zero-valent palladium complexes in a polar solvent such as N,N-dimethylformamide (DMF) or dimethylsulfoxide (DMSO).[187,188]

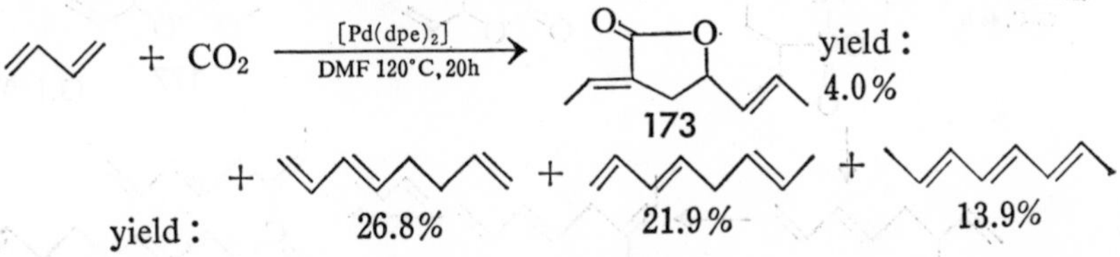

$$\text{(3.187)}$$

Although the yield of **173** is not high compared with ordinary butadiene dimers and their isomers, the turnover number of as high as about 17 was achieved. The zero-valent palladium complexes with bidentate tertiary phosphines, such as $Ph_2PCH_2CH_2PPh_2$ and $Me_2PCH_2CH_2PMe_2$, are better than those with monodentate phosphines as catalysts for the formation of **173**. Since the carboxylic acids **175** and **176** were detected in the system at an early stage of the reaction, a mechanism was proposed that involved the π-and σ-allyl type intermediates, followed by CO_2 insertion and the formation of acids **175** and **176**, (Eq. 3.188).[187,188]

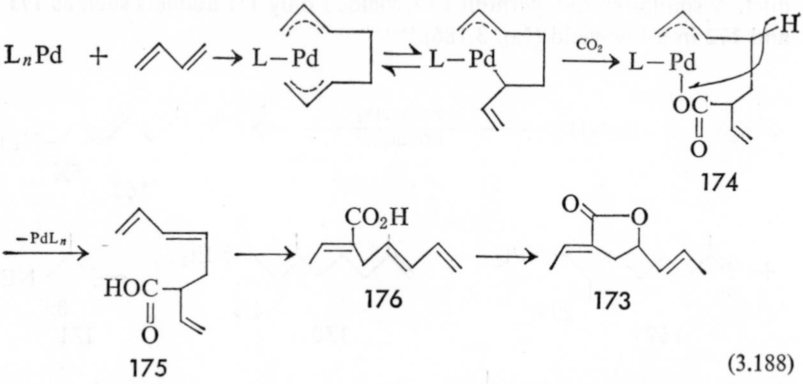

$$(3.188)$$

It is important to note that the nickel complex of the isomeric form of **174** has been isolated (**121**) by the reaction between σ- and π-complexes of butadiene dimer and carbon dioxide; it has been characterized by X-ray crystallography (Eq. 3.115).[132] When a similar reaction was carried out in a non-polar solvent such as benzene, other carboxylated compounds (**177, 178,** and **179**) were produced (Eq. 3.189).[189]

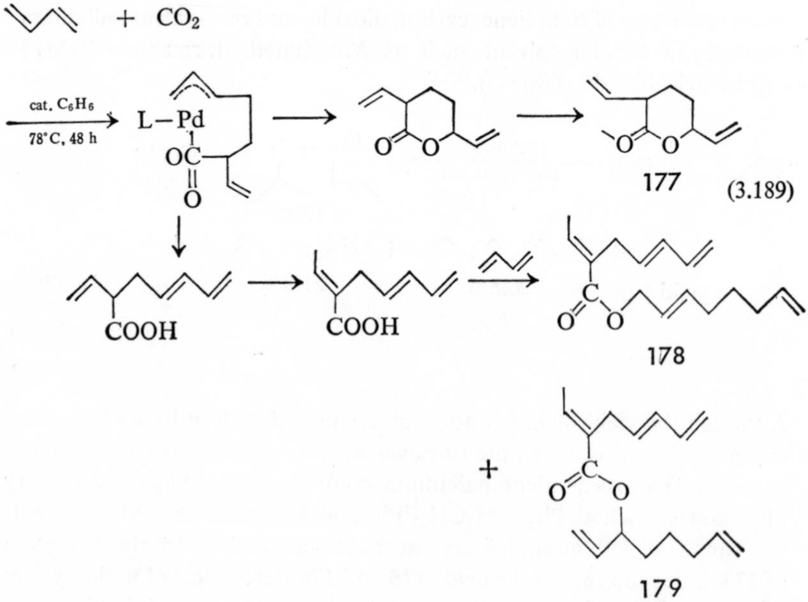

$$(3.189)$$

PdL$_4$ (L = P($cyclo$-C$_6$H$_{11}$)$_3$, PEt$_3$, and PPh$_3$), as well as a mixture of (η^3-2-methallylPdOCOMe)$_2$ and the appropriate phosphine ligand, can be used as an effective catalyst for this sytem. The addition of small quantities of water was found to accelerate the reaction.

Under similar conditions as those used for reaction (3.187) (80° C), isoprene underwent a tail-to-tail dimerization to give 2,7-dimethyl-1,3,7-octatriene as the main product, as shown in Eq. (3.183). In this reaction system, the formation of a very small amount of five-membered lactones was detected, indicating that the incorporation of CO_2 followed a reaction path similar to that shown for the butadiene system (Eq. 3.188).[190]

Oligomerization of alkylacetylenes with the catalyst system [Ni(1,5-$C_8H_{12})_2$]–$Ph_2P(CH_2)_nPPh_2$ in the presence of CO_2 yielded substituted 2–pyrones together with oligomeric hydrocarbons such as substituted benzene and cyclopentadiene (Eq. 3.190).[191,192]

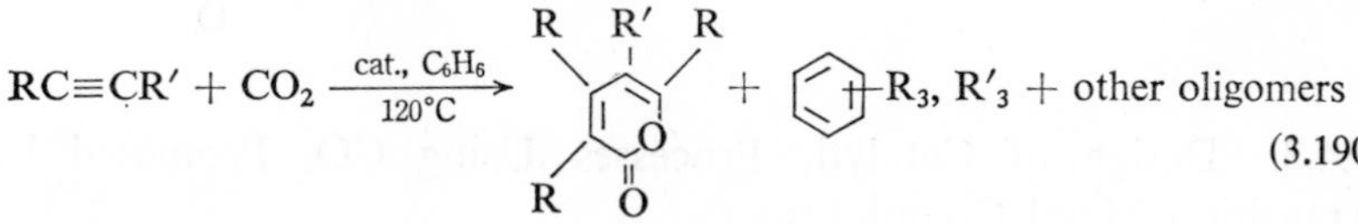

$$RC{\equiv}CR' + CO_2 \xrightarrow[120°C]{cat., C_6H_6}$$

$+ \text{—}R_3, R'_3 + \text{other oligomers}$

(3.190)

(180)

(R = *n*-Bu, R′ = H; R = Et, R′ = H; R = R′ = Et; R = R′ = *n*-Pr)

Among the diphosphines ($Ph_2P(CH_2)_nPPh_2$) examined, the one with $n = 4$ gave the most effective catalyst system in combination with [Ni(1,5-$C_8H_{12})_2$]. With this catalyst system, the highest yield of **180** (57 % based on the alkyne charged) was obtained for R = R′ = Et. The proposed mechanism for the formation of pyrones with this sytem, involves the insertion of CO_2 into the Ni–C bond of a nickelole intermediate, as shown in Eq. (3.191)[191,192]

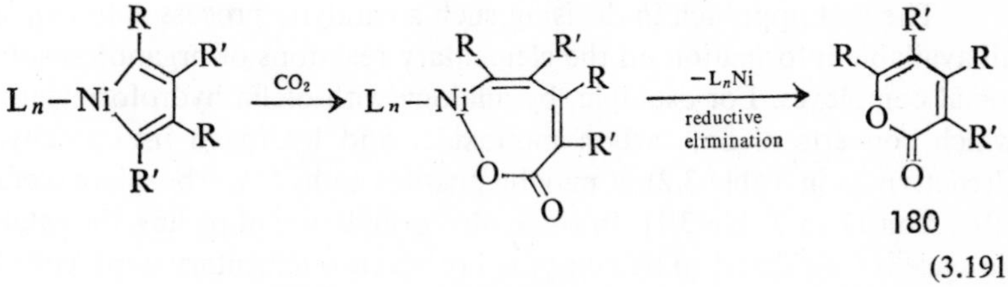

180

(3.191)

These reactions that cause the formation of lactones or carboxylic acid together with the dimerization of unsaturated hydrocarbons represent one type of catalytic reactions involving carbon dioxide, a type different from those described in Section 3.4.1. These catalytic reactions may be explained by the general seheme shown in Eq. (3.192), which proceeds by the formation of a π-σ or σ-σ-metallocycle intermediate followed by the insertion of CO_2 into the metal-carbon bond. The subsequent reductive elimination and β-hydrogen abstraction give lactones and unsaturated carboxylic acid, respectively (Eq. 3.213).

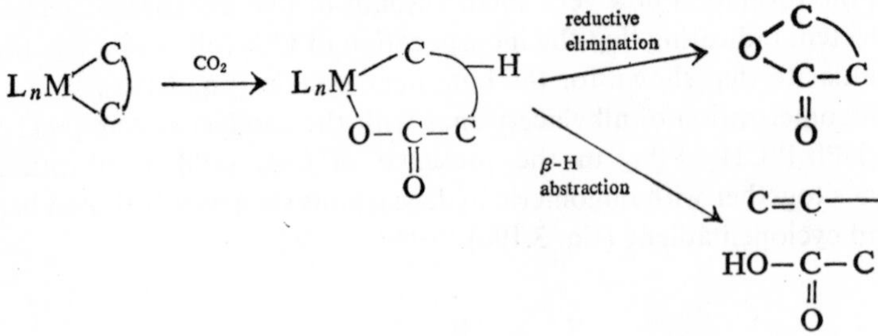

(3.192)

3.4.3. Design of Catalytic Processes Using CO_2 Promoted by Transition Metal Complexes

Among the various thermodynamically possible reactions involving carbon dioxide listed in Table 3.1, only a very limited number of processes employing transition metal complexes as catalysts have been developed so far, as described in Section 3.4.2, and the efficiency of these catalysts is still insufficient for practical applications. Although it is difficult to anticipate future uses of CO_2 involving transition metal catalysts at the present time, when the chemistry of interactions between carbon dioxide and transition metal cpmplexes is not well explored, we will attempt to indicate a few approaches which may be relevant to the further development of processes utilizing carbon dioxide.

The first approach in devising such a catalytic process is to employ all available information on the elementary reactions of organotransition metal complexes. For example, by analogy with olefin hydroformylation which converts olefin, carbon monoxide, and hydrogen into aldehyde (Reaction 28 in Table 3.2), it may be possible to hydrocarboxylate olefins (Reaction 17 in Table 3.1). In the hydroformylation of olefins, the catalytic cycle is considered to be composed of several elementary steps: coordination of the olefin to the cobalt hydride species, insertion of the olefin into the Co–H bond, further insertion of the coordinated carbon monoxide into the Co–C bond, oxidative addition of dihydrogen, and reductive elimination of acyl hydride (Fig. 3.13).[193]

Figure 3.14 shows a possible process for the hydrocarboxylation of olefins corresponding to hydroformylation. Olefin insertion into a M–H

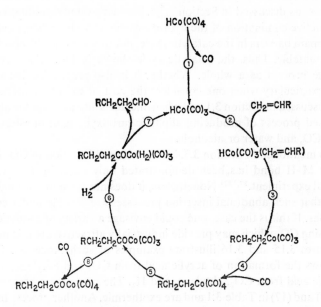

Fig. 3.13. A proposed mechanism for the olefin hydroformylation reaction. (Source: ref. 193. Reproduced by kind permission of the Chemical Society (London).)

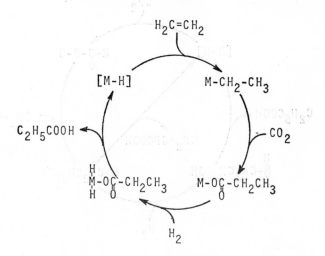

Fig. 3.14. A possible process for the hydrocarboxylation of olefins.

bond is a very well-known process, and CO_2 insertion into a M–C bond is also an established elementary process for some organotransition metal complexes, as discussed in Section 3.3.4. Further oxidative addition of H_2 and reductive elimination of the hydrido-carboxylato ligands may constitute the main barriers in the catalytic cycle, but they may not be absolutely unsurmountable. Thus, the catalytic cycle shown in Fig. 3.14 may be a realizable process as a whole. Whether it is also practicable is another matter, particulary when one considers the cost of the raw materials, as briefly discussed in Section 3.1 and when one takes into account the already established process of producing aliphatic carboxylic acids or esters from olefins, CO, and water or alcohols.

As mentioned in Section 3.3.4, the abnormal insertion of CO_2 into a M–C or M–H bond has been demonstrated only once, in a somewhat equivocal experiment.[57,122] Nonetheless, it does not seem unreasonable to assume that such abnormal insertion processes are feasible under certain conditions. If this is the case, one could envisage a variety of catalytic processes using CO_2 which may provide industrially attractive organic materials. Figures 3.15 and 3.16 illustrate examples of such processes. Figure 3.15 shows the formation of acrylic acid from CO_2 and ethylene, and of propionic acid from CO_2, ehtylene, and H_2. The processes correspond to Eqs. (15) and (17) in Table 3.1 and are exothermic. Another process involving the abnormal CO_2 insertion into a M–H bond is shown in Fig. 3.16,

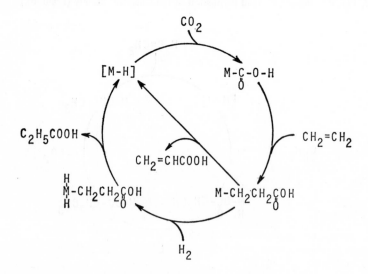

Fig. 3.15. Carboxylation and hydrocarboxylation of olefins involving the abnormal insertion of CO_2 into the M–H bond.

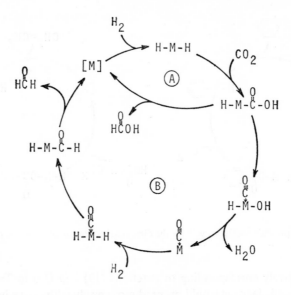

Fig. 3.16. Possible mechanisms for the hydrogenation of carbon dioxide.

cycle A. Figure 3.16 also illustrates a potential process that involves decarbonylation, followed by reductive elimination of the hydrido-hydroxo ligands, forming water (cycle B). The process is related to the water-gas shift reaction, and formation of carbon monoxide in the presence of hydrogen can lead to various products such as formaldehyde, methanol, and methane, obtained in the conversion of "water gas." Since commercial processes using water gas are well established, there may not be much practical importance in designing a catalytic process using CO_2 and H_2 unless we can obtain a specific product quite economically under energy-saving conditions.

The second approach to converting CO_2 into useful organic materials by means of transition metal complexes is to specifically activate organic compounds at a particular bond, such as the C–H, C–C, C–O, or C–N bond, and then to introduce CO_2 into the compounds. If the specific activation and cleavage of, for example, a C–H bond, can be promoted by a transition metal, a new species having M–C and M–H bonds may be formed into which CO_2 could be inserted. There is a growing member of reports of specific activation of aromatic,[194] olefinic,[194,195] and aliphatic[196] C–H bonds using transition metal complexes under mild conditions. The combination of C–H activation with CO_2 insertion may lead to a valuable catalytic process. Figure 3.17 indicates a few candidates for such catalytic carboxylation processes which could produce benzoic acid and acrylic

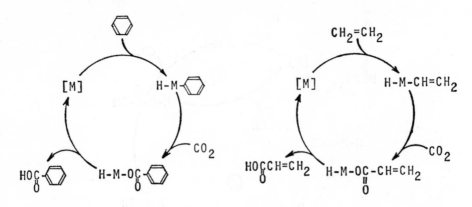

Fig. 3.17.   A possible catalytic cycle for the direct carboxylation of aromatic or olefinic hydrocarbons.

acid, respectively corresponding to reactions (19) and (15) in Table 3.1. Such C–H bond cleavage would proceed more easily with active hydrogen compounds, although in this case some means should be devised to prevent decarboxylation back to the original compound.

The third approach, so far unrealized but conceptually feasible and most interesting, is to use carbon dioxide as the oxidizing agent of an organic compound. A stoichiometric reaction in which CO_2 was partly reduced to CO and partly used to oxidize a tertiary phosphine is known, [73,96,97] as is a catalytic conversion of nitrous oxide into dinitrogen and a tertiary phosphine oxide.[117] Most of the current commercial processes using oxygen directly as an oxidizing agent are known to be radical processes, and consequently, the selectivities are low. Utilization of CO_2 as a mild oxidizing agent has not been achieved, but such an application might prove to be fruitful and deserves further development.

REFERENCES

1) P. D. Moore, *Nature*, **255**, 108 (1975).
2) The data on the standard enthalpy of the formation of each compound were cited from *Kagaku-Binran, Kiso-hen* (Japanese), p.965, Maruzen (1975).
3) R. P. A. Sneeden, *The Chemistry of Carboxylic Acids and Esters* (ed. by S. Patai), p. 137, Interscience-Publishers (1969).
4) E. J. Soloski and C. Tomborski, *J. Organometal. Chem.*, **157**, 373 (1978).
5) D. J. Cram, *Fundamentals of Carbanion Chemistry*, p. 123, Academic Press (1965).
6) J. F. Nobis and L. F. Moormeier, *Ind. Eng. Chem.*, **46**, 539 (1954); H. Gilman and J. W. Morton, Jr., *Org. Reactions*, **8**, 258 (1954).

7) M. Schlosser, *Angew. Chem.*, **76**, 124, 258 (1964); *Angew. Chem., Int. Ed. Engl.*, **3**, 287, 362 (1964).

8) A. S. Lindsey and H. Jeskey, *Chem. Rev.*, **57**, 583 (1957).

9) I. Hirao, *J. Synth. Org. Chem., Japan* (Japanese), **34**, 326 (1976).

10) J. I. Jones, *Chem. Ind.* (London), **1958**, 228.

11) K. Kudo and Y. Takezaki, *J. Chem. Soc. Japan, Ind. Chem. Section* (Japanese), **70**, 2147 (1967).

12) G. Bottaccio and G. P. Chiusoli, *Chem. Commun.*, **1966**, 618.

13) H. Mori, H. Yamamoto, and T. Kwan, *Chem. Pharm. Bull.*, **20**, 2440 (1972).

14) H. Mori, Y. Makino, H. Yamamoto, and T. Kwan, *Chem. Pharm. Bull.*, **21**, 915 (1973).

15) T. Kwan, H. Yamamoto, H. Mori, and H. Samejima, *Chemical Industry* (Japanese), **1972** 1618.

16) Y. Otsuji, M. Arakawa, N. Matsumura, and E. Haruki, *Chem. Lett.*, **1973**, 1193.

17) E. Haruki, M. Arakawa, N. Matsumura, Y. Otsuji, and E. Imoto, *Chem. Lett.*, **1974**, 427.

18) H. Finkbeiner and M. Stiles, *J. Am. Chem. Soc.*, **85**, 616 (1963).

19) For example, S. W. Pelletier, R. L. Chappell, P. C. Parthasarathy, and N. Lewin, *J. Org. Chem.*, **31**, 1747 (1966).

20) K. Ziegler, *Angew. Chem.*, **68**, 721 (1956); K. Ziegler, F. Krupp, K. Weyer, and W. Larbig, *Justus Liebigs Ann. Chem.*, **629**, 251 (1960).

21) H. Gilman and K. E. Marple, *Rec. Trav. Chim.*, **55**, 133 (1936).

22) H. Lehmkuhl, *Angew. Chem.*, **76**, 817 (1964).

23) S. Inoue and Y. Yokoo, *Bull. Chem. Soc. Japan*, **45**, 3651 (1972).

24) S. Inoue and N. Takeda, *Bull. Chem. Soc. Japan*, **50**, 984 (1977).

25) N. Takeda and S. Inoue, *Bull. Chem. Soc. Japan*, **51**, 3564 (1978).

26) G. A. Olah and J. A. Olah, *Friedel-Crafts and Related Reactions* (ed. G. A. Olah), vol. 3, p. 1257, Interscience-Publishers (1964).

27) H. Hopff and Th. Zimmermann, *Helv. Chim. Acta*, **47**, 1293 (1964).

28) M. Kumada, M. Ishikawa, and K. Yoshida, *Memoirs of the Fac. Eng. Osaka City Univ.*, **3**, 201 (1962); *Chem. Ab.*, **57**, 2238e (1962).

29) M. F. Lappert and B. Prokai, *Advances in Organometallic Chemistry*, **5**, 225 (1967).

30) R. T. Sanderson, *J. Am. Chem. Soc.*, **77**, 4531 (1955).

31) H. Breederveld, *Rec. Trav. Chim.*, **81**, 276 (1962).

32) G. Oertel, H. Malz, and H. Holtschmidt, *Chem. Ber.*, **97**, 891 (1964).

33) R. H. Cragg and M. F. Lappert, *J. Chem. Soc., A*, 82 (1966).

34) L. K. Peterson and K. I. Thé, *Can. J. Chem.*, **50**, 562 (1972).

35) J. Satgé, M. Lesbre, and M. Baudet, *Compt. Rend.*, **259**, 4733 (1964).

36) M. R.-Baudet and J. Satgé, *Bull. Chim. Soc. France*, **1969**, 1356.

37) R. F. Dalton and K. Jones, *J. Chem. Soc.*, *A*, **1970**, 590.

38) T. A. George, K. Jones, and M. F. Lappert, *J. Chem. Soc.*, **1965**, 2157.

39) A. J. Bloodworth, A. G. Davies, and S. C. Vasishtha, *J. Chem. Soc., C*, **1967**, 1309.

40) A. G. Davies and P. G. Harrison, *J. Chem. Soc., C*, **1967**, 1313.

41) J. Koketsu and Y. Ishii, *J. Chem. Soc., C*, **1971**, 511.

42) K. I. The, L. V. Griend, W. A. Whitla, and R. G. Cavell, *J. Am. Chem. Soc.*, **99**, 7379 (1977).

43) N. Yamazaki, *J. Synth. Org. Chem., Japan* (Japanese), **36**, **1978**, 684.

44) N. Yamazaki, F. Higashi, and T. Iguchi, *Tetrahedron Lett.*, **1974**, 1191; N. Yamazaki, T. Iguchi, and F. Higashi, *Tetrahedron*, **31**, 3031 (1975).

45) N. Yamazaki, T. Tomioka, and F. Higashi, *Synthesis*, **1975**, 384.

46) N. Yamazaki, T. Iguchi, and F. Higashi, *J. Polymer Sci., Polymer Chem. Ed.*, **13**, 785 (1975).

47) S. Inoue and Y. Yokoo, *J. Organometal. Chem.*, **39**, 11 (1972).

48) S. Inoue, M. Kobayashi, H. Koinuma, and T. Tsuruta, *Makromol. Chem.*, **155**, 61 (1972).

49) K. Kataoka and T. Tsuruta, *Polymer J.*, **9**, 595 (1977).

50) J. G. Noltes, *Rec. Trav. Chim.*, **84**, 126 (1965).
51) G. Bottaccio and G. P. Chiusoli, *Chem. and Ind.* (London), **1966**, 1457.
52) D. Grdenić and F. Zado, *J. Chem. Soc.*, **1962**, 521.
53) T. Tsuda and T. Saegusa, *Kagaku Sosetsu* (Japanese), No. 20, 211 (1978).
54) Y. Iwashita, *Chemical Industry* (Japanese), **23**, 1596 (1972).
55) M. E. Vol'pin, *Pure Appl. Chem.*, **30**, 607 (1972).
56) M. E. Vol'pin, *Z. Chem.*, **12**, 361 (1972).
57) M. E. Vol'pin and I. S. Kolomnikov, *Pure Appl. Chem.*, **33**, 567 (1973).
58) M. Kobayashi and S. Inoue, *Kagaku* (Japanese), **28**, 942 (1973).
59) M. Hidai, T. Hikida, and Y. Uchida, *J. Japan Petroleum Institute* (Japanese), **17**, 193 (1974).
60) S. Inoue, *J. Japan Petroleum Institute* (Japanese), **17**, 199 (1974).
61) S. Inoue, *Miriyo-Shigen to Gosei-Kagaku, Kagaku-Zokan,* **63** (Japanese), p. 29, Kagaku-Dojin(1974).
62) M. E. Vol'pin and I. S. Kolomnikov, *Organometallic Reactions*, **5**, 313 (1975).
63) T. Saegusa and T. Tsuda, *Chemical Industry* (Japanese), **26**, 562 (1975).
64) T. Ito and A. Yamamoto, *J. Synth. Org. Chem., Japan* (Japanese), **34**, 308 (1976).
65) I. S. Kolomnikov and M. Kh. Grigoryan, *Usp. Khim.*, **47**, 603 (1978); *Russ. Chem. Reviews*, **47**, 334 (1978).
66) A. D. Allen and C. V. Senoff, *Chem. Commun.*, **1965**, 621.
67) D. W. Turner and D. P. May, *J. Chem. Phys.*, **46**, 1156 (1967).
68) S. M. Vinogradova and G. L. Gutsev, *Koord. Khim.*, **4**, 992 (1978); *Chem. Abst.*, **89**, 118039k (1978).
69) J. Chatt, M. Kubota, G. J. Leigh, F. C. March, R. Mason, and D. J. Yarrow, *J. Chem. Soc., Chem. Commun.*, **1974**, 1033.
70) H. H. Karsch, *Chem. Ber.*, **110**, 2213 (1977).
71) A. Simon, G. Speier, and L. Marko, Proc. 18th Int. Conf. Coordination Chemistry, vol. 1, p. 154, Cracow-Zakopane, Poland (1970).
72) B. R. Flynn and L. Vaska, *J. Chem. Soc., Chem. Commun.*, **1974**, 703.
73) M. Aresta and C. F. Nobile, *Inorg. Chim. Acta*, **24**, L49 (1977).
74) T. Herskovitz and L. J. Guggenberger, *J. Am. Chem. Soc.*, **98**, 1615 (1976).
75) T. Herskovitz, *J. Am. Chem. Soc.*, **99**, 2391 (1977).
76) M. Aresta, C. F. Nobile, V. G. Albano, E. Forni, and M. Manassero, *J. Chem. Soc., Chem. Commun.*, **1975**, 637.
77) M. Aresta and C. F. Nobile, *J. Chem. Soc., Dalton Trans.*, **1977**, 708.
78) J. Vlčková and J. Barton, *J. Chem. Soc., Chem. Commun.*, **1973**, 306.
79) T. Yamamoto, M. Kubota, and A. Yamamoto, *Bull. Chem. Soc. Japan*, **53**, 680 (1980).
80) Y. Iwashita and A. Hayata, *J. Am. Chem. Soc.*, **91**, 2525 (1969).
81) M. E. Vol'pin, I. S. Kolomnikov, and T. S. Loobeva, *Izvest. Akad. Nauk SSSR, Ser. Khim.*, **1969**, 2084.
82) I. S. Kolomnikov, T. S. Belopotapova, T. V. Lysyak, and M. E. Vol'pin, *J. Organometal. Chem.*, **67**, C25 (1974).
83) P. W. Jolly, K. Jonas, C. Krüger, and Y.-H. Tsay, *J. Organometal. Chem.* **33**, 109 (1971).
84) C. D. M. Beverwijk and G. J. M. van der Kerk, *J. Organometal, Chem.* **49**, C59 (1973).
85) K. Nakamoto, *Infrared and Raman Spectra of Inorganic and Coordination Compounds*, 3rd. ed., p. 117, John Wiley & Sons (1978).
86) T. A. Stephenson, S. M. Morehouse, A. R. Powell, J. P. Heffer, and G. Wilkinson, *J. Chem. Soc.*, **1965**, 3632.
87) T. Kashiwagi, N. Yasuoka, T. Ueki, N. Kasai, M. Kakudo, S. Takahashi, and N. Hagihara, *Bull. Chem. Soc. Japan*, **41**, 301 (1968).
88) R. Mason and A. I. M. Rae, *J. Chem. Soc., A*, **1970**, 1768.
89) H. Felkin and P. J. Knowles, *J. Organometal. Chem.*, **37**, C14 (1972).
90) H. Felkin, P. J. Knowles, and B. Meunier, *J. Organometal. Chem.*, **146**, 151 (1978).
91) S. G. Davies and M. L. H. Green, *J. Chem. Soc., Dalton Trans.*, **1978**, 1510.

92) R. A. Forder, M. L. H. Green, R. F. MacKenzie, J. S. Poland, and K. Prout, *J. Chem. Soc., Chem. Commun.*, **1973**, 426.

93) B. Demerseman, G. Bouquet, and M. Bigorgne, *J. Organometal. Chem.*, **145**, 41 (1978).

94) G. O. Evans, W. F. Walter, D. R. Mills, and C. A. Streit, *J. Organometal. Chem.*, 144, C34 (1978).

95) P. Svoboda, T. S. Belopotapova, and J. Hetflejš, *J. Organometal. Chem.*, **65**, C37 (1974).

96) T. Ito and A. Yamamoto, *J. Chem. Soc., Dalton Trans.*, **1975**, 1398.

97) H.-F. Klein and H. H. Karsch, *Chem. Ber.*, **108**, 944 (1975).

98) S. Krogsrud, S. Komiya, T. Ito, J. A. Ibers, and A. Yamamoto, *Inorg. Chem.*, **15**, 2799 (1976).

99) S. Komiya and A. Yamamoto, *J. Organometal. Chem.*, **46**, C58 (1972).

100) F. Ozawa, T. Ito, and A. Yamamoto, *Chem. Lett.*, **1979**, 735.

101) T. Yoshida, D. L. Thorn, T. Okano, J. A. Ibers, and S. Otsuka, *J. Am. Chem. Soc.*, **101**, 4212 (1979).

102) T. Tsuda, Y. Chujo, and T. Saegusa, *J. Am. Chem. Soc.*, **120**, 431 (1980).

103) C. J. Nyman, C. E. Wymore, and G. Wilkinson, *J. Chem. Soc., A*, **1968**, 561.

104) P. J. Hayward, D. M. Blake, G. Wilkinson, and C. J. Nyman, *J. Am. Chem. Soc.*, **92**, 5873 (1970).

105) H. C. Clark, A. B. Goel, and C. S. Wong, *J. Organometal. Chem.*, **152**, C45 (1978).

106) M. C. Scrutton, *Inorganic Biochemistry*, (ed. G. L. Eichhorn), vol. 1, p. 401, Elsevier (1973).

107) S. Komiya and A. Yamamoto, *J. Organometal. Chem.*, **46**, C58 (1972); *Bull. Chem. Soc. Japan*, **49**, 784 (1976).

108) I. S. Kolomnikov, A. I. Gusev, G. C. Aleksandrov, T. S. Lobeeva, Yu. T. Struchkov, and M. E. Vol'pin, *J. Organometal. Chem.*, **59**, 350 (1973).

109) A. Immirzi and A. Musco, *Inorg. Chim. Acta.*, **22**, L35 (1977).

110) C. Floriani and G. Fachinetti, *J. Chem. Soc., Chem. Commun.*, **1974**, 615.

111) G. Fachinetti, C. Floriani, and P. F. Zanazzi, *J. Am. Chem. Soc.*, **100**, 7406 (1978).

112) T. Tsuda, S. Sanada, K. Ueda, and T. Saegusa, *Inorg. Chem.*, **15**, 2329 (1976).

113) T. Tsuda, Y. Chujo, and T. Saegusa, *J. Chem. Soc., Chem. Commun.*, **1975**, 963.

114) T. Tsuda, Y. Chujo, and T. Saegusa, *J. Am. Chem. Soc.*, **100**, 630 (1978).

115) T. Tsuda, Y. Chujo, and T. Saegusa, *J. Chem. Soc., Chem. Comm.*, **1976** 415.

116) L. S. Pu, A. Yamamoto, and S. Ikeda, *J. Am. Chem. Soc.*, **90**, 3896 (1968).

117) A. Yamamoto, S. Kitazume, L. S. Pu, and S. Ikeda, *J. Am. Chem. Soc.*, **93**, 371 (1971).

118) A. Misono, Y. Uchida, M. Hidai, and T. Kuse, *Chem. Commun.*, **1968**, 981.

119) V. D. Bianco, S. Doronzo, and M. Rossi, *J. Organometal. Chem.*, **35**, 337 (1972).

120) G. LaMonica, S. Ceinini, F. Porta, and M. Pizzotti, *J. Chem. Soc., Dalton Trans.*, **1976**, 1777.

121) G. Fachinetti, C.Floriani, A. Roselli,and S.Pucci, *J. Chem.Soc., Chem. Commun.*, **1978**, 269.

122) I. S. Kolomnikov, G. Stempowska, S. Tyrlik, and M. E. Vol'pin, *Zhur. Obshch. Khim.*, **42**, 1652 (1972).

123) U. Zucchini, E. Albizzati, and U. Giannini, *J. Organometal. Chem.*, **26**, 357 (1971).

124) I. S. Kolomnikov, T. S. Lobeeva, V. V. Gorbachevskaya, G.G. Aleksandrov, Y. T. Struchkov, and M. E. Vol'pin, *Chem. Commun.*, **1971**, 972.

125) I. S. Kolomnikov, T. S. Lobeeva, and M. E. Vol'pin, *Zhur. Obshch. Khim.*, **42**, 2232 (1972).

126) M. Kh. Grigoryan, I. S. Kolomnikov, E. G. Berkovich, T. V. Lysyak, V. B. Shur, and M. E. Vol'pin, *Izv. Akad. Nauk SSSR, Ser. Khim.*, **1978**, 1177.

127) G. N. Schrauzer and J. W. Sibert, *J. Am. Chem. Soc.*, **92**, 3509 (1970).

128) I. S. Komlonikov, A. O. Gusev, T. S. Belopotapova, M. Kh. Grigoryan, T. V. Lysyak, Yu. T. Struchkov, and M. E. Vol'pin, *J. Organometal. Chem.*, **69**, C10 (1974).

129) K. Maruyama, T. Ito, and A. Yamamoto, *Bull. Chem. Soc. Japan*, **52**, 849 (1979).
130) T. Yamamoto and A. Yamamoto, *Chem. Lett.*, **1978**, 615.
131) T. Tsuda, Y. Chujo, and T. Saegusa, *Synth. Comm.*, in press.
132) P. W. Jolly, S. Stobbe, G. Wilke, R. Goddard, C. Krüger, J. C. Sekutowski, and Y.-H. Tsay, *Angew. Chem., Int. Ed. Engl.*, **17**, 124 (1978).
133) A. D. English and T. Herskovitz, *J. Am. Chem. Soc.*, **99**, 1648 (1977).
134) C. D. Ittel, C. A. Tolman, A. D. English, and J. P. Jesson, *J. Am. Chem. Soc.*, **100**, 7577 (1978).
135) A. Miyashita and A. Yamamoto, *J. Organometal. Chem.*, **49**, C57 (1973); A. Miyashita and A. Yamamoto, *J. Organometal. Chem.*, **113**, 187 (1976).
136) T. Ikariya and A. Yamamoto, *J. Organometal. Chem.*, **72**, 145 (1974).
137) T. Tsuda, K. Ueda, and T. Saegusa, *J. Chem. Soc., Chem. Commun.*, **1974**, 380.
138) J. F. Normant, G. Cahitz, C. Chuit, and J. Villieras, *J. Organometal. Chem.*, **77**, 281 (1974).
139) G. Dubot, D. Mansuy, S. Lecolier, and J. F. Normant, *J. Organometal. Chem.*, **42**, C105 (1972).
140) G. Chandra, A. D. Jenkins, M. F. Lappert, and R. C. Srivastava, *J. Chem. Soc., A*, **1970**, 2550.
141) R. Choukoun and D. Gervais, *Synth. React. Inorg. Met.-Org. Chem.*, **8**, 137 (1978).
142) M. H. Chisholm and M. Extine, *J. Am. Chem. Soc.*, **96**, 6214 (1974).
143) M. H. Chisholm and M. Extine, *J. Chem. Soc., Chem. Commun.*, **1975**, 438.
144) M. H. Chisholm and M. Extine, *J. Am. Chem. Soc.*, **97**, 1623 (1975).
145) M. H. Chisholm and M. Extine, *J. Am. Chem. Soc.*, **99**, 782 (1977).
146) M. H. Chisholm and M. Extine, *J. Am. Chem. Soc.*, **99**, 792 (1977).
147) M. H. Chisholm, F. A. Cotton, and M. W. Extine, *Inorg. Chem.*, **17**, 2000 (1978).
148) M. H. Chisholm, F. A. Cotton, M. W. Extine, and D. C. Rideout, *Inorg. Chem.*, **17**, 3536 (1978).
149) M. H. Chisholm and M. Extine, *J. Am. Chem. Soc.*, **97**, 5625 (1975).
150) M. H. Chisholm, M. Extine, F. A. Cotton, and B. R. Stultz, *J. Am. Chem. Soc.* **98**, 4683 (1976).
151) M. H. Chisholm and W. W. Reichert, *Inorg. Chem.*, **17**, 767 (1978).
152) Y. Yoshida and S. Inoue, *Bull. Chem. Soc. Japan*, **51**, 559 (1978).
153) T. V. Ashworth, M. Nolte, and E. Singleton, *J. Organometal. Chem.*, **121**, C58 (1976).
154) T. Tsuda, H. Washita, K. Watanabe, M. Miwa, and T. Saegusa, *J. Chem. Soc., Chem. Commun.*, **1978**, 815.
155) J. N. Armor, *Inorg. Chem.*, **17**, 213 (1978).
156) F. Calderazzo, G. dell'Amico, R. Netti, and M. Pasquali, *Inorg. Chem.*, **17**, 471 (1978).
157) T. Tsuda, H. Washita, and T. Saegusa, *J. Chem. Soc., Chem. Commun.*, **1977**, 468.
158) E. Chaffee, T. P. Dasgupta, and G. M. Harris, *J. Am. Chem. Soc.*, **95**, 4169 (1973).
159) T. P. Dasgupta and G. M. Harris, *Inorg. Chem.*, **17**, 3304 (1978).
160) B. R. Flynn and L. Vaska, *J. Am. Chem. Soc.*, **95**, 5081 (1973).
161) T. Ito, H. Tsuchiya, and A. Yamamoto, *Chem. Lett.*, **1976**, 851.
162) R. J. Crutchley, J. Powell, R. Faggiani, and C. J. L. Lock, *Inorg. Chim. Acta.*, **24**, L15 (1977).
163) T. V. Ashworth and E. Singleton, *J. Chem. Soc., Chem. Commun.*, **1975**, 204.
164) T. Tsuda, S.-I. Sanada, and T. Saegusa, *J. Organometal. Chem.*, **116**, C10 (1976).
165) M. H. Chisholm, W. W. Reichert, F. A. Cotton, and C. A. Murillo, *J. Am. Chem. Soc.*, **99**, 1653 (1977).
166) M. H. Chisholm, F. A. Cotton, M. W. Extine, and W. W. Reichert, *J. Am. Chem. Soc.* **100**, 1727 (1977).
167) M. Hidai, T. Hikita, and Y. Uchida, *Chem. Lett.*, **1972**, 521.
168) A. J. Goodsel and G. Blyholder, *J. Am. Chem. Soc.*, **94**, 6725 (1972).
169) P. Haynes, L. H. Slaugh, and J. F. Kohnle, *Tetrahedron Lett.*, **1970**, 365.

170) K. Kudo, H. Phala, N. Sugita, and Y. Takezaki, *Chem. Lett.*, **1977**, 1495.
171) I. S. Kolomnikov, T. S. Lobeeva, and M. E. Vol'pin, *Izv. Akad. Nauk SSSR, Ser. Khim.*, **1972**, 2329.
172) Y. Inoue, Y. Sasaki, and H. Hashimoto, *J. Chem. Soc., Chem. Commun.*, **1975**, 718.
173) G. O. Evans and C. J. Newell, *Inorg. Chim. Acta.*, **31**, L387 (1978).
174) H. Koinuma, Y. Yoshida, and H. Hirai, *Chem. Lett.*, **1975**, 1223.
175) C.-H. Cheng and R. Eisenberg, *J. Am. Chem. Soc.*, **100**, 5968 (1978), and references cited therein.
176) Y. Inoue, H. Izumida, Y. Sasaki, and H. Hashimoto, *Chem. Lett.*, **1976**, 863.
177) K. Kudo, N. Sugita, and Y. Takezaki, *Nippon Kagakukaishi* (Japanese), **1977**, 302.
178) H. Koinuma, H. Kato, and H. Hirai, *Chem. Lett.*, **1977**, 517.
179) R. J. DePasquale, *J. Chem. Soc., Chem. Commun.*, **1973**, 157.
180) B. Jeżoska-Trzebiatowska and P. Sobota, *J. Organometal. Chem.*, **80**, C27 (1974).
181) B. Jeżoska-Trzebiatowska and P. Sobota, *J. Organometal. Chem.*, **76**, 43 (1974).
182) J. F. Kohnle, L. H. Slaugh, and K. L. Nakamaye, *J. Am. Chem. Soc.*, **91**, 5904 (1969).
183) A. Musco and A. Silvani, *J. Organometal. Chem.*, **88**, C41 (1975).
184) A. Musco, *J. Molecular Catalysis*, **1**, 443 (1975/76).
185) K. E. Atkins, W. E. Walker, and R. M. Manyik, *Chem. Commun.*, **1971**, 330.
186) M. Hidai, H. Mizuta, K. Hirai, and Y. Uchida, 25th Symposium on Organometallic Chemistry Japan, Osaka, 1978, Abstracts, p. 135.
187) Y. Sasaki, Y. Inoue, and H. Hashimoto, *J. Chem. Soc., Chem. Commun.*, **1976**, 605.
188) Y. Inoue, Y. Sasaki, and H. Hashimoto, *Bull. Chem. Soc. Japan*, **51**, 2375 (1978).
189) A. Musco, C. Perego, and V. Tartiari, *Inorg. Chim. Acta.*, **28**, L147 (1978).
190) S. Inoue, S. Sekiya, Y. Sasaki, and H. Hashimoto, *J. Synth. Org. Chem., Japan* (Japanese), **36**, 328 (1978).
191) Y. Inoue, Y. Itoh, and H. Hashimoto, *Chem. Lett.*, **1977**, 855.
192) Y. Inoue, Y. Itoh, and H. Hashimoto, *Chem. Lett.*, **1978**, 633.
193) C. A. Tolman, *Chem. Soc. Rev.*, **1**, 346 (1972).
194) G. W. Parshall, *Acc. Chem. Res.*, **8**, 113 (1975).
195) S. Komiya, T. Ito, M. Cowie, A. Yamamoto, and J. A. Ibers, *J. Am. Chem. Soc.*, **98**, 3874 (1976), and references cited therein.
196) A. E. Shilov and A. A. Shteinman, *Coord. Chem. Rev.*, **24**, 97 (1977).

4

Syntheses of Macromolecules
from Carbon Dioxide

Noboru YAMAZAKI[*1]
Department of Polymer Science, Tokyo Institute of Technology
Ohokayama, Meguro-ku, Tokyo 152, Japan

Fukuji HIGASHI[*1]
Faculty of Engineering, Tokyo University of Agriculture
and Technology, Koganei-shi, Tokyo 184, Japan

Shohei INOUE[*2]
Faculty of Engineering, the University of Tokyo
Bunkyo-ku, Tokyo 113, Japan

4.1. CONDENSATION POLYMERIZATIONS
INVOLVING CARBON DIOXIDE

4.1.1. Introduction

Condensation reactions involving carbon dioxide have usually been performed under drastic conditions such as high pressure and high temperature. As a result only low molecular weight polymers could be obtained when these reactions were employed for the preparation of polymers. Recently, however, new condensation reactions with carbon dioxide have been developed in the field of organic chemistry, some of which are applicable to the preparation of polymers.

We demonstrated a direct condensation reaction with carbon dioxide and amine under mild conditions using phosphorus compounds as condensing agents. This reaction has been successfully applied to the synthesis of polyurea with aryl diamines and yielded a very high molecular weight polymer. Soga *et al.* have also reported a new method for the preparation of polycarbonates from carbon dioxide, dihalides, and diols in the presence of crown ethers as catalysts.

This portion of Chapter 4 will limit its focus to polycondensation reactions involving carbon dioxide that produce polyureas and polycarbonates.

4.1.2. Polyureas

A. Reactions of *N*-phosphonium salts from phosphoric acid and its esters

Phosphites, especially diphenyl and triaryl phosphites, have been found to react with carboxylic acids in the presence of pyridine to give the acyloxy-*N*-phosphonium salts of pyridine (**1** and **2**, Scheme 4.1); dephenoxylation occurs at the same tine, producing the corresponding amides and esters on aminolysis and alcoholysis. The *N*-phosphonium salts were presumed to be present because of the stoichiometric relationship among the phosphite, pyridine, and carboxyl components.[1]

In these reactions, hydrolysis of diphenyl and triaryl phosphites to monoaryl phosphites and phenol was coupled by dehydration between carboxylic acids and amines or alcohols to the corresponding amides and esters. Therefore, the reaction was generalized as a hydrolysis-dehydration reaction (Scheme 4.1).

The proposed concept of the hydrolysis-dehydration reaction using phosphites was also shown to be applicable to reactions with other

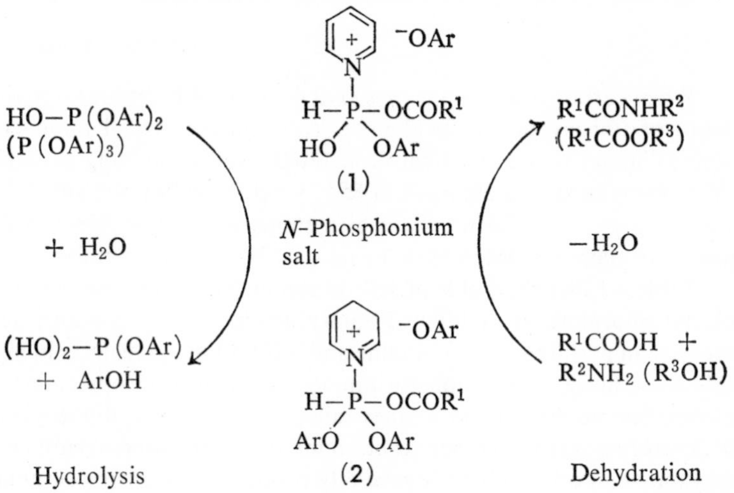

$$HO-P(OAr)_2 \quad (P(OAr)_3)$$

$$+ \; H_2O$$

$$(HO)_2-P(OAr) \\ + \; ArOH$$

Hydrolysis

$$\overset{+}{N}\text{—} \; {}^-OAr$$
$$H\text{—}\overset{|}{P}\text{—}OCOR^1$$
$$HO \quad OAr$$
$$(1)$$

N-Phosphonium salt

$$\overset{+}{N}\text{—} \; {}^-OAr$$
$$H\text{—}\overset{|}{P}\text{—}OCOR^1$$
$$ArO \quad OAr$$
$$(2)$$

$$R^1CONHR^2 \\ (R^1COOR^3)$$

$$- \; H_2O$$

$$R^1COOH \; + \\ R^2NH_2 \; (R^3OH)$$

Dehydration

Scheme 4.1

phosphorus compounds, such as phosphonites, phosphinites, and phosphonates (Eqs. 4.1–4.3).[2]

$$\overset{R}{\underset{R}{\diagdown}}P\text{-}OR + R^1COOH + RNH_2(R^3OH) \xrightarrow{Py} R^1CONHR^2(R^1COOR^3) + ROH$$

$$\overset{R}{\underset{R}{\diagdown}}P\text{-}OH \qquad (4.1)$$

$$R\text{-}P(OR)_2 + R^1COOH + R^2NH_2(R^3OH) \xrightarrow{Py} R^1CONHR^2(R^1COOR^3) + ROH$$

$$R\text{-}P(OH)(OR) \qquad (4.2)$$

$$\overset{O}{\overset{\|}{R\text{-}P(OR)_2}} + R^1COOH + R^2NH_2(R^3OH) \xrightarrow{Py} R^1CONHR^2(R^1COOR^3) + ROH$$

$$\underset{\underset{O}{\|}}{R\text{-}P(OR)(OH)} \qquad (4.3)$$

Since the chemical reactivity of carboxylic acids is similar to that of carbonic acid, as observed in amide and ester formation, the substitution of carbon dioxide for carboxylic acids in the coupling reaction with amines was attempted using phosphite in pyridine or imidazole, and it was found that ureas are, in fact, produced in good yields (Eq. 4.4).[3]

$$CO_2 + 2\ RNH_2 + HO\ P(OPh)_2 \xrightarrow[Py]{} RNHCONHR$$

$$+ (HO)_2\text{-}P(OPh) + PhOH \quad (4.4)$$

When carbon dioxide was passed at 40°C for 4 h through a pyridine solution of diphenyl phosphite (2 moles, twice the theoretical amount) and aniline (1 mole) in a cylindrical flask fitted with a bubbling glass filter, N,N'-diphenylurea was obtained in 85% yield (based on the aniline), together with phenol. The urea was also prepared in 93% yield in N,N-dimethylformamide (DMF) by using imidazole in lieu of pyridine.

Table 4.1 lists the yields of several symmetrical ureas prepared with diphenyl phosphite in pyridine. Primary aromatic amines with a lower basicity than pyridine, such as aniline and o-toluidine, gave ureas in higher yields than more basic aliphatic amines, such as isopropylamine and cyclohexylamine. Secondary amines failed to give ureas; diphenylamine (90%) was recovered from one reaction mixture, and N-methylaniline and diethylamine in other mixtures gave oily products which showed none of the C=O stretching characteristic of ureas in the infrared spectra.

Table 4.1.   Symmetrical ureas prepared from carbon dioxide and amines with diphenyl phosphite in pyridine.[†1]

Amine	Yield (%) of ureas[†2]
Aniline	85
o-Toluidine	74
Isopropylamine	23
Cyclohexylamine	27
N-Methylaniline	0
Diphenylamine	0
Diethylamine	0

[†1] $HO\text{-}P(OC_6H_5)_2$ = amine, 50 mmoles in 40 ml of pyridine; temp. 40°C; time, 4 h.
[†2] Based on the amine used.

Unsatisfactory results with aliphatic amines may be due to a retardation in the formation of carbamoyloxy N-phosphonium salts (3, Scheme 4.2); instead, the production of pyridine-insoluble and unreactive ammonium salts may predominate because of their higher basicity than pyridine. Steric hindrance around the nitrogen atom may account for the lack of reaction with secondary amines as found in an unsuccessful attempt at direct carbonylation of diphenylamine with phosgene.[4]

Tertiary amines facilitated the reaction of carbon dioxide with an amine (aniline) (Table 4.2). The yield of urea was affected by both the basicity of the tertiary amines and the steric hindrance around the nitrogen atom. Steric hindrance may be more influential on the reaction in pyridine derivatives (pK_a values in the range of 5.0–7.0), as reflected by unsatis-

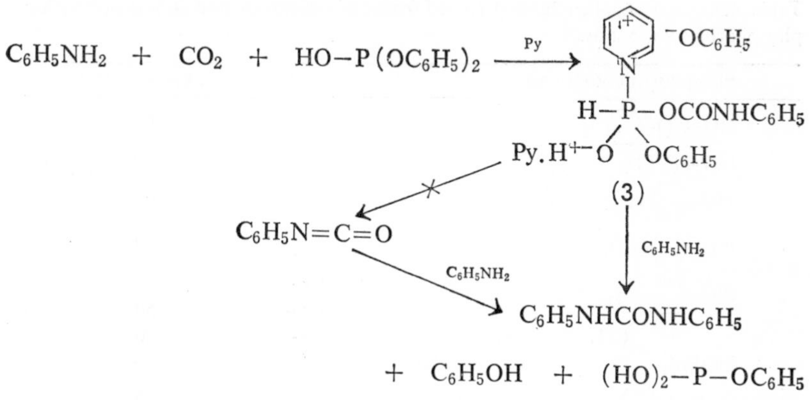

$$C_6H_5NH_2 \ + \ CO_2 \ + \ HO-P(OC_6H_5)_2 \ \xrightarrow{\text{Py}}$$

Scheme 4.2

Table 4.2. N,N'-Diphenyl urea prepared from carbon dioxide with aniline in various tertiary amines.[†1]

Tertiary amine	pK_a	Yield (%)[†2]
Imidazole	7.12	93
γ-Picoline	6.02	88
β-Picoline	5.52	84
Pyridine	5.23	85
2,4-Lutidine	6.99	74
α-Picoline	5.97	60
2,6-Lutidine	6.99	32
Triethylamine	10.87	70
None	—	4[†3]

[†1] $HO-P(OC_6H_5)_2$ = aniline, 50 mmoles in 40 ml of tertiary amines; imidazole, 100 mmoles in 40 ml of DMF; temp., 40°C; time, 4 h.
[†2] Base on aniline used.
[†3] The reaction was carried out in 40 ml of DMF.

factory results in α-picoline, in 2,4-lutidine, and, remarkably, in 2,6-lutidine. Unlike the poor results obtained in the amidation of carboxylic acids, triethylamine with high basicity gave a favorable result despite its considerable steric hindrance.

Urea production was also favored by the presence of phosphorus compounds other than diphenyl phosphite (Table 4.3). As in the coupling between carboxylic acids and amines by hydrolysis-dehydration, triaryl phosphites, diphenyl ethylphosphonite, phenyl diphenylphosphinite, and diphenyl n-butylphosphonate were effective for the reaction of carbon dioxide and amines. No urea was produced when these phenyl esters were replaced with alkyl esters. Both triphenyl and triethyl phosphates failed to promote the reaction.

Table 4.3. *N,N'*-Diphenylurea prepared from carbon dioxide and aniline with various phosphorus compounds.[†1]

Phosphorus compound		Yield (%)[†2]
HO–P(O–$\langle\!\!\langle\rangle\!\!\rangle$)$_2$		94
HO–P(OEt)$_2$		0
P(O–$\langle\!\!\langle\rangle\!\!\rangle$)$_3$		68
P(O–$\langle\!\!\langle\rangle\!\!\rangle$–Cl)$_3$		87
P(O–$\langle\!\!\langle\rangle\!\!\rangle$$\diagdown$CH$_3$)$_3$	o- m- p-	21 50 40
P(OBu)$_3$		0
Et–P(O–$\langle\!\!\langle\rangle\!\!\rangle$)$_2$		90
Et–P(OEt)$_2$		0
($\langle\rangle$)$_2$ P–O–$\langle\!\!\langle\rangle\!\!\rangle$		71
Bu–P(O)(O–$\langle\!\!\langle\rangle\!\!\rangle$)$_2$		15
Et–P(O)(OEt)$_2$		0
O=P(O–$\langle\!\!\langle\rangle\!\!\rangle$)$_3$		0
O=P(OEt)$_3$		0

[†1] Phosphorus compounds = aniline, 50 mmoles in 40 ml of DMF containing imidazole (100 mmoles); temp., 40° C; time, 4 h.
[†2] Based on aniline used.

Both phosphorus compounds and tertiary amines were found to be stoichiometrically involved in the reaction between carbon dioxide and aniline. We investigated the effect of the reaction conditions on the yield of urea in the reaction with aniline (2 moles) using diphenyl phosphite (1 mole) in pyridine. Better results were obtained at higher reaction temperatures, and the yield approached that theoretically predicted by Eq. (4.4), despite a decrease in the concentration of carbon dioxide in the reaction mixture at high temperatures. In addition, a higher concentration of reactants in the mixture resulted in a higher rate of urea production (Table 4.4). These results suggest that the initial reaction of carbon dioxide,

Table 4.4. Effect of the concentration of reactants on urea formation[†]

HO–P(OC$_6$H$_5$)$_2$ mmole	C$_6$H$_5$NH$_2$ mmole	*N,N'*-Diphenylurea Yield, m mole
12	24	4.7
24	48	11.7
48	96	23.1

[†] The reaction was carried out at 40° C for 4 h in 40 ml of pyridine.

phosphite, and aniline, which probably forms **3**, is rapid. So a rate-determining step is the aminolysis of **3** with amine into the urea.

The reaction of carbon dioxide with amines is thought to follow a hydrolysis-dehydration scheme similar to that in the amidation of carboxylic acids with amines. The reaction with CO_2 using diphenyl phosphite in pyridine, for example, may be represented as in Scheme 4.2. A carbamoyloxy N-phosphonium salt of pyridine (**3**) is initially formed by the reaction of diphenyl phosphite, carbon dioxide, and an amine (aniline) in pyridine, and a phenolate anion is released from the phosphite. Compound **3** is converted into the corresponding urea, monophenyl phosphite, and phenol by aminolysis with the amine. Another possible pathway for urea formation through the isocyanate from **3**, can be ruled out by the finding in a separate experiment that only urea, and no urethane, was obtained from the reaction in the presence of both phenol and aniline.

Similarly, the reactions using triaryl phosphites, diphenyl ethylphosphonite, phenyl diphenylphosphinite, and diphenyl *n*-butylphosphonate may proceed via the corresponding carbamoyloxy N-phosphonium salts, like **3**, produced by hydrolysis-dehydration followed by aminolysis.

B. Polyureas from carbon dioxide and diamines with various N-phosphonium salts of pyridines

The reactions of these N-phosphonium salts of pyridines with CO_2 have been extended to the preparation of polyureas from carbon dioxide and diamines at ordinary or low pressure and at low temperature.[5] Although the preparation of polyureas from carbon dioxide under drastic conditions (high temperature and high pressure)[6] or from carbon oxysulfide[7] has been reported, no preparative methods operative under moderate conditions have been described.

When CO_2 was passed into a mixture of diphenyl phosphite and diamines in pyridine, polyureas were obtained in quantitative yield, but their molecular weights were not satisfactorily high. An increased pressure of CO_2 enhanced the rate of polycondensation and the molecular weight of the resulting polymer. Therefore, polycondensation under a pressure of CO_2 was studied in detail.

$$(PhO)_2\text{-}P(OH) + NH_2\text{-}R\text{-}NH_2 \xrightarrow[Py]{CO_2} \{NH\text{-}R\text{-}NHCO\}_n$$

$$+ (HO)_2\text{-}P(OPh) + PhOH \qquad (4.5)$$

The polycondensation reaction (Eq. 4.5) was carried out by introducing carbon dioxide at a given initial pressure into a 200-m*l* autoclave containing phosphorus compounds and a diamine in pyridine. The CO_2 pressure was allowed to drop during polycondensation.

The polycondensation of 4,4'-diaminodiphenylmethane (MDA) with CO_2 using diphenyl phosphite in pyridine was carried out in order to ascertain the effect of the initial pressure of CO_2 on the molecular weight of the resulting polyurea (Fig. 4.1). Although a quantitative yield of polymer was obtained independent of the pressure, a maximum viscosity of the product was obtained when the reaction was carried out at a pressure around 20 atm. The reaction temperature also affected the molecular weight of the resulting polymer, and an optimum viscosity was observed at temperatures around 40° C.

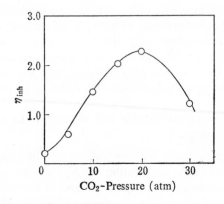

Fig. 4.1. Effect of pressure of carbon dioxide on the viscosity of polyurea from 4,4'-diaminodiphenylmethane (MDA) at 40°C.

The unfavorable effects of both high temperature and high pressure on the molecular weight may be caused by various reactions: depolymerization reactions, intermolecular or intramolecular exchange reactions of amino groups between polymer chains (Eq. 4.6),[8] and CO_2-catalyzed exchanges of amino groups (Eq. 4.7).[9]

$$NH_2-R-NH_2 + NH_2-CO-NH_2 \longrightarrow 1/n\{NH-R-NHCO\}_n + 2\,NH_3 \qquad (4.6)$$

$$R^1CONHR^2 + R^3NH_2 \xrightarrow{CO_2} R^1CONHR^3 + R^2NH_2 \qquad (4.7)$$

To confirm the occurrence of these side reactions, the relationship between the viscosity and the molecular weight (calculated from end groups) was investigated using polyurea from 2,2-bis-[p-(4-aminophenoxy) phenyl]propane (PDAP) since the polymer from MDA was insoluble in *m*-cresol (Table 4.5). End group analysis was conducted in *m*-cresol by means of potentiometric titration with *p*-toluenesulfonic acid. As with polyurea from MDA, the viscosity of the polymer from PDAP was lower at a high concentration of carbon dioxide (40 atm) than at 20 atm (Nos. 1 and 4, Table 4.5).

Table 4.5. Relationship between viscosity and molecular weight calculated by end group analysis of polyurea from 2,2-bis-[*p*-(4-aminophenoxy)phenyl]propane

No.	Polycondensation conditions	Further treatment[1]	η_{inh}[2]	$\bar{M}n \times 10^4$[3]
1	20 atm, 40°C, 4 h	—	2.14	5.41
2	—	CO_2 (40 atm)	1.87	5.54
3	—	CO_2 (40 atm) and phosphite	1.73	5.41
4	40 atm, 40°C, 4 h	—	1.43	3.20

[1] Polymer, 0.8 g; [HO–P(OC$_6$H$_5$)$_2$], 0.01 mole in 20 ml of pyridine; temp., 60°C; time, 4 h.
[2] Measured in HMPA at 30°C.
[3] Calculated from results of end group analysis by potentiometric titration in *m*-cresol.

When the original polymer ($\eta_{inh} = 2.14$) was further treated with CO_2 in pyridine in the absence or presence of diphenyl phosphite, the viscosity of the recovered polymer was reduced, but no practical changes in the molecular weight were observed (Nos. 2 and 3). These results suggest that the side reactions are mainly exchange reactions, such as those shown in Eqs. (4.6) and (4.7), rather than depolymerization, and that an intramolecular exchange reaction to form a cyclic polymer is possibly involved.

The undesirable low viscosity of the polymer obtained at high temperatures and pressures may be attributed to these exchange reactions that occur during polycondensation; they may be promoted by higher reaction temperatures, higher pressures of carbon dioxide, as well as the presence of carbon dioxide and diphenyl phosphite.

The results of the polycondensation of MDA and CO_2 (10 atm) in DMF containing various amounts of pyridine or in a variety of other solvents are given in Tables 4.6 and 4.7. Polymer was obtained in almost quantitative yields independent of the amount of pyridine, but the viscosity increased with the amount of pyridine, the highest value being obtained in pyridine alone. Highly polar solvents such as DMF or NMP, which readily dissolve polyureas, gave rather good results. Hexamethylphosphoric triamide (HMPA), however, was less effective for the reaction, despite its strong dissolving capability and marked efficiency under atmospheric

Table 4.6.   Polycondensation of carbon dioxide (10 atm) and 4,4'diaminodiphenylmethane (MDA) with diphenyl phosphite in the presence of various amounts of pyridine in DMF.†

Solvent (Py/DMF)	Polymer	
(m/l/ml)	Yield (%)	η_{inh}
5/45	97	0.18
10/40	98	0.24
20/30	100	0.48
30/20	100	1.14
50/ 0	100	1.45

† The reaction was carried out at 40°C for 4 h in 50 ml of solvent.

Table 4.7.   Polycondensation of carbon dioxide (10 atm) and MDA with diphenyl phosphite in various solvents.†

Solvent	Polymer	
	Yield (%)	η_{inh}
Pyridine	100	1.45
DMF	100	0.48
NMP	100	0.32
DMAc	97	0.24
HMPA	68	0.14
Toluene	84	0.14
Dioxane	86	0.14
Ethanol	0	—

† The reaction was carried out at 40°C for 4 h in pyridine or in various solvents containing pyridine (8 equiv.).

conditions;[10] the unsatisfactory results may have been a product of a reaction of the HMPA with the CO_2 under pressure. No polymer was obtained in ethanol.

The effect of tertiary amines on the polycondensation reaction of MDA with CO_2 (20 atm) in DMF is shown in Table 4.8. The yield and viscosity of the polymer were affected by the basicity of the amine, reaching a maximum when an amine with a pKa value around 5.5, such as β-picoline, was used. The finding of higher viscosity in α-picoline than in γ-picoline suggested that steric hindrance around the nitrogen atom may not have a marked influence on the polycondensation.

Several phosphorus compounds were used for the reaction of MDA with CO_2 (20 atm) in pyridine or in MDF (Table 4.9). Aryl esters of phosphorous acid, especially diphenyl phosphite, were remarkably effective. Triphenyl ester yielded polymer in the presence of imidazole, but not in pyridine. Phosphoric acid and its alkyl esters failed to give polymer, however. This lack of reactivity may be attributed to difficulty in forming an

Table 4.8.   Polycondensation of carbon dioxide (20 atm) and MDA in the presence of various tertiary amines.†

Tertiary amine	pK_a	Polymer	
		Yield (%)	η_{inh}
Pyridine	5.23	100	0.68
β-Picoline	5.52	100	0.85
α-Picoline	5.97	100	0.70
γ-Picoline	6.02	100	0.43
2,6-Lutidine	6.99	62	0.17
Imidazole	7.12	89	0.31
Triethylamine	10.87	76	0.18

† The reaction was carried out at 40°C for 4 h with diphenyl phosphite in DMF containing various tertiary amines (8 equiv).

Table 4.9.   Polycondensation of carbon dioxide (20 atm) and MDA with various phosphorus compounds.†

Phosphorus compound	Tertiary amine	Solvent	Polymer	
			Yield (%)	η_{inh}
$P(OH)_3$	Py	Py	0	—
$HOP(O-n-Bu)_2$	Py	Py	0	—
$HOP(O-\langle\overline{}\rangle)_2$	Py	Py	100	2.24
$P(O-\langle\overline{}\rangle)_3$	Py	Py	0	—
$P(O-\langle\overline{}\rangle)_3$	Imidazole	DMF	100	0.37
$P(O-\langle\overline{}\rangle-Cl)_3$	Imidazole	DMF	69	0.16
$P(O-n-Bu)_3$	Imidazole	DMF	0	—

† The reaction was carried out at 40°C for 4 h in pyridine or in DMF containing imidazole (8 equiv.).

N-phosphonium salt because of the low nucleophilicity of the alkoxy groups.

The polycondensation of several diamines with CO_2 was carried out at 40°C for 4 h under 20 atm pressure of carbon dioxide (Table 4.10). Aromatic diamines, from which polymers with good solubility in pyridine were formed, gave polymers of higher molecular weight, whereas polymers from 4,4'-diaminodiphenylsulfone and p-phenylenediamine were insoluble even in HMPA and showed low viscosity in sulfuric acid. In contrast, an aliphatic diamine with high basicity afforded polymer of low viscosity in a low yield because of retardation by the formation of pyridine-insoluble and unreactive ammonium carbamate.

The reaction of phenyl dichlorophosphite with CO_2 and amines via

Table 4.10.   Polycondensation of carbon dioxide (20 atm) and various diamines with diphenyl phosphite in pyridine.[†1]

Diamine	Yield (%)	η_{inh}[†2]
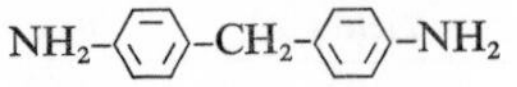 NH_2-⟨◯⟩-CH_2-⟨◯⟩-NH_2	100	2.24
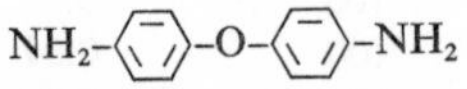 NH_2-⟨◯⟩-O-⟨◯⟩-NH_2	100	0.51
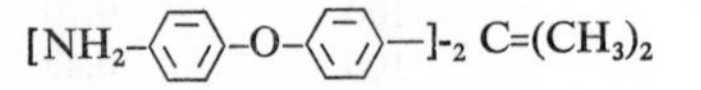 $[NH_2$-⟨◯⟩-O-⟨◯⟩-$]_2$ C=$(CH_3)_2$	100	2.14
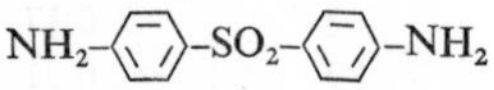 NH_2-⟨◯⟩-SO_2-⟨◯⟩-NH_2	86	0.14[†3]
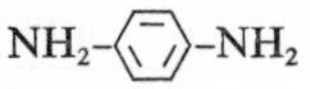 NH_2-⟨◯⟩-NH_2	100	0.09[†3]
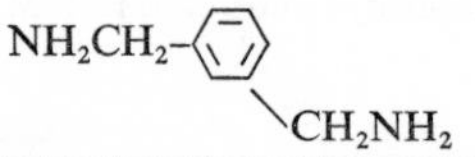 NH_2CH_2-⟨◯⟩ ⟍CH_2NH_2	46	0.13

[†1] The reaction was carried out at 40°C for 4 h.
[†2] Measured in HMPA at 30°C.
[†3] The polymers were insoluble in HMPA, and the viscosity was measured in sulfuric acid at 30°C.

the N-phosphonium salt was extended to the reactions of CO_2 with diamines in pyridine under mild conditions.[11]

$$PhO-PCl_2 + NH_2R-NH_2 \xrightarrow[Py]{CO_2} \{NH-R-NHCO\}_n + PhO-P=O \qquad (4.8)$$

The reaction (Eq. 4.8) was conducted in an autoclave at 60°C for 6 h under an initial pressure of 20 atm of carbon dioxide. Polymers from p-phenylenediamine and 4,4'-diaminodiphenylsulfone precipitated from the reaction mixture, but those from other diamines resulted in viscous solutions as the polycondensation proceeded.

The results of polycondensation of several aromatic diamines with carbon dioxide are shown in Table 4.11. When the polycondensation was carried out for 6 h or less at ordinary pressure, polymers with high molecular weights were obtained from MDA, 4,4'-diaminodiphenyl ether, and PDAP. On the other hand, 4,4'-diaminodiphenylsulfone and p-phenylenediamine, whose polymers have a low solubility in pyridine, gave polymers with low viscosity.

Unlike the results obtained with diphenyl phosphite, where the molecular weight depended significantly on the pressure, increasing the pressure of carbon dioxide in this case did not affect the molecular weight of the resulting polymer.

Polyureas of MDA maintained at 60°C for 6 h were completely soluble in HMPA, whereas the polymers obtained by heating for more

Table 4.11. Polyureas from carbon dioxide and diamines with phenyl dichlorophosphite in pyridine at 60°C under atmospheric conditions.†1

Diamine	Time, h	Yield (%)	η_{inh} †2
$NH_2-C_6H_4-CH_2-C_6H_4-NH_2$	2	38	0.17
	4	76	0.27
	6	84	0.43
	8	100	0.38
$NH_2-C_6H_4-O-C_6H_4-NH_2$	6	92	0.52
$NH_2-C_6H_4-SO_2-C_6H_4-NH_2$	6	100	0.07†3
$[NH_2-C_6H_4-O-C_6H_4-]_2\ C{=}(CH_3)_2$	6	100	0.43 (0.44)†4
$NH_2-C_6H_4-NH_2$	6	71	0.05

†1 $C_6H_5O-PCl_2$ = diamine, 10 mmole; pyridine, 30 ml.
†2 Measured at 30°C in HMPA.
†3 Polymers were insoluble in HMPA, and their viscosities were measured in sulfuric acid at 30°C.
†4 Reaction was carried out under a CO_2 pressure of 20 atm.

than 6 h contained some insoluble solids, and consequently, the solution viscosity was reduced. These insoluble products may be formed by a side reaction, such as cross-linking, at a later stage of the reaction. A side reaction involving incorporation of the phosphorus moiety into the polymer backbone is unlikely, because no phosphoric acid was detected when the polymer from MDA was oxidized with a mixture of sulfuric and nitric acids.

4.1.3. Polycarbonates

The reaction of CO_2 with α,ω-dibromo compounds and the potassium salts of diols was promoted by crown ethers and yielded polycarbonates with the structure (shown in Eq. (4.9).[12]

$$n\,XCH_2-C_6H_4-CH_2X + n\,KO-R-OK \xrightarrow[\text{crown ethers}]{CO_2}$$

$$\{OCOCH_2-C_6H_4-CH_2OCOO-R\}_n + 2n\,KX \qquad (4.9)$$

$$R = -C_6H_4-,\ -CH_2-C_6H_4-CH_2-,\ -C_6H_4-C(CH_3)_2-C_6H_4-$$

$$X = Cl,\ Br.$$

A number of crown ethers were employed for the reaction, and the ethers having an 18-member ring, especially 18-crown-6 and dicyclohexyl-

18-crown-6, were found to give the best results. Dibenzo-18-crown-6 was relatively less effective in the reaction.

The reaction was significantly affected by solvents. Ether solvents such as THF, dioxane, and diglyme gave polymer in high yields, and aprotic polar solvents such as DMF retarded the reaction. The reaction in dioxane at a temperature around 100°C gave a maximum yield of polymer of the highest molecular weight.

In a similar manner, crown ethers facilitated the reaction between α, ω-dibromide and potassium carbonate, producing polycarbonates in moderate yields (Eq. 4.10).[13]

$$\text{BrCH}_2\text{-}\langle\_\rangle\text{-CH}_2\text{Br} + \text{K}_2\text{CO}_3 \xrightarrow{\text{crown ethers}}$$

$$\text{BrCH}_2\text{-}\langle\_\rangle\text{-CH}_2\{\text{OCOO-CH}_2\text{-}\langle\_\rangle\text{-CH}_2\}_n\text{-Br} \qquad (4.10)$$

Reactions between p-xylylene dibromide (PXDB) and potassium carbonate were carried out with various solvents and crown ethers (Table 4.12). Crown ethers having an 18-member ring, a ring size suitable to the potassium cation, were the most useful for the reaction. Like the reaction between carbon dioxide, α, ω-dihalide, and potassium salts of diols (Eq. 4.9), the reaction proceeded best in ether solvents; among these ethers, diglyme gave the highest polymer yield as well as the highest molecular weight. The amount of crown ethers affected the yield and the molecular weight of the polymer. Both values increased rapidly and reached asymptotic values at a ratio of 18-crown-6-ether/potassium of about 0.1.

Table 4.12. Reactions between p-xylylene dibromide and K_2CO_3 in the presence of various crown ethers.

Ethers	Solvent	Conversion (%)
15-Crown-5	Benzene	24.9
18-Crown-6	Toluene	62.5
Dicyclohexyl-18-crown-6	Toluene	69.2
Dicyclohexyl-18-crown-6	Diglyme	73.2
Dibenzo-18-crown-6	Diglyme	58.7
Dicyclohexyl-24-crown-8	Benzene	34.0
$\text{C}_{17}\text{H}_{35}\text{COO}(\text{C}_2\text{H}_4\text{O})_{25}\text{H}$	Benzene	Trace
$\text{C}_{17}\text{H}_{35}\text{COO}(\text{C}_2\text{H}_4\text{O})_4\text{H}$	Benzene	0
None	Benzene	0

Several aliphatic α, ω-dibromides were tested in the reaction with potassium carbonate in the presence of 18-crown-6-ether (Eq. 4.11).

$$\text{Br--(CH}_2)_m\text{-Br} + \text{K}_2\text{CO}_3 \xrightarrow{\text{crown ether}} \{\text{OCOO--(CH}_2)_m\}_n- \quad (4.11)$$

The bromides with more than four methylene groups gave polymers with high molecular weights, up to 33,000. Only cyclic carbonates, and no polymers, were obtained from the dibromides with shorter chains.

4.2. RING-OPENING AND ADDITION POLYMERIZATIONS INVOLVING CARBON DIOXIDE

Addition copolymerization involving carbon dioxide and unsaturated or cyclic compounds offer both fundamental and practical aspects of considerable interest. They may be regarded as catalytic processes since the reacting species at the growing polymer end react with the monomer to regenerate the same reactive species. Addition copolymerization employing carbon dioxide as a monomer, therefore, presents an intriguing possibility in the search for new techniques for the catalytic fixation of carbon dioxide, examples of which are rather limited, as seen in the previous chapters. Polymer synthesis using CO_2 as a monomer leads to the formation of products with a large amount of oxygen in the main chain, and high molecular-weight polymers with polar groups in the main chain are potentially useful as functional material for various fields.

It should be noted that the discovery of the copolymerization of carbon dioxide and epoxide by our group has provided new impetus to the growing interest in the organic chemistry of carbon dioxide. In this section we will review recent progress in the field of addition copolymerization of carbon dioxide and cyclic or unsaturated compounds.

4.2.1. Copolymerization with epoxide and other cyclic ethers

The discovery of an alternating copolymerization of carbon dioxide and epoxide that yields an aliphatic polycarbonate of high molecular weight was the first demonstration of a higher polymer synthesis involving the direct use of carbon dioxide (Eq. 4.12).[14,15]

$$\text{R}^1\text{R}^2\text{C}\underset{\text{O}}{\overset{\diagdown\diagup}{\text{--}}}\text{CR}^3\text{R}^4 + \text{CO}_2 \longrightarrow (\text{-R}^1\text{R}^2\text{C--R}^3\text{R}^4\text{C-O-}\underset{\overset{\|}{\text{O}}}{\text{C}}\text{-O--})_x \quad (4.12)$$

By using an appropriate catalyst, the reaction proceeds under mild conditions such as room temperature and moderate pressure. Typically, propylene oxide or ethylene oxide gives a copolymer with a molecular weight of 50,000–150,000 in the reaction under a CO_2 pressure of 40–50 kg/cm². Physical and other properties of the carbon dioxide-epoxide copolymers have been reported.[16–18]

A. Catalyst

The catalyst system originally found to be effective was an equimolar reaction product between diethylzinc and a compound with two active hydrogens, such as water or primary amine.[15,19] The combination of diethylzinc with dihydric phenol,[20,21] aromatic hydroxycarboxylic acid,[22] or aromatic dicarboxylic acid[22] is even more effective as a catalyst system. The reaction mechanism of some of these catalyst systems has been studied in detail and was described earlier.[23]

In general, the copolymerization of carbon dioxide and epoxide with an organozinc catalyst system is thought to proceed by the repetition of two elementary reactions, (4.13) and (4.14).

$$-C-C-O-ZnX + CO_2 \longrightarrow -C-C-O-\underset{\underset{O}{\|}}{C}-O-ZnX \qquad (4.13)$$

$$-C-C-O-\underset{\underset{O}{\|}}{C}-O--ZnX + \underset{\underset{O}{\diagdown\diagup}}{C-C} \longrightarrow -C-C-O-\underset{\underset{O}{\|}}{C}-O-C-C-O-ZnX \qquad (4.14)$$

Group X varies depending upon the nature of the compound with active hydrogens that is coupled to diethylzinc. Because systems composed of compounds with one active hydrogen (e.g., methanol, secondary amine, monohydric phenol, and aromatic monocarboxylic acid) coupled to diethylzinc exhibit little or no catalytic activity for the copolymerization, the presence in group X of a structure such as $(-Y-Zn-Y-Zn-)_x$ [where $Y = O$, NR, OC_6H_4O (m,p), OC_6H_4COO (m,p), or $OOCC_6H_4COO$ (m,p)] is thought to be very advantageous for catalysis. For example, the reaction product of diethylzinc and water has a $(-O-Zn-O-Zn-)_x$ linkage (Eq. 4.15), while in the diethylzinc-methanol system no such group is present (Eq. 4.16).

$$n\,(C_2H_5)_2Zn + n\,H_2O \xrightarrow{-(2n-1)C_2H_6} C_2H_5-Zn(-O-Zn-)_{n-1}OH \qquad (4.15)$$

$$(C_2H_5)_2Zn + 2CH_3OH \xrightarrow{-2C_2H_6} (CH_3O)_2Zn \qquad (4.16)$$

The situation is similar for dihydric and monohydric phenols (Eqs. 4.17, 4.18).

$$(C_2H_5)_2Zn + C_6H_4(OH)_2(m) \longrightarrow C_2H_5(-Zn-O-C_6H_4-O(m)-)_xH \quad (4.17)$$
resorcinol

$$(C_2H_5)_2Zn + 2\,C_6H_5OH \longrightarrow (C_6H_5-O-)_2Zn \quad (4.18)$$

With the diethylzinc-resorcinol system (4.17), the copolymerization reaction takes place at the zinc-resorcinolate linkage of the catalyst, but not at the zinc-ethyl bond, since the end group of the copolymer is always a resorcinol group.[21]

Of the two elementary steps, reaction (4.13) is a typical reaction of carbon dioxide involving the addition of a metal alkoxide. It should be noted here that, in the absence of CO_2, epoxide is known to homopolymerize (to a polyether) with similar catalyst systems. Therefore, it may be concluded that CO_2 is much more reactive with zinc alkoxide than is epoxide. The key step by which the whole copolymerization reaction proceeds catalytically is reaction (4.14), in which zinc alkoxide is regenerrated. In this step, epoxide appears more reactive than CO_2. The factors governing this apparent reversal of reactivity have not yet been elucidated.

The nature of group X bound to the zinc atom is considered to affect 1) the nucleophilicity of the zinc alkoxide or zinc alkylcarbonate, 2) the ability of the zinc atom to be coordinated by carbon dioxide or epoxide, and 3) the strength and the extent of the association of the catalyst system.

Building on the information derived from these catalyst systems and the proposed mechanism, further studies have been made to develop other and better catalyst systems. If the diethylyzinc-water system, a typical catalyst, is aged in the presence of CO_2 prior to its use for copolymerization, its catalytic activity is enhanced.[24] The reaction between CO_2 and the reaction product of diethylzinc-water (Eq. 4.15) is thought to produce a zinc carbonate structure in the catalyst system (Eq. 4.19).

$$C_2H_5(ZnO)_nH + CO_2 \longrightarrow C_2H_5(ZnCO_3)_k(ZnO)_lH \quad (4.19)$$

Although a polyether linkage can be formed under relatively low pressure (e.g., 1.26 kg/cm^2) with the usual diethylzinc-water catalyst system, the aged system gives a higher yield of the copolymer. The combination of diethylzinc and trihydric phenol or aminophenol has also been found to possess high catalytic activity for the copolymerization.[25]

Although most of the above-mentioned catalyst systems are heterogeneous, the diethylzinc-pyrogallol system is homogeneous, and the structure shown in Formula (4.1) has been suggested to be the active species.[25]

Intramolecular coordination is thought to prevent intermolecular association, keeping the system homogeneous and resulting in enhanced

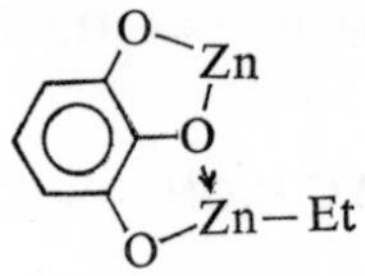

Formula (4.1)

catalytic activity. These and other diethylzinc-polyhydric phenol systems have been investigated in detail.[26-28] It has been concluded that the $-Zn-O \rightarrow Zn$-linkage and the coordinative unsaturation of the zinc atom in the $O-Zn-O$ linkage are important for the catalytic activity. The diethylzinc-o-phenylenediamine and diethylzinc-thioresorcinol systems are also effective as catalysts for carbon dioxide-propylene oxide copolymerization.[29]

An organozinc catalyst supported on alumina also facilitates the copolymerization of carbon dioxide and propylene oxide.[30] The reaction between diethylzinc and the surface hydroxyl groups of γ-alumina gives an active species with the structure shown in Eq. (4.20).

$$Al\text{-}OH + Zn(C_2H_5)_2 \xrightarrow{-C_2H_6} Al\text{-}O\text{-}Zn\text{-}C_2H_5 \qquad (4.20)$$

A related active catalyst system is the reaction product between diethylzinc and the polymer of p-hydroxystyrene in an equimolar ratio (Eq. 4.21).[31]

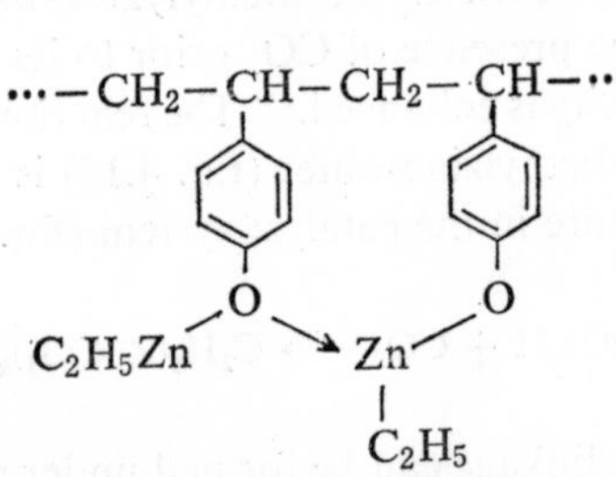

$$(4.21)$$

End group analysis of the copolymer indicates that, in contrast to the diethylzinc-resorcinol system[21], it is the zinc-ethyl group in this catalyst system that is involved in the copolymerization reaction.

The mechanism suggested for the copolymerization of carbon dioxide and epoxide (Eqs. 4.13 and 4.14) points to the possible catalytic activity of metal carbonate or carboxylate. In fact, zinc carbonate is effective as a catalyst,[15] though the activity is not high. Some metal acetates can catalyze the copolymerization of CO_2 and propylene oxide.[32] Zn $(OAc)_2$ and $Co(OAc)_2$ give alternating copolymer; $Mg(OAc)_2$, $Cr(OAc)_3$, and $Ni(OAc)_2$ give random copolymer; and $Sn(OAc)_2$ gives homopolymer of the epoxide. When the copolymerization was carried out with cobalt (II) acetate as the catalyst, the acetic acid added to the mixture caused a chain transfer reaction.[33]

Metal amides such as ethylzinc diphenylamide and diethylaluminum diethylamide are effective as catalysts for the copolymerization of CO_2 and cyclohexene oxide, although the product is not an alternating polymer,[34] while a titanium amide gave a hydroxycarbamate in a quantitative yield.[35] Since the titanium amide is known to react rapidly with carbon dioxide to form a titanium carbamate, reaction (4.22) may be regarded as a model for one of the elementary steps (Eq. 4.14) of the copolymerization reaction.

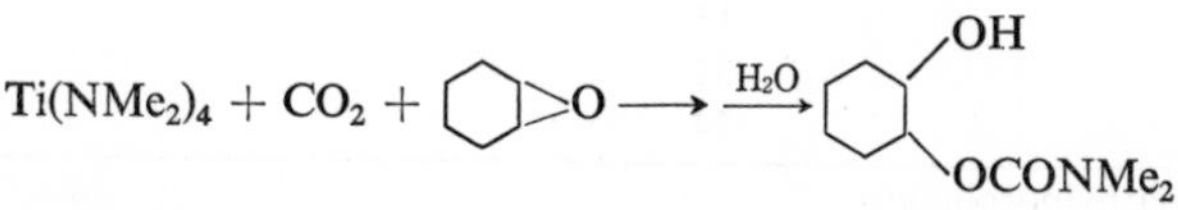

$$Ti(NMe_2)_4 + CO_2 + \underset{}{\bigcirc}\!\!\!\triangleright\!\!O \longrightarrow \overset{H_2O}{\longrightarrow} \quad \bigcirc\!\!\!\underset{OCONMe_2}{\overset{OH}{}} \qquad (4.22)$$

Organoaluminum systems are known as effective catalysts for the homopolymerization of epoxide, but they tend to give the homopolymer even in the presence of carbon dioxide. However, the addition of a Lewis base, such as triphenylphosphine, to triethylaluminum improves the catalytic activity for copolymerization.[36] Tetraphenylporphinatoaluminum methoxide (Formula 4.2) is a unique catalyst for the copolymerization of CO_2 and propylene oxide, where the molecular weight distribution of the copolymer is very narrow.[37] Some organomagnesium systems, such as the diethylmagnesium-primary amine system, are also effective.[38]

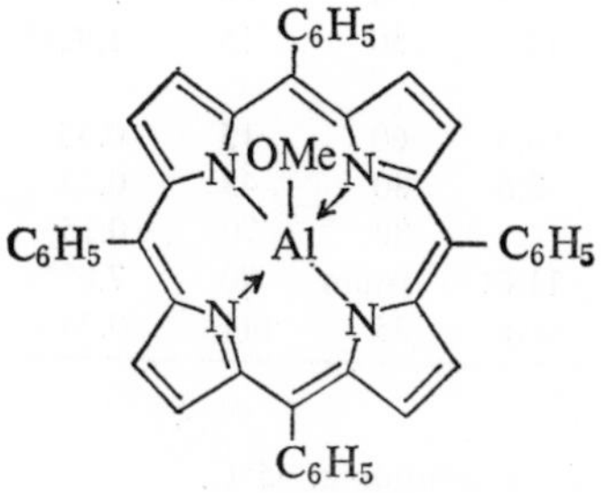

Formula (4.2)

Catalyst systems for the copolymerization of CO_2 and epoxide, mainly propylene oxide, are listed in Table 4.13. Table 4.14 summarizes some examples of the copolymerization of CO_2 and propylene oxide.

Table 4.13. Catalyst systems for the copolymerization of carbon dioxide and epoxide.

	Ref.		Ref.
$ZnEt_2$–H_2O	15, 19	$AlEt_3$–PPh_3	36
$ZnEt_2$–H_2O–CO_2 (aged)	24	MeOAlTPP†	37
$ZnEt_2$–RNH_2	15, 19	Et_2AlNEt_2	34
$ZnEt_2$–$C_6H_4(OH)_2(m,p)$	20,21		
$ZnEt_2$–$C_6H_4(COOH)_2(m,p)$	22	$MgEt_2$–H_2O	38
$ZnEt_2$–$C_6H_4(OH)(COOH)(m,p)$	22	$MgEt_2$–RNH_2	38
$ZnEt_2$–$C_6H_3(OH)_3$	25	$MgEt_2$–$C_6H_4(OH)(COOH)(m)$	38
$ZnEt_2$–$C_6H_4(OH)(NH_2)(o)$	25		
$ZnEt_2$–$C_6H_4(NH_2)_2(o)$	29	$Mg(OAc)_2$	32
$ZnEt_2$–$C_6H_4(SH)_2(m)$	29		
$ZnEt_2$–poly (p-OH-styrene)	31	$Co(OAc)_2$	32
$ZnEt_2$–γ-alumina	30	$Ni(OAc)_2$	32
$Zn(CO_3)_2$	15		
$Zn(OAc)_2$	32		
$EtZnNPh_2$	34		

† TPP: Tetraphenylporphinato

Table 4.14. Copolymerization of carbon dioxide and propylene oxide†[1]

Catalyst	Mono-mer (g)	Temp. (°C)	Time (h)	Copolymer yield†[2] (g)	$[\eta]$†[3] (dl/g)	Mn (x 10^3)	Ref.
$ZnEt_2$–water (1:1)	17.3	room	70	4.74	1.00		19
$ZnEt_2$–water–CO_2 (aged)	3.1†[4]	30	48	0.73	1.27		24
$ZnEt_2$–α-phenethylamine (1:1)	26.2	40	68	6.01	1.14		19
$ZnEt_2$–resorcinol (1:1)	17.3	35	48	8.24	0.61	47	21
$ZnEt_2$–m-hydroxybenzoic acid (1:1)	12.5	35	44	15.76	1.21	111	22
$ZnEt_2$–pyrogallol (2:1)	5.8	35	44	5.9†[5]	1.90		25
$ZnEt_2$–poly (p-hydroxystyrene) (1:1)	17.3	50	15	1.93†[6]	1.05	93	31
$ZnEt_2$–alumina	4.3	60	15	0.55		28	30
$Zn(OAc)_2$	2.6	80	43	0.23		20	32
$Co(OAc)_2$	2.6	80	70	0.15		25	32
$AlEt_3$–PPh_3	11.6†[7]	room	90	7.0†[8]			36
$MgEt_2$–water (1:1)	8.6	35	144	0.24			38

†[1] CO_2, 40–50 atm.
†[2] Alternating copolymer
†[3] Intrinsic viscosity of benzene solution at 35°C.
†[4] CO_2, 1.26 atm.
†[5] 48% to propylene oxide.
†[6] 2 mole/mole diethylzinc.
†[7] CO_2, 5 atm.
†[8] Not alternating; (C=O)–O unit, 22 mole %.

Although the conditions of the reactions are not identical, the data are useful for a comparison of the catalyst systems.

B. Monomer

The copolymerization reaction using the diethylzinc-water catalyst system is generally applicable to various epoxides, nonsubstituted, monosubstituted, and disubstituted, as summarized in Table 4.15. However, no good catalyst system is yet available for the copolymerization of epichlorohydrin and carbon dioxide.

Table 4.15. Cyclic ethers for copolymerization with carbon dioxide by organozinc catalyst system.

Copolymerizable	Not Copolymerizable

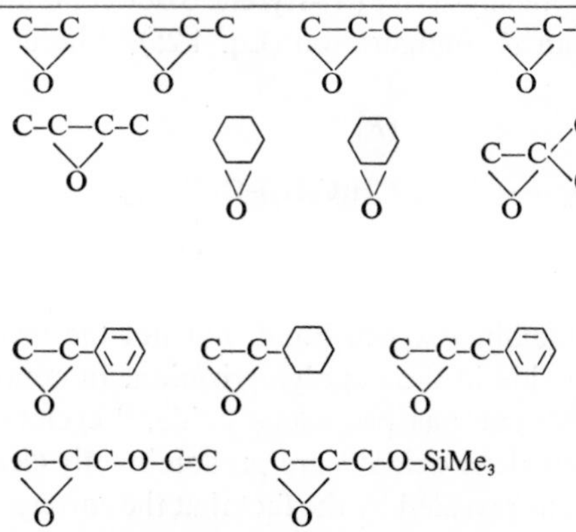

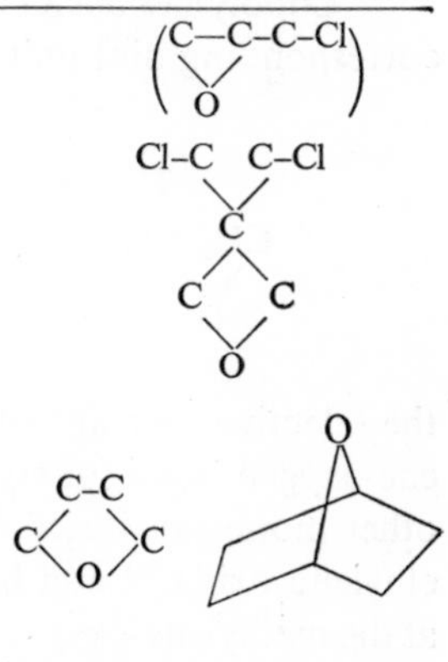

Cyclic ethers with a four-membered ring (3,3 - bischloromethyloxetane) or a five-membered ring (tetrahydrofuran, 1,4-epoxycyclohexane) can not be copolymerized with CO $_2$by the organozinc system. Oxetane, a nonsubstituted four-membered cyclic ether, does yield a copolymer (intrinsic viscosity $[\eta] = 0.4$) with carbon dioxide, though not an alternating polymer, when a triethylaluminum-water-acetylacetone system is used as the catalyst (Eq. 4.23).[39]

$$\begin{array}{l} CH_2\text{-}CH_2 \\ \;|\qquad\;| \\ CH_2\text{-}O \end{array} + CO_2 \longrightarrow \{(CH_2)_3\text{-}O\}_{0.8}\{(CH_2)_3\text{-}O\text{-}\underset{\underset{O}{\|}}{C}\text{-}O\}_{0.2} \qquad (4.23)$$

C. Stereochemical aspects

In the copolymerization of CO_2 and epoxide with the diethylzinc-water catalyst system, an inversion in the configuration takes place at the carbon atom of the epoxide ring where the ring is cleaved. Thus, *cis*-cyclohexene oxide is converted to a diol structure with a *trans* form when it is incorporated into the copolymer (Eq. 4.24).[40]

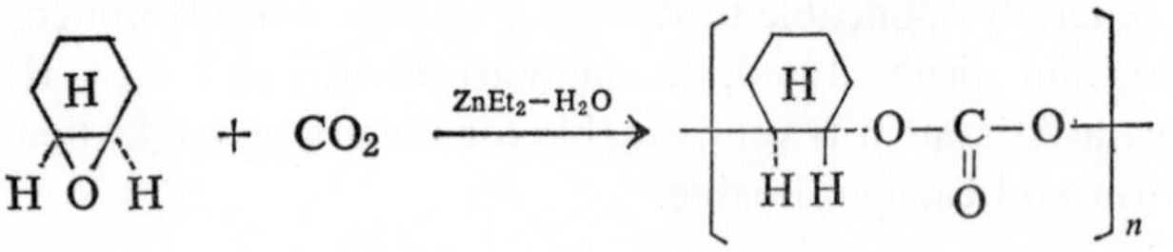

$$
\begin{array}{c}
\text{(structure)}
\end{array}
\quad + \quad CO_2 \quad \xrightarrow{ZnEt_2-H_2O} \quad
\begin{array}{c}
\text{(structure)}
\end{array}
\tag{4.24}
$$

Similarly, in the copolymer from CO_2 and *cis*-2-butene oxide the 2,3-butanediol unit is in a *trans* form.[41]

Copolymer of CO_2 and optically active (*S*)-styrene oxide contains the corresponding diol unit with an *R* configuration (Eq. 4.25),[42] indicating

$$
\underset{\overset{*}{O}}{\overset{(S)}{CH_2-CH-C_6H_5}} + CO_2 \longrightarrow \underset{O}{\overset{(R)}{\{CH_2-CH(C_6H_5)-O-C-O\}_x}}
\tag{4.25}
$$

the selective cleavage of the methine-oxygen bond, but not the methylene-oxygen bond, of styrene oxide in the copolymerization. In contrast, other monosubstituted epoxides such as propylene oxide,[43] cyclohexylethylene oxide,[44] and benzylethylene oxide[45] are predominantly cleaved at the methylene-oxygen bond, as revealed by the fact that the corresponding diol unit in the copolymer has the same configuration as that of the epoxide. Equation (4.26) is an example.

$$
\underset{\overset{*}{O}}{\overset{(R)}{CH_2-CH-CH_3}} + CO_2 \longrightarrow \underset{O}{\overset{(R)}{\{CH_2-CH(CH_3)-O-C-O\}_x}}
\tag{4.26}
$$

In the copolymerization of carbon dioxide and 2-butene oxide with the diethylzinc-water catalyst system, the *cis* isomer of the epoxide is much more highly reactive than the *trans* isomer (Table 4.16).[41] The *trans* isomer gives a very low yield, if any. On the other hand, in the homopolymerization of 2-butene oxide with the same catalyst, *cis* as well as *trans* isomers are polymerized in similar yields.[46] Such selectivity toward geometrical isomers is undoubtedly related to the difference in the readiness of the isomers to coordinate with the catalyst. The varied findings in the selec-

Table 4.16. Copolymerization of carbon dioxide and butene oxide (BO).[1,41]

Monomer[2]			Polymer			
A	B	(Ratio)	Yield (g)	Carbonate unit (%)	Epoxide unit A:B	$[\eta]$
cis-2-BO	—		5.7	50	—	2.0
trans-2-BO	—		0.08	50	—	—
cis-2-BO	1–BO	(1:1)	4.50	50	0.67:1	2.9
trans-2-BO	1–BO	(1:1)	0.93	50	0.04:1	0.77

[1] $ZnEt_2$–H_2O catalyst; 35°C; CO_2, 40 kg/cm^2; in dioxane.
[2] 0.1 mole

tivity in the homopolymerization of epoxide and in the copolymerization with carbon dioxide are thought to result from a modification of the catalyst system by the reaction with CO_2 (e.g., Eq. 4.19) under the conditions of the copolymerization.

As with other carbon dioxide-epoxide copolymerization reactions, the copolymerization of CO_2 and 2-butene oxide is accompanied by the formation of a cyclic carbonate (Eq. 4.27).

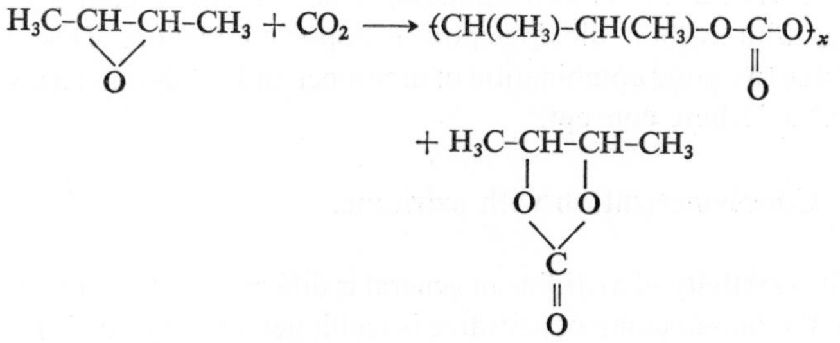

$$H_3C-CH-CH-CH_3 + CO_2 \longrightarrow \{CH(CH_3)-CH(CH_3)-O-C-O\}_x$$

$$+ H_3C-CH-CH-CH_3 \qquad (4.27)$$

Both *cis* and *trans* epoxides react to give comparable amounts of the corresponding cyclic carbonates (inverted). The considerable differences in the selectivity of the geometrical isomers in the copolymerization and in the cyclic adduct formation indicate that the cyclic adduct is not formed via the copolymer. However, cyclic carbonate is not the precursor of the copolymer because propylene carbonate is not polymerized under the relevant conditions. Thus, the copolymer and the cyclic adduct are formed by mechanisms that are distinct from one another.

In the copolymerization of CO_2 and racemic monosubstituted epoxide, isomeric structures of the copolymer molecule occur in the arrangement of the enantiomeric units from the epoxide. The stereoregularity of, for example, the carbon dioxide-racemic propylene oxide copolymer has not yet been elucidated, however. Related to this is the observation that, in the ternary copolymerization of carbon dioxide, (*R*)-propylene oxide, and styrene oxide with a diethylzinc-water catalyst, both R and S enantiomers of styrene oxide were incorporated into the carbon dioxide copolymer chain with (R)-propylene oxide.[47]

4.2.2.   Copolymerization with episulfide

Organometallic catalyst systems have been extended to the copolymerization of CO_2 and propylene sulfide (Eq. 4.28).[48]

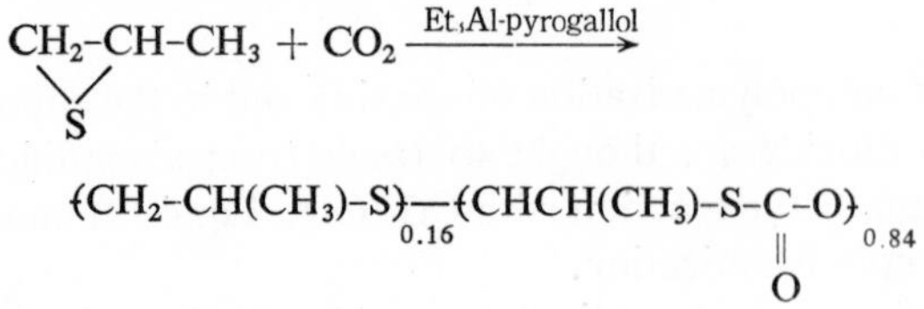

$$CH_2\text{-}CH\text{-}CH_3 + CO_2 \xrightarrow{\text{Et}_3\text{Al-pyrogallol}}$$

$$\{CH_2\text{-}CH(CH_3)\text{-}S\}_{0.16}\text{---}\{CHCH(CH_3)\text{-}S\text{-}\underset{\|}{C}\text{-}O\}_{0.84} \quad (4.28)$$
$$O$$

It is interesting to note in this reaction that the triethylaluminum-pyrogallol (2:1) system yields a copolymer with a higher content of oxycarbonyl-thio units than does the corresponding zinc system, in contrast to the situation with carbon dioxide-epoxide copolymerization. This is considered due to a good combination of monomer and catalyst in terms of the hard-soft-acid-base concept.

4.2.3.   Copolymerization with aziridine.

The reactivity of aziridine in general is different from that of epoxide because the ring-opening of aziridine is facilitated only by acidic reagents, while the epoxide ring is cleaved by both acidic and basic reagents.

Carbon dioxide is copolymerized with N-phenylethylenimine in the presence of a weak protonic acid like phenol or acetic acid as catalyst (Eq. 4.29).[49]

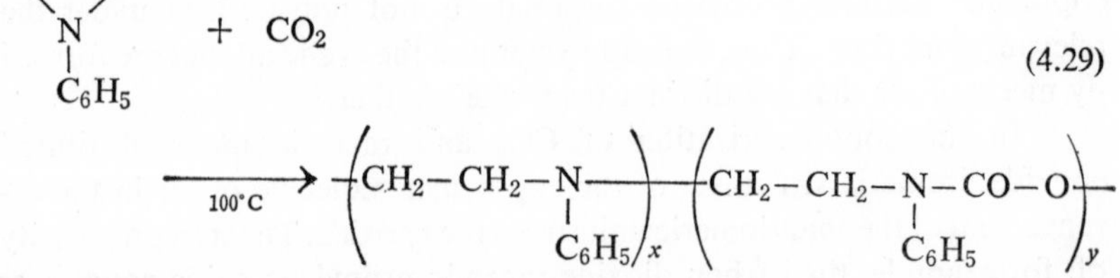

$$CH_2\text{-}CH_2$$
$$\diagdown N \diagup \quad + \quad CO_2$$
$$\underset{C_6H_5}{|}$$

$$\xrightarrow[100°C]{} \left(CH_2\text{-}CH_2\text{-}\underset{C_6H_5}{\overset{}{N}}\right)_x \left(CH_2\text{-}CH_2\text{-}\underset{C_6H_5}{\overset{}{N}}\text{-}CO\text{-}O\right)_y \quad (4.29)$$

The maximum content of carbamate (urethane) linkage in the copolymer is 20%. The carbamate linkage could be formed by the ring-opening of a cyclic carbamate (an oxazolidone), a one-to-one cyclic adduct of carbon dioxide and *N*-phenylethylenimine, but this possibility is excluded by the fact that the oxazolidone, 3-phenyloxazolidone-2, is not copolymerized with *N*-phenylethylenimine under the relevant conditions.

    N-Phenylethylenimine is known to be polymerized with an acid catalyst by a cationic mechanism (Eqs. 4.30, 4.31).

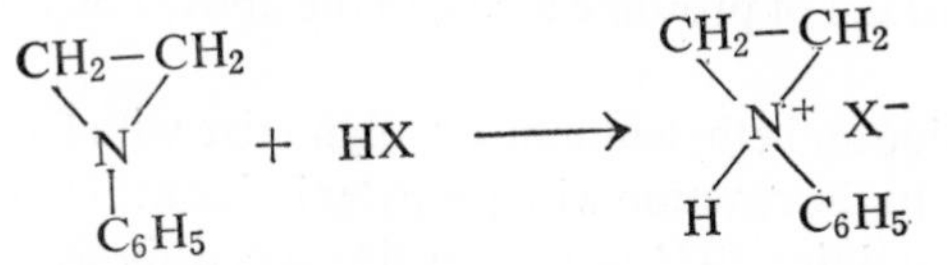

$$
\begin{array}{c}
CH_2-CH_2 \\
\diagdown \diagup \\
N \\
| \\
C_6H_5
\end{array}
\quad + \quad HX \quad \longrightarrow \quad
\begin{array}{c}
CH_2-CH_2 \\
\diagdown \diagup \\
N^+\ X^- \\
\diagup\ \diagdown \\
H\quad C_6H_5
\end{array}
\qquad (4.30)
$$

$$
\begin{array}{c}
CH_2-CH_2 \\
\diagdown \diagup \\
\sim\!\!\!N^+ \\
| \\
C_6H_5
\end{array}
\ + \
\begin{array}{c}
CH_2-CH_2 \\
\diagdown \diagup \\
N \\
| \\
C_6H_5
\end{array}
\ \longrightarrow \
\begin{array}{c}
\qquad\qquad\qquad CH_2-CH_2 \\
\qquad\qquad\qquad \diagdown \diagup \\
\sim\!\!\!N-CH_2-CH_2-N^+\ X^- \\
| \qquad\qquad\qquad | \\
C_6H_5 \qquad\qquad C_6H_5
\end{array}
$$

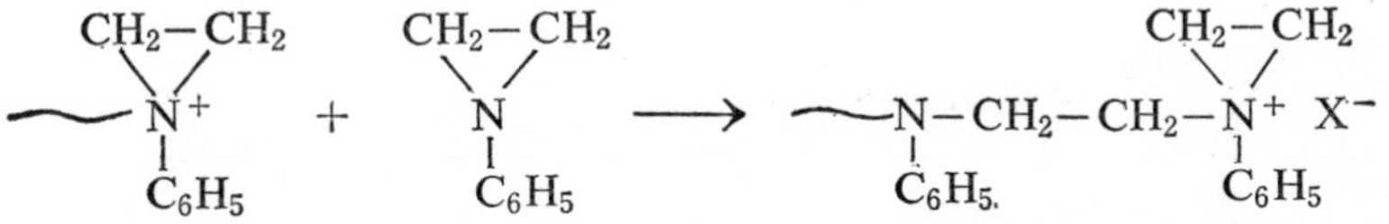

$$(4.31)$$

The mechanism in Eq. (4.32) has been proposed for the reaction between such a cationic growing end and carbon dioxide.

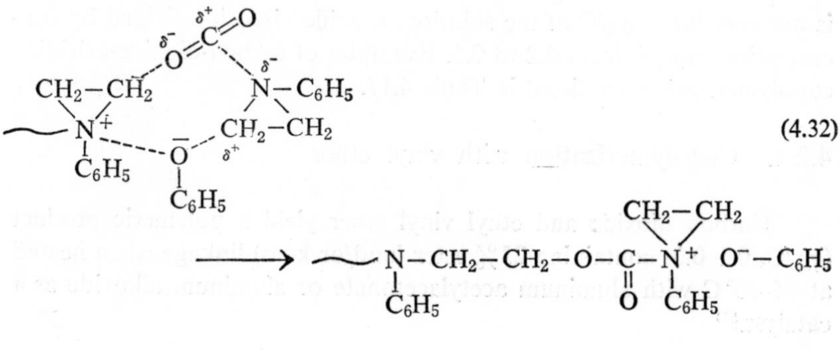

$$
\begin{array}{c}
\qquad\qquad\quad \overset{\delta^+}{\underset{\ }{C}}{\overset{\displaystyle =O}{}} \\
\qquad\quad \overset{\delta^-}{O}\diagup\quad\diagdown\overset{\delta^-}{N}-C_6H_5 \\
CH_2-CH_2 \qquad\quad | \\
\sim\!\!\!N^+\cdots\cdots\underset{\delta^-}{O}\diagup CH_2-CH_2 \\
|\qquad\qquad | \quad\ \overset{\delta^+}{} \\
C_6H_5 \qquad C_6H_5
\end{array}
\qquad (4.32)
$$

$$
\longrightarrow \quad
\begin{array}{c}
\qquad\qquad\qquad\qquad\qquad CH_2-CH_2 \\
\qquad\qquad\qquad\qquad\qquad \diagdown \diagup \\
\sim\!\!\!N-CH_2-CH_2-O-C-N^+\ O^--C_6H_5 \\
| \qquad\qquad\qquad\quad \| \quad | \\
C_6H_5 \qquad\qquad\qquad O\ \ C_6H_5
\end{array}
$$

As shown, the interaction between N-phenylethylenimine and CO_2 raises the electron density of an oxygen atom in the latter, resulting in enhanced susceptibility to attack by an immonium ion. The interaction between the aziridine and CO_2 can be observed in the UV spectrum of the mixture.

Various metallic compounds such as aluminum butoxide, titanium (IV) butoxide, zinc chloride, manganese(II) chloride, and manganese(II) acetylacetonate are effective for the copolymerization of CO_2 and N-phenylethylenimine.[50,51] When manganese(II) dichloride tetrahydrate is used as catalyst, a fraction of the copolymer obtained has a carbamate unit content as high as 80%. The metallic compounds are considered to behave as Lewis acids, instead of protonic acids, in the above reaction mechanism (Eq. 4.32).

Propylenimine and ethylenimine copolymerize with CO_2 without any added catalyst.[52] In the reaction with propylenimine at 100° C, the product is a copolymer containing 50% carbamate linkage, while at −20° C a cyclic carbamate is also formed (Eq. 4.33).

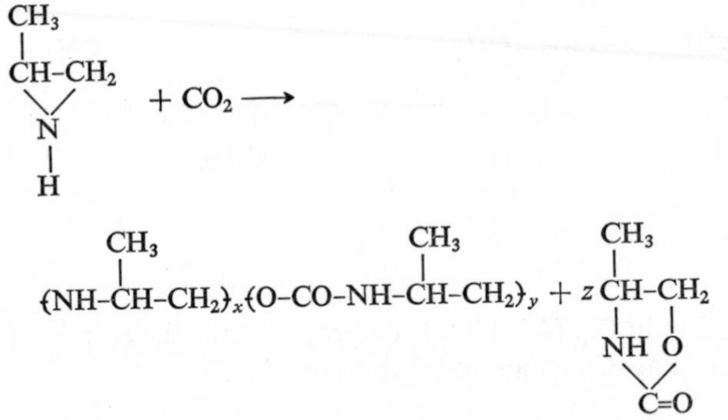

$$
\begin{array}{c}
CH_3 \\
| \\
CH-CH_2 \\
\diagdown \diagup \\
N \\
| \\
H
\end{array}
\; + CO_2 \longrightarrow
$$

$$
\{NH-\underset{\underset{CH_3}{|}}{CH}-CH_2\}_x\{O-CO-NH-\underset{\underset{CH_3}{|}}{CH}-CH_2\}_y + z\;
\begin{array}{c}
CH_3 \\
| \\
CH-CH_2 \\
| \quad\; | \\
NH \;\; O \\
\diagdown \diagup \\
C=O
\end{array}
\qquad (4.33)
$$

The molecular weight of these aziridine-carbon dioxide copolymers is not very high; η_{sp}/C of the solution, specific viscosity divided by concentration, ranges from 0.2 to 0.5. Examples of carbon dioxide-aziridine copolymerization are listed in Table 4.17.

4.2.4. Copolymerization with vinyl ether

Carbon dioxide and ethyl vinyl ether yield a polymeric product (η_{sp}/C, 0.1–0.3) containing 23% ester (and/or keto) linkage when heated at 65–80° C with aluminum acetylacetonate or aluminum alkoxide as a catalyst.[53]

Table 4.17. Copolymerization of carbon dioxide and aziridine.

Aziridine	Quantity	CO_2	Catalyst	Temp. (°C)	Yield (g)	η_{sp}/C (dl/g)	Content of carbamate bond (mole %)	Ref.
N-Phenylethylenimine	2.0 g	25.1 g	phenol	100	0.28	0.21[1]	41	49
	0.3 ml	20.8/imine	Al(OBu)$_3$	100	0.29		19	51
	0.3 ml	20.8/imine	Ti(OBu)$_4$	100	0.20		7.5	51
	0.3 ml	20.8/imine	MnCl$_2$.4H$_2$O	100	0.28		47	51
	0.3 ml	20.0/imine	Mn(acac)$_2$	100	0.29		59	51
Propylenimine	0.57 g	1.111	—	0	0.20	0.11[2]	8	52
	0.57 g	1.111	—	40	0.27	0.30[2]	26	52
	0.57 g	1.111	—	100	0.76	0.52[2]	47	52

[1] Specific viscosity divided by concentration, in 90% formic acid, 35°C.
[2] In water, 30°C.

Methyl vinyl ether gives a copolymer (molecular weight, 1,300) with CO_2 when heated at $80°C$ without any added catalyst. The carbonyl group in the copolymer, as observed by the infrared spectrum, disappears on treatment with sodium borohydride. Thus, keto groups, but not ester groups, are present, and the structure in Eq. (4.34) has been suggested for the copolymer.

$$CH_2=CH + CO_2 \longrightarrow \{CH-CH_2\}_x\{C-CH_2-CH-O-CH-CH_2\}, \qquad (4.34)$$
$$\underset{OCH_3}{|} \qquad \underset{OCH_3}{|} \quad \underset{O}{\|} \quad \underset{OCH_3}{|} \; \underset{OCH_3}{|}$$

In order to form a keto group, one of the two carbon-oxygen bonds of CO_2 must be cleaved, and a mechanism involving the formation of a β-lactone has been proposed (Eq. 4.35).

$$\begin{array}{l} OCH_3 \\ | \\ CH=CH_2 \end{array} + CO_2 \longleftrightarrow \begin{array}{c} OCH_3 \\ | \\ {}^{\oplus}CH=CH_2{}^{\ominus} \\ | \\ O=C=O \\ {}_{\ominus}\;{}_{\oplus} \end{array} \longrightarrow \left[\begin{array}{c} OCH_3 \\ | \\ CH-CH_2 \\ | \quad / \\ O-C=O \end{array} \right]$$

$$\sim CH_2-\overset{OCH_3}{\underset{|}{CH}}{}^{\oplus} + \begin{array}{c} OCH_3 \\ | \\ CH-CH_2 \\ | \quad / \\ O-C=O \\ {}_{\ominus}\;{}_{\oplus} \end{array} \longrightarrow \sim CH_2-\overset{OCH_3}{\underset{|}{CH}}-O-\overset{OCH_3}{\underset{|}{CH}}-CH_2-\underset{\|}{\overset{}{C}}{}^{\oplus}$$
$$O$$

$$\xrightarrow[\quad CH=CH_2 \quad]{\overset{OCH_3}{|}} \sim CH_2-\overset{OCH_3}{\underset{|}{CH}}-O-\overset{OCH_3}{\underset{|}{CH}}-CH_2-\underset{\underset{O}{\|}}{C}-CH_2-\overset{OCH_3}{\underset{|}{CH}}{}^{\oplus} \Longrightarrow \qquad (4.35)$$

4.2.5. Copolymerization with diene

Carbon dioxide and butadiene, when heated at $80–100°C$, yield a rubbery product insoluble in organic solvents.[55] The product, which shows carbonyl infrared absorption, on hydrolysis gives products with hydroxyl and carboxyl groups. Thus, the product contains ester groups formed by the copolymerization of these monomers. Isoprene or 2, 3-dimethylbutadiene also copolymerizes with CO_2 to give a viscous product. Infrared carbonyl absorption of the product almost disappears on treatment with sodium borohydride, indicating the presence of keto groups in the copolymer. No details of the structure of the copolymer or the reaction mechanism are yet known.

4.2.6. Copolymerization with acrylonitrile

Carbon dioxide and acrylonitrile, when heated at 120–160°C in the presence of triethylenediamine, give a polymeric product (molecular weight, 1,500–2,200) showing infrared absorption apparently due to carbamate (urethane) linkage.[56] In the presence of other tertiary amines such as triethylamine and pyridine, no polymerization takes place. From the analytical data, Formula (4.3) has been proposed as the structure of the copolymer.

$$\left(\!\!\begin{array}{l} CH_2-CH_2-O-\overset{\overset{\textstyle O}{\|}}{C}-N\!\!\underset{CH_2-CH_2}{\overset{CH_2-CH_2}{<}}\!\!N-\overset{\overset{\textstyle }{}}{\underset{\overset{\|}{O}}{C}}-O-CH_2-\underset{\overset{|}{CN}}{CH}\end{array}\!\!\right)_{\!(x)}\!\!\left(\!\!CH_2-\underset{\overset{|}{CN}}{CH}\!\!\right)_{\!(1-x)}$$

(Formula 4.3)

Initiation

$$:N\!\!\underset{CH_2-CH_2}{\overset{CH_2-CH_2}{<}}\!\!CH_2-CH_2-N: + \overset{\overset{\textstyle O}{\|}}{C}\!=\!O \longrightarrow \left[\ ^-O-\overset{\overset{\textstyle }{}}{\underset{\overset{\|}{O}}{C}}-\overset{+}{N}\!\!\underset{CH_2-CH_2}{\overset{CH_2-CH_2}{<}}\!\!\overset{+}{N}-\overset{\overset{\textstyle O}{\|}}{C}-O^- \right]$$

$CH_2=CH-CN$

$CH_2=CH-CN$ Polyacylonitrile

$$\left[\ ^-O\!\!-\!\!\overset{\overset{\textstyle O}{\|}}{\underset{+CH_2-CH_2}{C}}\!\!\underset{CH_2-CH_2}{\overset{+}{N}}\!\!\underset{CH_2-CH_2}{\overset{CH_2-CH_2}{<}}\!\!N-\overset{}{\underset{\overset{\|}{O}}{C}}-O-CH_2-\underset{\overset{|}{CN}}{CH}^- \right]$$

(4.36)

$$\left[\ ^+CH_2-CH_2-O-\overset{\overset{\textstyle O}{\|}}{C}-N\!\!\underset{CH_2-CH_2}{\overset{CH_2-CH_2}{<}}\!\!N-\overset{}{\underset{\overset{\|}{O}}{C}}-O-CH_2-\underset{\overset{|}{CN}}{CH}^- \right]\ (\text{IV})$$

Propagation $CH_2=CH-CN$

Alternating 1 : 2 : 1 terpolymer

$$\left(\!\!\begin{array}{l} CH_2-CH_2-O-\overset{\overset{\textstyle O}{\|}}{C}-N\!\!\underset{CH_2-CH_2}{\overset{CH_2-CH_2}{<}}\!\!N-\overset{}{\underset{\overset{\|}{O}}{C}}-O-CH_2-\underset{\overset{|}{CN}}{CH}\end{array}\!\!\right)_{\!(x)}\!\!\left(\!\!CH_2-\underset{\overset{|}{CN}}{CH}\!\!\right)_{\!(1-x)}$$

The carbamic salt from triethylenediamine and carbon dioxide, after reacting with acrylonitrile, is thought to undergo ring-opening (Eq. 4.36) and rearrangement to form zwitter ions (Formula 4.4) which combine with each other, directly or after the addition of some acrylonitrile molecules, to produce the copolymer.

4.2.7. Ternary copolymerization of carbon dioxide, cyclic phosphonite, and acrylic compounds

Heating at 130–150° C a mixture of carbon dioxide, ethylene phenylphosphonite, and acrylic ester or acrylonitrile yields a ternary 1:1:1 copolymer with molecular weight about 2,000.[57] An example is shown in Eq. (4.37).

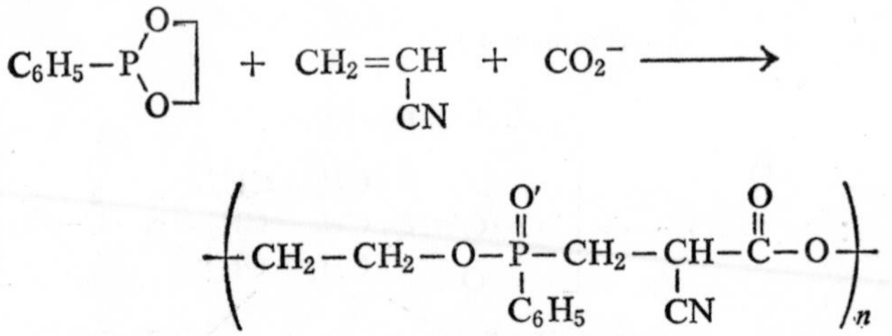

$$\begin{array}{c} C_6H_5-P \overset{O-}{\underset{O-}{\Big]}} \;+\; CH_2\!\!=\!\!\underset{CN}{CH} \;+\; CO_2^- \longrightarrow \end{array}$$

$$\left(\!-CH_2-CH_2-O-\underset{\underset{C_6H_5}{|}}{\overset{\overset{O'}{\|}}{P}}-CH_2-\underset{\underset{CN}{|}}{CH}-\overset{\overset{O}{\|}}{C}-O\!-\!\right)_n \qquad (4.37)$$

A mechanism has been proposed that involves the formation of a zwitter ion from the nucleophilic cyclic phosphonite and the electrophilic acrylic compound, followed by the reaction with carbon dioxide (Eq. 4.38).

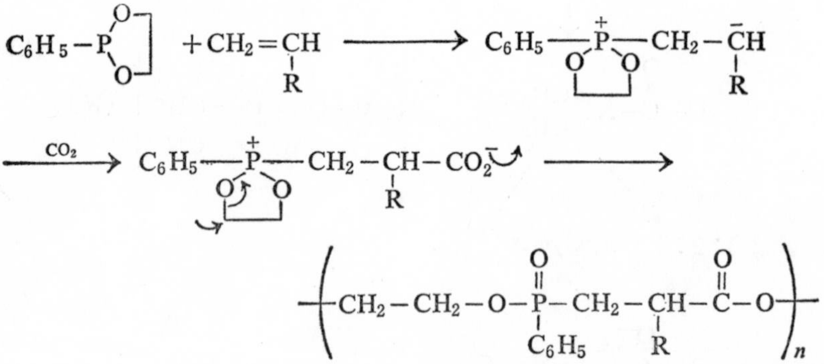

$$C_6H_5-P \overset{O-}{\underset{O-}{\Big]}} \;+CH_2\!\!=\!\!\underset{R}{CH} \longrightarrow C_6H_5-\overset{+}{\underset{O\;\;O}{P}}-CH_2-\underset{R}{\overset{-}{CH}}$$

$$\overset{CO_2}{\longrightarrow} C_6H_5-\overset{+}{\underset{O\;\;O}{P}}-CH_2-\underset{R}{CH}-CO_2^- \longrightarrow$$

$$\left(\!-CH_2-CH_2-O-\underset{\underset{C_6H_5}{|}}{\overset{\overset{O}{\|}}{P}}-CH_2-\underset{\underset{R}{|}}{CH}-\overset{\overset{O}{\|}}{C}-O\!-\!\right)_n$$

(R=CN, COOCH₃)

$$(4.38)$$

References

1) N. Yamazaki and F. Higashi, *Tetrahedron*, **30**, 1323 (1974).
2) N. Yamazaki, M. Niwano, J. Kawabata, and F. Higashi, *Tetrahedron*, **31**, 665 (1975).
3) N. Yamazaki, T. Iguchi, and F. Higashi, *Tetrahedron Lett.*, 1191 (1974); *Tetrahedron*, **31**, 3031 (1975).
4) E. Erdman and H. van der Smissen, *Liebigs Ann. Chem.*, **361**, 32 (1908).
5) N. Yamazaki, T. Iguchi, and F. Higashi, *J. Polym. Sci. Polym. Chem. Ed.*, **13**, 785 (1975).
6) G. D. Buckley and N. H. Ray, U. S. Patent 2,550,767 (1951).
7) G. J. M. van der Kerk, H. G. J. Overmars, and G. M. van der Want, *Rec. Trav. Chim.*, **74**, 1301 (1955).
8) H. Iiyama, M. Asakura, and K. Kimoto, *Kogyo Kagaku Zasshi* (Japanese), **68**, 240 (1965).
9) Y. Otsuji, N. Matsumura, and E. Imoto, *Bull. Chem. Soc. Jpn.*, **41**, 1485 (1968).
10) N. Yamazaki, F. Higashi, and T. Iguchi, *J. Polym. Sci. Polym. Lett. Ed.*, **12**, 517 (1974).
11) N. Yamazaki, T. Tomioka, and F. Higashi, *J. Polym. Sci. Polym. Lett. Ed.*, **14**, 55 (1976).
12) K. Soga, Y. Toshida, S. Hosoda, and S. Ikeda, *Makromol. Chem.*, **179**, 2379 (1978).
13) K. Soga, S. Hosoda, and S. Ikeda, *J. Polym. Sci. Polym. Chem. Ed.*, **17**, 517 (1979).
14) S. Inoue, H. Koinuma, and T. Tsuruta, *Polymer Letters*, **7**, 287 (1969).
15) S. Inoue, H. Koinuma, and T. Tsuruta, *Makromol. Chem.*, **130**, 210 (1969).
16) S. Inoue, *Chemtech.*, **6**, 588 (1976).
17) S. Inoue, T. Tsuruta, T. Takada, N. Miyazaki, M. Kambe, and T. Takaoka, *Appl. Polym. Symp.*, **26**, 257 (1975).
18) K. Udipi and J. K. Gillham, *J. Appl. Polym. Sci.*, **18**, 1575 (1974).
19) S. Inoue, M. Kobayashi, H. Koinuma, and T. Tsuruta, *Makromol. Chem.*, **155**, 61 (1972).
20) M. Kobayashi, S. Inoue, and T. Tsuruta, *Macromolecules*, **4**, 658 (1971).
21) M. Kobayashi, Y. -L. Tang, T. Tsuruta, and S. Inoue, *Makromol. Chem.*, **169**, 69 (1973).
22) M. Kobayashi, S. Inoue, and T. Tsuruta, *J. Polym. Sci., Polym. Chem. Ed.*, **11**, 2383 (1973).
23) S. Inoue, *Progr. Polym. Sci. Japan*, **8**, 1 (1975).
24) M. Rätzch and W. Haubold, *Faserforsch. Textiltech.*, **28**, 15 (1977).
25) W. Kuran, S. Pasynkiewicz, J. Skupinska, and A. Rokicki, *Makromol. Chem.*, **177**, 11 (1976).
26) W. Kuran, S. Pasynkiewicz, and J. Skupinska, *Makromol. Chem.*, **177**, 1283 (1976).
27) W. Kuran, S. Pasynkiewicz, and J. Skupinska, *Makromol. Chem.*, **178**, 47 (1977).
28) W. Kuran, S. Pasynkiewicz, and J. Skupinska, *Makromol. Chem.*, **1 78**, 2149 (1977).
29) W. Kuran A. Rokicki, E. Wilinska, *Makromol. Chem.*, **180**, 361 (1979).
30) K. Soga, K. Hyakkoku, K. Izumi, and S. Ikeda, *J. Polym. Sci., Polym. Chem. Ed.*, **16**, 2383 (1978).
31) M. Nishimura, M. Kasai, and E. Tsuchida, *Makromol. Chem.*, **179**, 1913 (1978).
32) K. Soga, K. Uenishi, S. Hosoda, and S. Ikeda, *Makromol. Chem.*, **178**, 893 (1977).
33) K. Soga, K. Hyakkoku, and S. Ikeda, *Makromol. Chem.*, **179**, 2837 (1978).
34) Y. Yoshida and S. Inoue, *Polymer J.*, in press.
35) Y. Yoshida and S. Inoue, *Bull. Chem. Soc., Japan*, **51**, 559 (1978).
36) H. Koinuma and H. Hirai, *Makromol. Chem.*, **178**, 1283 (1977).
37) N. Takeda and S. Inoue, *Makromol. Chem.*, **179**, 1377 (1978).

38) M. Kobayashi, S. Inoue, and T. Tsuruta, unpublished data cited in S. Inoue, *Progr. Polym. Sci. Japan,* **8**, 8 (1975).
39) H. Koinuma and H. Hirai, *Makromol. Chem.,* **178**, 241 (1977).
40) S. Inoue, H. Koinuma, Y. Yokoo, and T. Tsuruta, *Makromol. Chem.,* **143**, 97 (1971).
41) S. Inoue, K. Matsumoto, and Y. Yoshida, *Makromol. Chem.,* in press
42) T. Hirano, S. Inoue, and T. Tsuruta, *Makromol. Chem.,* **176**, 1913 (1975).
43) S. Inoue, H. Koinuma, and T. Tsuruta, *Polymer J.,* **2**, 220 (1971).
44) T. Hirano, S. Inoue, and T. Tsuruta, *Makromol. Chem.,* **177**, 3237 (1976).
45) T. Hirano, S. Inoue, and T. Tsuruta, *Makromol. Chem.,* **177**, 3245 (1976).
46) E. J. Vandenberg, *J. Polym. Sci.,* A-1, **7**, 529 (1969).
47) S. Inoue, T. Hirano, and T. Tsuruta, *Polymer J.,* **9**, 101 (1977).
48) W. Kuran, A. Rokicki, and W. Wielgopolan, *Makromol. Chem.,* **179**, 2545 (1978).
49) T. Kagiya and T. Matsuda, *Polymer J.,* **2**, 398 (1971).
50) K. Soga and S. Ikeda, *J. Polym. Sci., Polym. Lett. Ed.,* **11**, 479 (1973).
51) K. Soga, S. Hosoda, and S. Ikeda, *J. Polym. Sci., Polym. Chem. Ed.,* **12**, 729 (1974).
52) K. Soga, W. -Y. Chang, and S. Ikeda, *J. Polym. Sci., Polym. Chem. Ed.,* **12**, 121 (1974).
53) K. Soga, M. Sato, Y. Tazaki, and S. Ikeda, *J. Polym. Sci., Polym. Lett. Ed.,* **13**, 265 (1975).
54) K. Soga, M. Sato, S. Hosoda, and S. Ikeda, *J. Polym. Sci., Polym. Lett. Ed.,* **13**, 543 (1975).
55) K. Soga, S. Hosoda, and S. Ikeda, *Makromol. Chem.,* **176**, 1907 (1975).
56) W. -Y. Chang, *Proc. Natl. Sci. Council, Rep. of China,* **2**, 170 (1978).
57) T. Saegusa, S. Kobayashi, and Y. Kimura, *Macromolecules,* **10**, 68 (1977).

5

Biological Carboxylations

Kozi ASADA

The Research Institute for Food Science, Kyoto University

Uji-shi, Kyoto 611, Japan

5.1. Introduction

Almost all the organic compounds on our planet are derived from substances produced by photosynthetic CO_2 fixation in plants. Even the by-products of fossil oil and coal have their origins in photosynthetic processes that took place in plants several hundred million years ago (the Carboniferous period). Thus, present and past photosynthetic CO_2 fixation is the ultimate source of food and energy for human beings and for other organisms that are not able to use CO_2 as a carbon source. All organisms can be divided into two groups, depending on their source of carbon: heterotrophs that require organic compounds for nutrients, and autotrophs that are able to use CO_2 as their sole carbon source. Among the latter, those organisms that acquire energy for CO_2 fixation through photochemical reactions are called photoautotrophs. Other organisms that obtain energy for CO_2 fixation through chemical reactions such as $H_2S + 2 O_2 \rightarrow H_2SO_4$ (*Thiobacillus*), $Fe^{2+} \rightarrow Fe^{3+}$ (*Ferrobacillus*), $H_2 \rightarrow 2 H^+$ (*Hydrogenomonas*), and $NO_2^- \rightarrow NO_3^-$ (*Nitrobacter*) are referred to as chemoautotrophs. Photoautotrophs are found among organisms at different evolutionary stages, prokaryotes (photosynthetic bacteria), algae, and plants, but chemoautotrophs are limited only to primitive prokaryotes.

On our planet, the carbon cycle between CO_2 fixation by autotrophs and CO_2 release by the respiration of heterotrophs has maintained the CO_2 level in the atmosphere at about 330 ppm (Fig. 5.1). However, recent industrial consumption of fossil fuels has resulted in a small but gradual increase in the CO_2 concentration of the atmosphere (about 50 ppm since 1850[1]), despite the large buffering capacity of bicarbonates in the sea. Another buffer for CO_2 in the atmosphere is photosynthesis by plants because the CO_2 fixation rate is not saturated at the present concentration of CO_2 in the air. The annual amount of CO_2 fixed by plants, algae, and photosynthetic bacteria has been estimated to be approximately 10^{17} g,

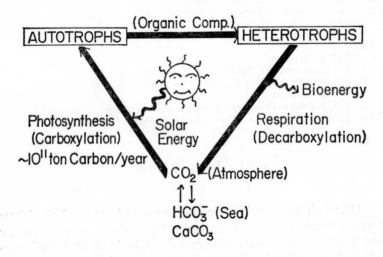

Fig. 5.1. The carbon cycle on the earth.

almost all of which is counterbalanced by respiration and combustion. The contribution of chemoautotrophs to the production of organic compounds on Earth is negligible compared to that of the photoautotrophs.

After the first report of photosynthetic CO_2 fixation in plants by de Saussure in 1804, CO_2-fixing organisms were believed to be limited to the autotrophs. However, in 1935 Wood and Werkman[2] found that sizable amounts of CO_2 are fixed in a heterotroph, *Propioniobacterium*, growing on glycerol and since then many CO_2 fixation reactions have been found in heterotrophs. The amount of CO_2 fixed by heterotrophs is very small compared to CO_2 released by respiration, and in most cases CO_2 fixation by heterotrophs does not function as a carbon source. For example, in the rat, only 0.01 % of the body carbon is derived from CO_2. However, in an anaerobic bacterium, *Clostridium kluyveri*, a quarter of the bacterial carbon is the result of CO_2 fixation.[3]

Because heterotrophs evolve CO_2 by respiration, metabolic disturbances caused by a CO_2 deficiency are rarely observed under normal conditions. However, many CO_2 effects have been found by controlling the CO_2 concentration in heterotrophs. A typical example is the growth of the heterotrophic bacterium, *Escherichia coli*, which is affected by environmental CO_2 concentration; growth stops when CO_2 is removed.[4] Similar CO_2-induced effects have been found in diverse organisms: initiation of seed germination, fruit-body formation in *Basidiomycetes*, attraction of mosquitoes, formation of potato tubers, termination of diapause in insects, and sexual differentiation of coelenterates. These are only a few of the interesting effects, and further examples will undoubtedly be observed when careful control of the CO_2 concentration is maintained in other biological systems.

How these CO_2-induced biological effects occur is not well understood, but the following three mechanisms may contribute to the effects. The first is the production of metabolites by carboxylation. An important biosynthetic pathway uses carboxylation reactions in, for example, the biosynthesis of malonyl CoA, the key intermediate in fatty acid biosynthesis, and the biosynthesis of carbamoylphosphate, the starting compound of the ornithine cycle. In addition, several carboxylation reactions that yield di- and tricarboxylic acids regulate the tricarboxylic acid cycle. Thus, non-photosynthetic CO_2 fixation has important physiological functions in the heterotrophs and in the autotrophs as well. The second is the modification of enzyme proteins by CO_2 as examplified by its effect on hemoglobin. The activities of many other enzymes are also influenced by CO_2. The third mechanism is the pH effect from the hydration of CO_2 and the dehydration of bicarbonate. In summary CO_2 has at least three biological functions:

1) as the carbon source in autotrophs 2) in the biosynthesis of metabolites by carboxylation in both autotrophs and heterotrophs and 3) in the regulation of organisms through pH control and enzyme modification.

This chapter will first describe the properties of the hydration and dehydration of carbon dioxide and the function of the enzyme that catalyzes these processes, carbonic anhydrase. These processes are important to the transport of carbon dioxide in organisms at the tissue, cell or organelle level, to the buffering effects of carbon dioxide in the cells, and to carboxylations in which carbon dioxide is utilized as CO_2 or HCO_3^-. We will then turn to a summary of carboxylation reactions by autotrophs and heterotrophs and a discussion of their physiological functions. In the concluding sections we will describe photosynthetic CO_2 fixation and its regulation and the modification of enzyme proteins by carbon dioxide.

5.2. Hydration and Dehydration of Carbon Dioxide

5.2.1. Effect of pH on the Hydration of Carbon Dioxide

Air-equilibrated water near neutral pH contains a concentration of carbon dioxide (Table 5.1) that is similar to that of molecular oxygen (0.25 mM 25°C) although the concentration of CO_2 (0.033%) is much lower than that of O_2 (21%) in the atmosphere. Hydrophobic properties of molecular oxygen are responsible for its low solubility in water, which has a high dielectric constant; oxygen is more soluble in non-polar solvents than in water. In contrast carbon dioxide has higher solubility in water than oxygen has; the Bunsen absorption coefficients of CO_2 and O_2 are 0.76 and 0.028 ml gas/ml water at 25°C, respectively. Further, carbon dioxide dissolved in water is hydrated to yield H_2CO_3, and accompanying the increased pH in the medium, the hydrated carbon dioxide releases H^+ to yield bicarbonate and carbonate. These properties explain why carbon dioxide is utilized by organisms as effectively as oxygen despite the low concentration of CO_2 in the atmosphere. In this section we will discuss the hydration of CO_2 and the dissociation of hydrated carbon dioxide at various pH's.

Carbon dioxide in the gas phase dissolves in water and the dissolved CO_2 is hydrated to form carbonic acid and bicarbonate (Eqs. 5.1 and 5.2).

$$CO_2 + H_2O \underset{k_{-1}}{\overset{k_1}{\rightleftharpoons}} H_2CO_3 \qquad (5.1)$$

$$CO_2 + H_2O \underset{k_{-2}}{\overset{k_2}{\rightleftarrows}} H^+ + HCO_3^- \qquad (5.2)$$

Above pH 7.0 the reaction of dissolved CO_2 with OH^- participates in the formation of bicarbonate; however, its contribution is very low in the range of physiological pH and is only 1.2% at pH 7.0. At sufficiently high pH values, reaction (5.3) becomes important (Section 5.2.2).

$$CO_2 + OH^- \underset{k_{-3}}{\overset{k_3}{\rightleftarrows}} HCO_3^- \qquad (5.3)$$

Carbonic acid dissociates to bicarbonate, then to carbonate (Eqs. 5.4 and 5.5).

$$H_2CO_3 \underset{k_{-4}}{\overset{k_4}{\rightleftarrows}} H^+ + HCO_3^- \qquad (5.4)$$

$$HCO_3^- \underset{k_{-5}}{\overset{k_5}{\rightleftarrows}} H^+ + CO_3^{2-} \qquad (5.5)$$

The following dissociation or equilibrium constant is the value at 25°C. The equilibrium of CO_2 hydration (Eq. 5.1) lies so far to the left[5];

$$K_H = \frac{[H_2CO_3]}{[CO_2][H_2O]} = 2.56 \times 10^{-3} \qquad (5.6)$$

at low pH, therefore, that most of the dissolved carbon dioxide occurs as CO_2, and H_2CO_3 is a very minor molecular species. Because of the low concentration of H_2CO_3, the first dissociation of carbonic acid (Eq. 5.4) is defined as in Eq. (5.7).

$$K_{a1} = \frac{[H^+][HCO_3^-]}{[CO_2 + H_2CO_3]} = 4.3 \times 10^{-7}, pK_{a1} = 6.35 \qquad (5.7)$$

The second dissociation (Eq. 5.5) is given in Eq. (5.8).

$$K_{a2} = \frac{[H^+][CO_3^{2-}]}{[HCO_3^-]} = 5.6 \times 10^{-11}, pK_{a2} = 10.32 \qquad (5.8)$$

Using the Henderson-Hasselbach equation (5.9 and 5.10) we can estimate the molar ratio of CO_2, H_2CO_3, HCO_3^-, CO_3^{2-} at various pH's (Fig. 5.2).

$$pH = pK_{a1} + \log \frac{[HCO_3^-]}{[H_2CO_3 + CO_2]} \qquad (5.9)$$

$$pH = pK_{a2} + \log \frac{[CO_3^{2-}]}{[HCO_3^-]} \qquad (5.10)$$

At 1 atmosphere of pressure, 0.76 ml of CO_2 dissolves per ml of water at 25° C; therefore, air-equilibrated water dissolves 11.2 μM CO_2. Table 5.1 shows the concentration of each molecular species of carbon dioxide in air-equilibrated water containing enough buffer at each pH. At low pH, CO_2 is the major molecular species. In the neutral pH region, the amount

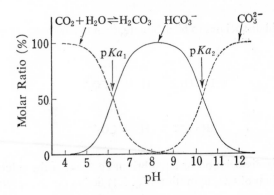

Fig. 5.2.   Molar ratio of CO_2, H_2CO_3, HCO_3^-, and CO_3^{2-} in water at various pH's (25° C). Molar ratio of H_2CO_3 is 0.26% of the concentration of $H_2CO_3 + CO_2$.

Table 5.1.   Concentrations of CO_2, H_2CO_3, HCO_3^-, and CO_3^{2-} in water equilibrated with air (330 ppm CO_2) at various pH's (25° C).

pH	CO_2 (μM)	H_2CO_3 (μM)	HCO_3^- (μM)	CO_3^{2-} (μM)	Total "CO_2" (mM)
3.5	11.2	0.029			0.0112
4.0	11.2	0.029	0.05		0.0113
4.5	11.2	0.029	0.16		0.0114
5.0	11.2	0.029	0.50		0.0117
5.5	11.2	0.029	1.58		0.0128
6.0	11.2	0.029	5.01		0.0162
6.5	11.2	0.029	15.8		0.0270
7.0	11.2	0.029	50.1	0.02	0.0613
7.5	11.2	0.029	158	0.24	0.169
8.0	11.2	0.029	501	2.4	0.515
8.5	11.2	0.029	1580	24	1.62
9.0	11.2	0.029	5010	240	5.26
9.5	11.2	0.029	15800	2400	18.2
10.0	11.2	0.029	50100	24000	74.1
10.5	11.2	0.029	158000	240000	398

of HCO_3^- increases in addition to CO_2, and at high pH, CO_3^{2-} is the major species. The total concentration of carbon dioxide increases with the increase in pH. However, the concentrations of dissolved CO_2 and H_2CO_3 are not affected by pH although the molar ratio of CO_2 and H_2CO_3 diminishes with increased pH. Thus, the pH in cells affects not only the molar ratio of CO_2, H_2CO_3, HCO_3^-, and CO_3^{2-} but also the content of carbon dioxide, and, conversely, carbon dioxide affects the pH in the cells.

5.2.2. Rate of Hydration and Dehydration of Carbon Dioxide

The concentrations of the four molecular species of carbon dioxide at various pH's given in Fig. 5.2 and Table 5.1 are the values at equilibrium. However, the rate of interconversion of these molecular species is also very important in biological respects. Carbon dioxide is fixed either as CO_2 or HCO_3^- by carboxylases (Table 5.2). The form of carbon dioxide released by decarboxylases and the form transported in tissues, cells, or organelles is also either CO_2 or HCO_3^-. Furthermore, enzyme proteins are modified only by CO_2 (Section 5.4.2.D). Thus, the rate of hydration and dehydration of carbon dioxide is intimately related to these processes.

Table 5.2. Substrates of carbon dioxide for carboxylases.

Carboxylase (equation number)	Substrate	References
RuBP carboxylase (5.17)	CO_2	20
PEP carboxykinase (5.19)	CO_2	21
PEP carboxytransphosphorylase (5.20)	CO_2	21
Malic enzyme (5.31)†	CO_2	23
Isocitrate dehydrogenase (5.32)	CO_2	23
6-Phosphogluconate dehydrogenase (5.33)	CO_2	24
CO_2-reductase (5.27)	CO_2	25
Prothrombin-precursor carboxylating enzyme (5.35)	CO_2	26
PEP carboxylase (5.18)	HCO_3^-	27
Pyruvate carboxylase (5.36)	HCO_3^-	28
Acetyl CoA carboxylase (5.37)	HCO_3^-	28
Propionyl CoA carboxylase (5.38)	HCO_3^-	28
β-Methylcrotonyl CoA carboxylase (5.39)	HCO_3^-	28
Geranoyl CoA carboxylase (5.40)	HCO_3^-	28
Urea carboxylase (5.41)	HCO_3^-	28
Carbamate kinase (5.44)	HCO_3^-	29
Carbamoylphosphate synthase (5.45)	HCO_3^-	30

† The enzyme from maize uses HCO_3^- as the substrate.[22]

Because reaction (5.3) contributes little in the region of physiological pH, reactions (5.1) and (5.2) are the most important in the hydration or dehydration of carbon dioxide in the cells. The dissociation rates of H_2CO_3

(reaction 5.4) and of HCO_3^-(reaction 5.5) are very rapid ($k_{-4} = {\sim}5 \times 10^{10}$ $M^{-1}s^{-1}$, $k_4 = {\sim}10^7s^{-1}$), so reactions (5.1) and (5.2) are the rate-limiting steps in the formation of HCO_3^- and CO_3^{2-}. The hydration and dehydration of carbon dioxide involved in establishing equilibrium at neutral pH can be formulated as in Eq. (5.11).

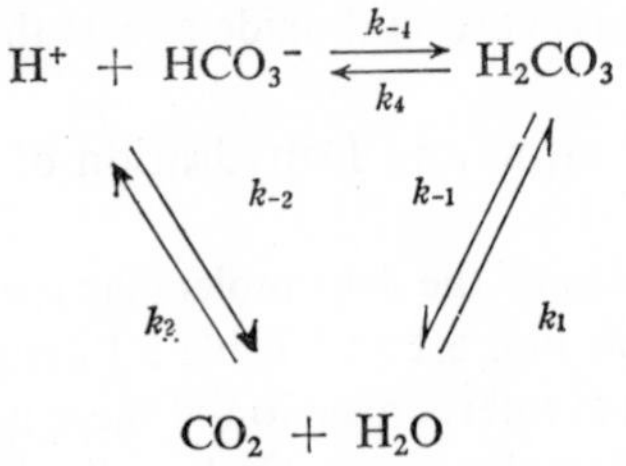

$$H^+ + HCO_3^- \underset{k_4}{\overset{k_{-4}}{\rightleftarrows}} H_2CO_3$$

$$(5.11)$$

$$CO_2 + H_2O$$

Therefore, the over-all hydration rate of CO_2 to bicarbonate is the balance of the hydration of CO_2 and the dehydration of H_2CO_3 (Eq. 5.12):

$$-\frac{d[CO_2]}{dt} = k_{1+2}[CO_2] - k_{-1-2}[HCO_3^-][H^+] \qquad (5.12)$$

where

$$k_{1+2} = k_1 + k_2, \quad k_{-1-2} = \frac{k_{-1}}{K_A} + k_{-2}, \text{ and}$$

$$K_A = \frac{[H^+][HCO_3^-]}{[H_2CO_3]} = 2.5 \times 10^{-4}.$$

The hydration (k_{1+2}) and dehydration (k_{-1-2}) rates were determined from the change of pH upon mixing CO_2 solution (for hydration) or bicarbonate solution (for dehydration) with a buffer-indicator solution by a stopped-flow apparatus.[6] The following rates are the values at 25° C:

$$k_{1+2} = 3.75 \times 10^{-2} \text{ s}^{-1},$$

$$k_{-1-2} = 5.5 \times 10^{4} M^{-1}s^{-1},$$

$$k_{-1} = 30 \text{ s}^{-1}.$$

The hydration of CO_2 is a slow reaction due to the extensive electronic rearrangement between CO_2 and H_2CO_3 or HCO_3^-. Its half-hydration time calculated from k_{1+2} is 18.5 s. On the other hand, the dehydration of HCO_3^- (H_2CO_3) depends on the concentration of H^+, and its rate is high at low pH. Its pseudofirst-order rate, $k_{-1-2}[H^+]$, is 5.5×10^{-1} s^{-1} (pH 5),

5.5×10^{-2} s^{-1} (pH 6), and 5.5×10^{-3} s^{-1} (pH 7), and the half-dehydration time is 1.3 s (pH 5), 13 s (pH 6), and 130 s (pH 7). Thus, at about pH 6.2, the rate constant for the hydration of CO_2, k_{1+2}, is equal to that for the dehydration of HCO_3^- (H_2CO_3), $k_{-1-2}[H^+]$, and CO_2 and HCO_3^- (H_2CO_3) are present in the same concentration. The hydration and dehydration rate constants for carbon dioxide shown above are several orders of magnitude lower than catalytic turnover rate constants of enzymes (10^2–10^3 s^{-1}/mol of enzyme). Carbonic anhydrase, which will be discussed in the next section, plays a role in adjusting the rate of the hydration and dehydration of carbon dioxide to that of the carboxylase or decarboxylase catalysis.

At high pH, the reaction of CO_2 with OH^- (reaction 5.3) predominates, and its rate constant, k_3, is 8.5×10^3 M^{-1} s^{-1}.[5] The pseudofirst-order rate constant, $k_3[OH^-]$, is 0.0085 s^{-1} (pH 8), 0.085 s^{-1} (pH 9), 0.85 s^{-1} (pH 10), and 8.5 s^{-1} (pH 11); it exceeds the hydration constant of CO_2, k_{1+2}, above pH 8.7. The reverse rate constant, k_{-3}, is estimated to be 2×10^{-4} s^{-1}.[5]

5.2.3. Carbonic Anhydrase

A. Physiological Functions of Carbonic Anhydrase

Carbonic anhydrase (carbonate hydro-lyase) is a zinc-containing enzyme that has been found in all organisms tested so far—bacteria, algae, mammals, and plants. This enzyme catalyzes reaction (5.2), and the universal distribution of the enzyme indicates the importance of the rapid interconversion between CO_2 and its hydrated form, even for unicellular organisms.

The requirement for carbonic anhydrase in algae and bacteria can be better defined from the observation that the biosynthesis of carbonic anhydrase appeared to be repressed when cells were cultured in high concentrations of CO_2,[7–9] revealing that carbonic anhydrase is essential only at low CO_2 concentrations for these unicellular organisms; at high CO_2 concentrations, molecular species of carbon dioxide for CO_2 fixation may be supplied in sufficient concentrations without carbonic anhydrase. Photosynthetic CO_2 fixation by carbonic anhydrase-deficient algal cells that have been cultured in 5% CO_2 is very low in air, but under the same conditions, carbonic anhydrase-containing cells that have been cultured in air have a high capacity for CO_2 fixation. The carbonic anhydrase-deficient algae can fix carbon dioxide at an ordinary rate only in 5% CO_2.[9]

It has been shown that carbon dioxide can be transported through algal cell membranes[10] and the envelope of spinach chloroplasts[11] only in the form of CO_2. The carboxylase for photosynthetic CO_2 fixation, ribulose bisphosphate carboxylase (section 5.4.2.B), can utilize carbon dioxide only in the form of CO_2. These observations indicate that carbon dioxide from outside the cells is transported to the site of the carboxylase in the chloroplast stroma and is fixed by the carboxylase only in the form of CO_2 (Fig. 5.3). In photosynthetic organisms, the rate of conversion of HCO_3^- to CO_2 should be the same as the rate of carboxylase turnover, and the enhanced dehydration catalyzed by carbonic anhydrase is essential for the transport and fixation of carbon dioxide. This is particularly so in the region of physiological pH, where the molar ratio of CO_2 is low (Fig. 5.2, Table 5.1). This is why carbonic anhydrase-deficient cells cannot fix carbon dioxide in air.

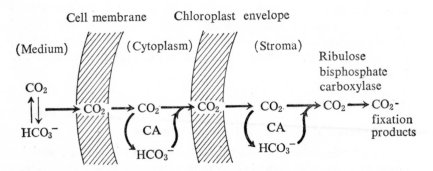

Fig. 5.3.   Transport and fixation of carbon dioxide in an algal cell. CA: carbonic anhydrase.

In contrast to photosynthetic organisms, heterotrophs excrete carbon dioxide that is produced by cellular metabolism. In mammals, carbon dioxide is transported to the lungs in the carbamate form of hemoglobin. Carbamino derivatives of hemoglobin are formed from CO_2 only, not from HCO_3^-, as a result of its binding to the basic form of the amino group of the NH_2-terminal residues (Section 5.4.2.D). Thus, the rate at which CO_2 is supplied to the hemoglobin in erythrocytes from the carbon dioxide pool produced by cellular metabolism is the key step for the excretion of carbon dioxide. Carbonic anhydrase in the erythrocytes enhances the rate of interconversion between CO_2 and HCO_3^-; in its absence the excretion of carbon dioxide is retarded and death results. Carbonic anhydrase also functions to release carbon dioxide from carbamino derivatives of hemoglobin in the lungs.

These two examples indicate the importance of carbonic anhydrase in facilitating the transport of carbon dioxide in cells or tissues. It has also been suggested that, in animal tissues, carbonic anhydrase plays a role in the transfer and accumulation of H^+ and HCO_3^-.[12] Enhancement of the hydration of CO_2 may increase the reaction rate of the biotin-dependent carboxylases which can utilize carbon dioxide only in the form of HCO_3^-. Carbonic anhydrase may also impede local changes of pH in the cells or organelles by promoting equilibrium between CO_2 and HCO_3^-.[12] Thus, the rapid establishment of equilibrium between CO_2 and HCO_3^- is indispensable for the cellular activities of all organisms. Let us now turn to a description of the catalytic properties and reaction mechanism of this enzyme.

B. Catalytic Properties of Carbonic Anhydrase

Carbonic anhydrase catalyzes reaction (5.2), the hydration of CO_2 and the dehydration of HCO_3^-.

$$CO_2 + H_2O \rightleftharpoons H^+ + HCO_3^- \tag{5.2}$$

H_2CO_3 is neither substrate for the dehydration nor a product of the hydration. When H_2CO_3 is the substrate or product, the observed catalytic rates exceed the diffusion-controlled limit for a low molar ratio of H_2CO_3 (Eq. 5.6).[13]

Over-all catalysis of hydration and dehydration by carbonic anhydrase (E) follows Michaelis-Menten kinetics; carbonic anhydrase-carbon dioxide complex (E-"CO_2") is first formed, then HCO_3^- or CO_2 is produced from the E-"CO_2" complex (Eq. 5.13).

$$E + CO_2 + H_2O \text{ or } HCO_3^- + H^+ \underset{k_{-13}}{\overset{k_{13}}{\rightleftharpoons}} E\text{-"}CO_2\text{"}$$

$$\overset{k_{cat}}{\longrightarrow} E + HCO_3^- + H^+ \text{ or } CO_2 + H_2O \tag{5.13}$$

The Michaelis-Menten parameters of carbonic anhydrase have been determined for mammalian and plant enzymes in detail. Figure 5.4 indicates the Michaelis constant, K_m ($=(k_{-13} + k_{cat})/k_{13}$), and the turnover number, k_{cat}, of the bovine enzyme at various pHs.[14,15] Similar values have been reported for a plant enzyme[16]. Although pH has little effect on the K_m values for hydration and dehydration, the turnover numbers of the enzyme are greatly influenced by pH. The basic form of the enzyme is apparently active in hydration, while the acidic form is required for dehydration, both with a pK_a of around 7. This inverse relationship between the pH-rate profiles for hydration and dehydration reflects the pH-dependence of the equilibrium in reaction (5.2).

The turnover number of carbonic anhydrase (k_{cat}) is one of the highest values known among enzymes. Furthermore, because of the high

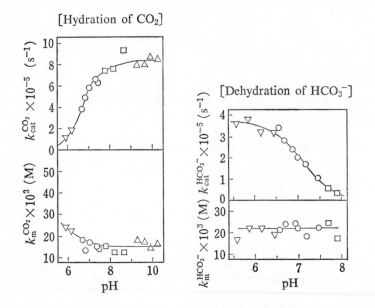

[Hydration of CO₂]

[Dehydration of HCO₃⁻]

Fig. 5.4. The pH-dependence of k_{cat} and K_m in the bovine carbonic anhydrase-catalyzed hydration of CO_2 and dehydration of HCO_3^-.[15]
Source: ref. 15. Reproduced by kind permission of the American Chemical Society.

content of carbonic anhydrase in cells and organelles, the catalyzed rates of hydration and dehydration of "CO_2" are much higher than the spontaneous rates. Spinach chloroplasts contain about 2 mM of carbonic anhydrase,[17] and 1% of the soluble protein of parsley leaves is carbonic anhydrase. The ratio k_{cat}/K_m gives the second-order reaction rate constant between "CO_2" and carbonic anhydrase, k_{enz}, which is 10^7-10^8 $M^{-1}s^{-1}$ at neutral pH. In chloroplasts, which have a concentration of carbonic anhydrase of 2×10^{-3} M, the first-order reaction rate constant for the hydration or dehydration of "CO_2", k_{enz} [carbonic anhydrase], is around 10^4 s^{-1}. Thus, in chloroplasts, the hydration or dehydration rate of "CO_2" is six orders of magnitude higher than the uncatalyzed spontaneous rates of hydration (k_{1+2}, reaction 5.11; k_3 [OH^-] (pH 8), reaction 5.3) and of dehydration (k_{-1-2}[H^+] (pH 6), reaction 5.11). The high catalytic rates and high concentration of carbonic anhydrase must serve the physiological needs discussed in the previous section.

In addition to the reversible hydration of CO_2, carbonic anhydrase can catalyze the reversible hydration of aldehydes and the hydrolysis of esters. Two typical reactions are shown in Eq. (5.14).

$$R'O-C\begin{matrix}R\\\diagdown C=O\\\diagup\\O\end{matrix} + H_2O \rightleftarrows R'O-C\begin{matrix}R\\\diagdown C^{\cdots OH}\\\diagup\\O\end{matrix}OH$$

Alkyl pyruvate

$$O_2N-\!\!\left\langle\!\!\begin{matrix}\\\_\end{matrix}\!\!\right\rangle\!\!-F + H_2O \longrightarrow O_2N-\!\!\left\langle\!\!\begin{matrix}\\\_\end{matrix}\!\!\right\rangle\!\!-O^- + 2\,H^+ + F^- \qquad (5.14)$$

$$\underset{NO_2}{} \qquad\qquad\qquad \underset{NO_2}{}$$

2,4-Dinitro fluorobenzene

These reactions involving carbonic anhydrase were discovered by Pocker and Sarkanv.[14] Although the physiological function of these reactions is unknown at present, the findings have contributed to a better understanding of the reaction mechanism. The reaction rate constants (k_{enz}) for these organic substrates are two to seven orders of magnitude lower than that for carbon dioxide.[14]

C. Catalytic mechanism of carbonic anhydrase

Carbonic anhydrase has a molecular weight near 30,000 and 1 atom of zinc per molecule. Several isozymes occur (3 isozymes in human erythrocytes, 2 isozymes in bovine erythrocytes), and spinach contains an oligomeric form of the enzyme (a hexamer composed of identical subunits with a molecular weight near 30,000, the molecular weight of the entire molecule, 160,000–180,000).[16] Their catalytic properties differ considerably from each other; for example, the turnover numbers of human isozymes A and B are indistinguishable, but that of isozyme C is fivefold higher than those of isozymes A and B. The structure of the enzyme has been determined by X-ray crystallography,[14,18] and the active center was found to be at the bottom of a crevice (1.5 nm in depth), as shown schematically in Fig. 5.5. The zinc atom is the catalytic site and is tetrahedrally coordinated to 3 residues of histidine (His–93, –95, and –118 in human isozyme C) and to 1 water molecule. Near the Zn atom in the crevice there is another histidine residue (His-63) that is not a ligand of the metal and that has a pK_a of 7.12.

As shown in Fig. 5.4, the basic form of carbonic anhydrase is the active species of the enzyme for the hydration of CO_2, and the acidic form effects the dehydration of HCO_3^-. The active site group with a pK_a of about 7 has been thought to be a zinc-coordinated water molecule, but the participation of the histidine-63 residue in the catalytic cycle is also possible.[15] Figure 5.6 shows that, in the hydration of CO_2, the electrophilic carbon atom of CO_2 is attacked by the zinc-bound hydroxide as a nucleo-

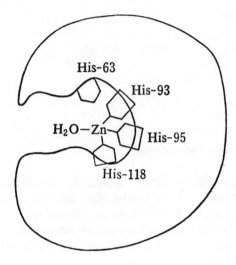

Fig. 5.5.   The active center of human carbonic anhydrase (acidic form).

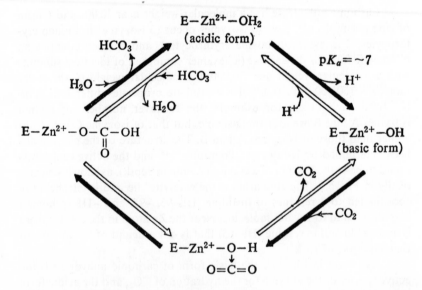

Fig. 5.6.   Catalytic mechanism of the hydration of CO_2 (■■■▶) and the dehydration of HCO_3^- (▭▭▷) by carbonic anhydrase (E).

phile, and an intermediate is formed. HCO_3^- is subsequently released from the intermediate, and the enzyme-zinc is coordinated to water. The release of a proton from the acidic form of carbonic anhydrase is the next step, and the basic form is regenerated. The dehydration of HCO_3^- is the reverse of the hydration of CO_2, and the binding of HCO_3^- to the acidic form of the enzyme is the start of the catalytic cycle.

The turnover number of carbonic anhydrase in the catalysis of the reversible hydration of carbon dioxide is very high, near 10^6 s^{-1}. This implies that the proton is transferred to and from the Zn-coordinated water molecule at a sufficient rate to attain this or higher frequency. However, the rate of protonation of the active site exceeds the diffusion-controlled rate if only a free proton (hydronium ion) in the medium transfers to the active site. This difficulty is avoidable if we assume the contribution of buffer molecules to the ionization process of the enzyme as proton donors and acceptors.[13] This has been confirmed experimentally; the catalytic rate of carbonic anhydrase is low, when the concentration of buffer in the reaction medium is sufficiently low, because of a slow rate in attaining equilibrium between the active site Zn-water molecule and the free H$^+$ in the medium.[19]

5.3. CARBOXYLATION REACTIONS

5.3.1. Classification of Biological Carboxylations

Biological carboxylation reactions can be categorized in terms of several features. One classification depends on the formation of either carboxylate or carbamate (Eqs. 5.15, 5.16).

$$R-\overset{|}{\underset{|}{C}}-H + CO_2 \text{ or } HCO_3^- \longrightarrow R-\overset{|}{\underset{|}{C}}-COO^- \qquad (5.15)$$

$$R-\overset{|}{\underset{|}{N}}-H + CO_2 \text{ or } HCO_3^- \longrightarrow R-\overset{|}{\underset{|}{N}}-COO^- \qquad (5.16)$$

$$CO_2\text{-acceptors} \qquad\qquad\qquad \text{carboxylated porducts}$$

Most carboxylations yield carboxylates. A C–N bond is formed only in the production of carbamoylphosphate by carbamate kinase and by carbamoylphosphate synthases (reactions 5.44–5.46) and of allophanate by urea carboxylase (reaction 5.41). However, in all biotin-dependent carboxylations (reactions 5.36–5.41), carbon dioxide binds first to the 1'-N of the ureido of biotin, forming a C–N bond, then the carboxyl group is transferred to a CO_2 acceptor, resulting in the formation of a C–C bond (except in the case of urea carboxylase).

As stated in previous chapters, the chemical, electrochemical, and radiochemical activation of carbon dioxide to CO_2^+ or CO_2^- causes the formation of a C–C bond. In biological systems, however, the activation of carbon dioxide itself and the release of free active carbon dioxide have not been observed. Enzymatic carboxylations involve the binding of the electrophilic carbon atom of carbon dioxide to the nucleophilic carbon or nitrogen atom of the CO_2 acceptor molecule on the enzyme; these reactions are similar to the binding of CO_2 to a nucleophilic Grignard reagent. With respect to C–C or C–N bond formation, CO_2 is a superior carboxylation species because the carbon atom of CO_2 is more electrophilic than that of HCO_3^-, but both are active species for carboxylating enzymes, and biological carboxylation reactions can be divided according to the form used. Strictly speaking, presently available method cannot distinguish among H_2CO_3, HCO_3^-, and CO_3^{2-} as the active species for carboxylases, but because HCO_3^- is a major species in neutral pH's, carbonic acid, bicarbonate, and carbonate are represented by HCO_3^-. All of the biotin enzymes utilize HCO_3^-, but spontaneous and $NAD(P)H_2$-dependent carboxylating enzymes use CO_2, except for phosphoenolpyruvate carboxylase (Table 5.2).

Chemically, CO_2 is more suspectible to nucleophilic attack by a CO_2 acceptor than HCO_3^- is. However, anionic properties of HCO_3^- may help to increase the interaction with enzymes, and the concentration of HCO_3^- may help to increase the interaction with enzymes, and the concentration of HCO_3^- is much higher than that of CO_2 above pH 7.0 (Table 5.1). In HCO_3^--utilizing carboxylases, the electrophilic activation of HCO_3^- must be mediated by the enzyme and its cofactor, especially by Mg^{2+}.

5.3.2. Bioenergetic Classification of Carboxylations

Carboxylations may also be classified from a bioenergetic viewpoint as spontaneous or energy-requiring carboxylations. In the former, carbon dioxide binds spontaneously to the CO_2 acceptor without any apparent requirement for extra energy, and in the latter, energy in the form of ATP, $NAD(P)H_2$, or an other reducing agent is necessary for the binding of carbon dioxide. In the following section, data on biological carboxylations are summarized, including CO_2 acceptor, carboxylated products, cofactors, and participating enzymes. The numbers given under the enzymes are numbering of enzymes recommended by the Nomenclature Committee of the International Union of Biochemistry (EC number). When the form of carbon dioxide used by a carboxylase has been determined, it is shown as CO_2 or HCO_3^-; when it has not been determined, it is designated as "CO_2." In the following reactions, $\longrightarrow$ shows that the equilibrium of the reaction lies far to the right, that is, to carboxylation; $\rightleftharpoons$ indicates that decarboxylation occurs depending on the substrate concentration. $\circledP$ refers to $-\overset{\underset{|}{O}}{\underset{|}{\overset{|}{P}}}-O^-$, $O-\circledP$ to phosphate ester, and P_i to orthophosphate.

A. Spontaneous Carboxylations

$$
\begin{array}{l}
CH_2O\circledP \\
\overset{|}{C}=O \\
\overset{|}{H\overset{|}{C}OH} \\
\overset{|}{H\overset{|}{C}OH} \\
\overset{|}{CH_2O\circledP}
\end{array}
+ CO_2 \xrightarrow[\substack{RuBP\ carboxylase \\ (4.1.1.39)}]{Mg^{2+}}
\begin{array}{l}
COOH \\
2\ \overset{|}{H\overset{|}{C}OH} \\
\overset{|}{CH_2O\circledP}
\end{array}
\qquad (5.17)
$$

Ribulose 1,5–bisphosphate   3–Phosphoglycerate
(RuBP)   (PGA)

$$
\begin{array}{l}
CH_2 \\
\overset{||}{C}-O\circledP \\
\overset{|}{COOH}
\end{array}
+ HCO_3^- \xrightarrow[\substack{PEP\ carboxylase \\ (4.1.1.31)}]{Mg^{2+}}
\begin{array}{l}
COOH \\
\overset{|}{CH_2} \\
\overset{|}{C}=O \\
\overset{|}{COOH}
\end{array}
+ P_i
\qquad (5.18)
$$

Phosphoenolpyruvate (PEP)   Oxaloacetate (OAA)

$$
PEP + \text{nucleotide diphosphate} + CO_2 \underset{\substack{PEP\ carboxykinase \\ (4.1.1.32)}}{\overset{Mg^{2+}\ or\ Mn^{2+}}{\rightleftharpoons}}
$$

$$
OAA + \text{nucleotide triphosphate} \qquad (5.19)
$$

$$
PEP + \text{orthophosphate } (P_i) + CO_2 \underset{\substack{PEP\ carboxytransphosphorylase \\ (4.1.1.38)}}{\overset{Mg^{2+}\ or\ Mn^{2+}}{\rightleftharpoons}}
$$

$$
OAA + \text{pyrophosphate} \qquad (5.20)
$$

$$
\begin{array}{c}
NH_2 \\
\fbox{} \\
N\quad N\text{–ribose–}\circledP
\end{array}
+ \text{``}CO_2\text{''} \underset{\substack{phosphoribosyl\ amino \\ imidazole\ carboxylase \\ (4.1.1.21)}}{\rightleftharpoons}
\begin{array}{c}
COO^-\ NH_2 \\
\fbox{} \\
N\quad N\text{–ribose–}\circledP
\end{array}
\qquad (5.21)
$$

5′-Phosphoribosyl-5-   5′-Phosphoribosyl
amino imidazole   5-amino-4-imidazole
                                carboxylate

Reaction (5.17) is the major carboxylation reaction in all photosynthetic organisms, including plants, algae, and photosynthetic bacteria.

Chemoautotrophs also contain this enzyme to fix CO_2, and thus, RuBP carboxylase is the key enzyme for the production of organic compounds in autotrophs. The reaction mechanism will be discussed in detail in Section 5.4.2.B. PEP carboxylase participates in the carboxylation of C_4-plants and is found also in non-photosynthetic organisms. Reactions (5.18), (5.19), and (5.20) differ in the substance that reacts with the phosphate group of the CO_2 acceptor, phosphoenolpyruvate: H_2O, nucleotide diphosphate, and orthophosphate, respectively. The oxaloacetate (OAA) produced by these three reactions may control the tricarboxylic acid cycle. Reaction (5.20) participates in the CO_2 fixation by the heterotroph, *Propioniobacterium*. The fifth spontaneous carboxylation (5.21) produces an intermediate of purine nucleotides. The reaction mechanism (5.19 and 5.20) proposed by Utter and Kolenbrander[31] is shown in Fig. 5.7. As in the carboxylation of a Grignard reagent, the CO_2 acceptor has a nucleophilic enolate form, and the electrophilic carbon atom of CO_2 binds to the acceptor. The same holds true for RuBP carboxylase (Fig. 5.17).

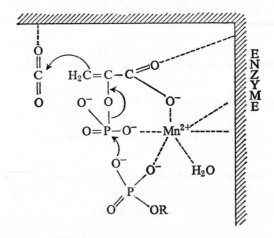

Fig. 5.7. Reaction intermediate with PEP carboxykinase (R = nucleotide diphosphate) and PEP carboxytransphosphorylase (R = H).[31]

In these spontaneous carboxylations, CO_2 acceptors have the potential to bind CO_2 and apparently no extra energy is required. Energy in the form of ATP and $NADPH_2$ has been consumed in the formation of the CO_2 acceptor. These carboxylases have no prosthetic group, but metal ions such as Mg^{2+} and Mn^{2+} are an essential cofactor. Except for PEP carboxylase, CO_2 is a substrate for the carboxylation.

B. Carboxylations Requiring Energy

a) Carboxylations requiring reduced ferredoxin

$$
\begin{array}{c}
CH_3 \\
| \\
C{=}O \\
| \\
CoA
\end{array}
+ \text{red. Fd} + \text{``}CO_2\text{''} \xrightarrow[\substack{\text{pyruvate synthase} \\ (1.2.7.1)}]{}
\begin{array}{c}
CH_3 \\
| \\
C{=}O \\
| \\
COO^-
\end{array}
+ CoA + Fd \qquad (5.22)
$$

Acetyl CoA   Reduced ferredoxin   Pyruvate   Oxidized ferredoxin

$$
\begin{array}{c}
COOH \\
| \\
CH_2 \\
| \\
CH_2 \\
| \\
C{=}O \\
| \\
CoA
\end{array}
+ \text{red. Fd} + \text{``}CO_2\text{''} \xrightarrow[\substack{\alpha\text{-ketoglutarate} \\ \text{synthase} \\ (1.2.7.3)}]{}
\begin{array}{c}
COOH \\
| \\
CH_2 \\
| \\
CH_2 \\
| \\
C{=}O \\
| \\
COOH
\end{array}
+ CoA + Fd \qquad (5.23)
$$

Succinyl CoA   α-Ketoglutarate

$$
\begin{array}{c}
CH_3 \\
| \\
CH_2 \\
| \\
C{=}O \\
| \\
CoA
\end{array}
+ \text{red. Fd} + \text{``}CO_2\text{''} \xrightarrow[\substack{\alpha\text{-ketobutyrate} \\ \text{synthase} \\ (1.2.7.2)}]{}
\begin{array}{c}
CH_3 \\
| \\
CH_2 \\
| \\
CH_2 \\
| \\
C{=}O \\
| \\
COOH
\end{array}
+ CoA + Fd \qquad (5.24)
$$

Propionyl CoA   α-Ketobutyrate

$$
\begin{array}{c}
H_3C \ \ CH_3 \\
\diagdown \diagup \\
CH \\
| \\
C{=}O \\
| \\
CoA
\end{array}
+ \text{red. Fd} + \text{``}CO_2\text{''} \xrightarrow[\substack{\alpha\text{-ketovalerate} \\ \text{synthase}}]{}
\begin{array}{c}
H_3C \ \ CH_3 \\
\diagdown \diagup \\
CH \\
| \\
C{=}O \\
| \\
COOH
\end{array}
+ CoA + Fd
$$
$$(5.25)$$

Isobutyryl CoA   α-Ketovalerate

$$
\begin{array}{c}
CoA \\
| \\
C{=}O \\
| \\
CH_2 \\
|
\end{array}
+ \text{red. Fd} + \text{``}CO_2\text{''} \xrightarrow[\substack{\text{phenylpyruvate} \\ \text{synthase}}]{}
\begin{array}{c}
COOH \\
| \\
C{=}O \\
| \\
CH_2 \\
|
\end{array}
+ CoA + Fd \qquad (5.26)
$$

Phenylacetyl CoA   Phenylpyruvate

$$
\text{Red. Fd} + H^+ + CO_2 \underset{CO_2\text{-reductase}}{\rightleftarrows} HCOO^- + Fd \qquad (5.27)
$$

These low-potential, iron sulfur protein-dependent carboxylations are limited to anaerobic bacteria. The redox potential of ferredoxin in these bacteria (-440 mV) is lower by about 100 mV than that of NAD (P)H$_2$. The carboxylations of acyl CoA (Eqs. 5.22–5.26) produce α-keto acids; these reactions participate in the CO_2 fixation by an anaerobic sulfur photosynthetic bacterium, *Chlorobium*. The detailed mechanism of CO_2 fixation by this bacterium will be described in Section 5.4.8. In the anaerobe *Clostridium pasteurianum*, CO_2 is reduced directly to formate (Eq. 5.27). This reaction does not produce a C–C bond, but the product works as a C$_1$-unit in biosynthesis. This CO_2 reductase is a Mo-Fe-S prote-in[32] and differs from pyridine nucleotide-dependent formate dehydroge-nase (Eq. 5.34) in the reductant of carbon dioxide and in the prosthetic group of the enzymes.

In contrast, *Clostridium kluyveri* reduces carbon dioxide to formate via a simple carbon cycle composed of reaction (5.27) and pyruvate-formate lyase (Fig. 5.8).[33] A quarter of the carbon of this anaerobic bacterium is the result of this carboxylation.[3] The CO_2 acceptor, acetyl CoA, is re-generated, and this cycle appears to be the most primitive cycle of carbon dioxide assimilation, although no C–C bond is formed. This cycle mimics the carbon cycle for CO_2 fixation in the autotrophs in the binding of CO_2 to the CO_2 acceptor and the generation of the CO_2 acceptor with part of the fixation products.

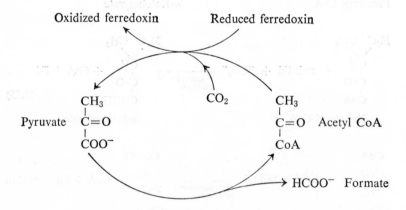

Fig. 5.8.   The simplest CO_2 fixation carbon cycle, yielding formate in *Clostridium kluy-veri*.[33]

It is interesting to note that chloroplasts that are formed in potato tubers under low temperature and dim light lack normal ribulose bisphosphate carboxylase. The first carboxylating product in this chloroplast is formate, a precursor of solanidine, a bitter compound formed in greening potato.[34] It has not yet been determined whether the reductant of carbon dioxide is reduced ferredoxin (Eq. 5.27) or pyridine nucleotide (Eq. 5.34).

b) Carboxylations requring reduced cytochrome and protein

$$\text{Cytochrome } b_1(\text{Fe}^{2+}) + \text{``CO}_2\text{''} \xrightleftharpoons[\substack{\text{formate dehydrogenase} \\ \text{(cytochrome)} \\ (1.2.2.1)}]{} \text{HCOO}^-$$

$$+ \text{ cytochrome } b_1(\text{Fe}^{3+}) \qquad (5.28)$$

$$\text{Cytochrome } b_1(\text{Fe}^{2+}) + \text{acetate} + \text{``CO}_2\text{''}$$

$$\xrightleftharpoons[\substack{\text{pyruvate dehydrogenase} \\ \text{(cytochrome)} \\ (1.2.2.2)}]{} \begin{array}{c} \text{CH}_3 \\ | \\ \text{C=O} \\ | \\ \text{COO}^- \end{array} + \text{cytochrome } b_1(\text{Fe}^{3+}) \qquad (5.29)$$

$$\text{5,10–Methylenetetrahydrofolate} + \text{NH}_3 + \text{AH}_2 + \text{``CO}_2\text{''}$$

$$\xrightleftharpoons[\substack{\text{glycine synthase} \\ (2.1.2.10)}]{} \begin{array}{c} \text{CH}_2\text{NH}_2 \\ | \\ \text{COOH} \end{array} + \text{tetrahydrofolate} + \text{A} \qquad (5.30)$$

$$\text{Glycine}$$

Cytochrome-dependent formate and pyruvate dehydrogenases are bacterial enzymes with equilibria and physiological functions that favor decarboxylation rather than carboxylation. Pyruvate dehydrogenase is a flavoprotein.[35] For glycine synthase, the reductant for the carboxylation (AH_2 in Eq. 5.30) is a low molecular weight (17,000), heat-stable protein, and its thiol group is an electron donor.[36] The disulfide of oxidized protein (A) is reduced by $NADH_2$ in a reaction catalyzed by a reductase. Glycine synthease has been found in mammalian mitochondria and also in an anaerobic bacterium, *Peptococcus glycinophilus* which ferments glycine to acetate, CO_2, and NH_3.[37]

c) Carboxylations requiring reduced pyridine nucleotides

$$\begin{array}{c} \text{CH}_3 \\ | \\ \text{C=O} \\ | \\ \text{COO}^- \end{array} + \text{NAD(P)H}_2 + \text{CO}_2 \text{ or } \text{HCO}_3^- \xrightleftharpoons[\substack{\text{malic enzyme} \\ (1.1.1.38\text{—}40)}]{\substack{\text{Mn}^{2+}, \text{Co}^{2+}, \\ \text{or Mg}^{2+}}} \begin{array}{c} \text{COO}^- \\ | \\ \text{HCOH} \\ | \\ \text{CH}_2 \\ | \\ \text{COO}^- \end{array} + \text{NADP}$$

$$(5.31)$$

$$\text{Pyruvate} \qquad\qquad\qquad\qquad\qquad\qquad \text{Malate}$$

$$\begin{array}{l}
\text{COO}^- \\
\text{CH}_2 \\
\text{CH}_2 \\
\text{C=O} \\
\text{COO}^-
\end{array} + \text{NAD(P)H}_2 + \text{CO}_2 \underset{\substack{\text{isocitrate} \\ \text{dehydrogenase} \\ (1.1.1.41-42)}}{\overset{\text{Mn}^{2+}}{\rightleftharpoons}}
\begin{array}{l}
\text{COO}^- \\
\text{HCOH} \\
\text{HC–COO}^- \\
\text{CH}_2 \\
\text{COO}^-
\end{array} + \text{NADP} \qquad (5.32)$$

α-Ketoglutarate   Isocitrate

$$\begin{array}{l}
\text{CH}_2\text{OH} \\
\text{C=O} \\
\text{HCOH} \\
\text{HCOH} \\
\\
\text{CH}_2\text{O}\textcircled{P}
\end{array} + \text{NAD(P)H}_2 + \text{CO}_2 \underset{\substack{\text{6-phosphogluconate} \\ \text{dehydrogenase} \\ (1.1.1.44)}}{\overset{\text{Mg}^{2+}}{\rightleftharpoons}}
\begin{array}{l}
\text{COO}^- \\
\text{HCOH} \\
\text{HOCH} \\
\text{HCOH} \\
\text{HCOH} \\
\text{CH}_2\text{O}\textcircled{P}
\end{array} + \text{NAD(P)} \quad (5.33)$$

Ribulose-5-phosphate   6–Phosphogluconate

$$\text{NADH}_2 + \text{``CO}_2\text{''} \underset{\substack{\text{formate dehydrogenase} \\ (1.2.1.2)}}{\rightleftharpoons} \text{HCOO}^- + \text{NAP} \qquad (5.34)$$

Formate

$$\begin{array}{cc}
\text{COOH} & \text{COOH} \\
\text{CH}_2 & \text{CH}_2 \\
\text{CH}_2 & \text{CH}_2
\end{array}$$

Ala–Asn–Lys–Gly–Phe–Leu–NH–CH–CO–NH–CH–CO–Val······+NAD(P)H$_2$
NH$_2$–terminal region of the precursor of prothrombin

$$\begin{array}{cc}
\text{HOOC COOH} & \text{HOOC COOH} \\
\diagdown\diagup & \diagdown\diagup \\
\text{CH} & \text{CH} \\
\text{CH}_2 & \text{CH}_2
\end{array}$$

+ 2 CO$_2$ $\xrightarrow[\text{O}_2]{\text{Vitamin K}}$ Ala–Asn–Lys–Gly–Phe–Leu–NH–CH–CO–NH–CH–CO–Val
           Prothrombin

$$\cdots\cdots + \text{NAD(P)} \qquad (5.35)$$

Malic enzyme donates malate to the tricarboxylic acid cycle, thus controlling the cycle. Because several organic acids in the cycle are drained off as the precursors of biosynthetic pathways, a supply of the cycle intermediates is necessary to maintain the catabolic activity of the cycle. In addition to malic enzyme, pyruvate carboxylase (5.36) and several other enzymes play a role in the supply of cycle intermediates. Isocitrate and 6-phosphogluconate dehydrogenases are the enzymes that participate in the tricarboxylic acid cycle and the phosphogluconate pathway, respectively, and under normal conditions work as decarboxylases. Formate dehydrogenase has been found in bacteria and plant tissues, especially in seedlings, and in contrast to CO_2 reductase (Eq. 5.27), the equilibrium lies in favor of decarboxylation.

Reaction (5.35) represents the carboxylation of protein. Glutamic acid residues in the NH_2-terminal region of the precursor of prothrombin, one of the four blood-clotting zymogens, are carboxylated to form γ-carboxyglutamic acid. In addition, glutamic acid residues 15, 17, 20, 21, 26, 27, 30, and 33 are also carboxylated. This modification of glutamic acid residues in the precursor is a Vitamin K-dependent reaction in microsomes and is required for the physiological activation of prothrombin. The carboxylation appears to be carried out post-translationally and is a unique example of the post-modification of enzyme protein by carboxylation. The carboxylase binds to the microsomes, and Vitamin K, $NAD(P)H_2$, and oxygen are required for the carboxylation.[38] However, when reduced Vitamin K (hydroquinone form) is added, $NAD(P)H_2$ is omittable. The interaction of O_2 with reduced Vitamin K appears to be essential for the abstraction of hydrogen from the γ-carbon of the glutamate residue and/or the activation of CO_2 to CO_2^-.[39]

d) Carboxylations requiring ATP

i) Biotin carboxylases

$$
\begin{array}{c}
CH_3 \\
| \\
C{=}O \\
| \\
COO^-
\end{array}
+ ATP + HCO_3^-
\underset{\substack{\text{pyruvate} \\ \text{carboxylase} \\ (6.4.1.1.)}}{\overset{\substack{Mg^{2+} \text{ or } Mn^{2+}, \\ \text{biotin, acetyl CoA}}}{\rightleftharpoons}}
\begin{array}{c}
COO^- \\
| \\
CH_2 \\
| \\
C{=}O \\
| \\
COO^-
\end{array}
+ ADP + P_i \quad (5.36)
$$

Pyruvate   Oxaloacetate

$$
\begin{array}{c}
CH_3 \\
| \\
C{=}O \\
| \\
CoA
\end{array}
+ ATP + HCO_3^-
\underset{\substack{\text{acetyl CoA} \\ \text{carboxylase} \\ (6.4.1.2)}}{\overset{\substack{Mg^{2+} \text{ or } Mn^{2+}, \\ \text{biotin}}}{\longrightarrow}}
\begin{array}{c}
COO^- \\
| \\
CH_2 \\
| \\
C{=}O \\
| \\
CoA
\end{array}
+ ADP + P_i \quad (5.37)
$$

Acetyl CoA   Malonyl CoA

$$
\begin{array}{c}
CH_3 \\
| \\
CH_2 \\
| \\
C{=}O \\
| \\
CoA
\end{array}
+ ATP + HCO_3^-
\underset{\substack{\text{propionyl CoA} \\ \text{carboxylase} \\ (6.4.1.3)}}{\overset{\substack{Mg^{2+} \text{ or } Mn^{2+}, \\ \text{biotin}}}{\rightleftharpoons}}
\begin{array}{c}
COO^- \\
| \\
HC{-}CH_3 \\
| \\
C{=}O \\
| \\
CoA
\end{array}
+ ADP + P_i \quad (5.38)
$$

Propionyl CoA   Methyl malonyl CoA

$$
\begin{array}{c}
\text{H}_3\text{C} \ \ \text{CH}_3 \\
\diagdown \diagup \\
\text{C} \\
\| \\
\text{CH} \\
| \\
\text{C=O} \\
| \\
\text{CoA}
\end{array}
\ + \ \text{ATP} \ + \ \text{HCO}_3^-
\ \xrightarrow[\substack{\beta\text{-methylcrotonyl} \\ \text{CoA carboxylase} \\ (6.\,4.\,1.\,4)}]{\substack{\text{Mg}^{2+} \text{ or Mn}^{2+}, \\ \text{biotin}}}
\begin{array}{c}
\text{H}_3\text{C} \ \ \text{CH}_2\text{COO}^- \\
\diagdown \diagup \\
\text{C} \\
\| \\
\text{CH} \\
| \\
\text{C=O} \\
| \\
\text{CoA}
\end{array}
\ + \ \text{ADP} \ + \ \text{P}_i
\qquad (5.39)
$$

β-Methylcrotonyl CoA $\qquad\qquad\qquad\qquad\qquad$ β-Methylglutaconyl CoA

$$
\begin{array}{c}
\text{H}_3\text{C} \ \ \text{CH}_3 \\
\diagdown \diagup \\
\text{C} \\
\| \\
\text{CH} \\
| \\
\text{CH}_2 \\
| \\
\text{CH}_2 \\
| \\
\text{C–CH}_3 \\
\| \\
\text{HC–C–CoA} \\
\| \\
\text{O}
\end{array}
\ + \ \text{ATP} \ + \ \text{HCO}_3^-
\ \xrightleftharpoons[\substack{\text{geranoyl CoA} \\ \text{carboxylase} \\ (6.\,4.\,1.\,5)}]{\text{Mg}^{2+}, \ \text{biotin}}
\begin{array}{c}
\text{H}_3\text{C} \ \ \text{CH}_3 \\
\diagdown \diagup \\
\text{C} \\
\| \\
\text{CH} \\
| \\
\text{CH}_2 \\
| \\
\text{CH}_2 \\
| \\
\text{C–CH}_2\text{–COO}^- \\
\| \\
\text{HC–C–CoA} \\
\| \\
\text{O}
\end{array}
\ + \ \text{ADP} \ + \ \text{P}_i
\qquad (5.40)
$$

Geranoyl CoA $\qquad\qquad\qquad\qquad\qquad$ Carboxy geranoyl CoA

$$
\text{O=C} \diagup^{\text{NH}_2}_{\diagdown \text{NH}_2}
\ + \ \text{ATP} \ + \ \text{HCO}_3^-
\ \xrightleftharpoons[\substack{\text{urea} \\ \text{carboxylase} \\ (6.\,3.\,4.\,6)}]{\text{Mg}^{2+}, \ K^+, \ \text{biotin}}
\ \text{O=C} \diagup^{\text{NH}_2}_{\diagdown \text{NHCOO}^-}
\ + \ \text{ADP} \ + \ \text{P}_i
\qquad (5.41)
$$

Urea $\qquad\qquad\qquad\qquad\qquad\qquad\qquad$ Allophanate

The carboxylases shown in Eqs. (5.36)–(5.41) are biotin enzymes. These biotin-dependent reactions are composed of two partial reactions (Eqs. 5.42 5.43).[28]

$$
\text{CCP–biotin} \ + \ \text{ATP} \ + \ \text{HCO}_3^-
\ \xrightleftharpoons[\substack{\text{biotin} \\ \text{carboxylsae} \\ (6.\,3.\,4.\,14)}]{\text{Mn}^{2+} \text{ or Mg}^{2+}}
\ \text{CCP–biotin-COO}^- \ + \ \text{ADP} \ + \ \text{P}_i
\qquad (5.42)
$$

$$
\text{CCP–biotin–COO}^- \ + \ \text{acceptor} \ \xrightleftharpoons[\substack{\text{carboxyl} \\ \text{transferase}}]{} \ \text{carboxylated products}
$$

$$
+ \ \text{CCP–biotin}
\qquad (5.43)
$$

In the first half-reaction biotin bound to carboxyl carrier protein (CCP) is carboxylated by biotin carboxylase (Fig. 5.9). This carboxylation is dependent on Mg^{2+} or Mn^{2+} and is driven by ATP. All biotin dependent carboxylations utilize HCO_3^- as substrate rather than CO_2. Carboxyl transfer from CCP-carboxybiotin to the acceptor molecule is catalyzed by

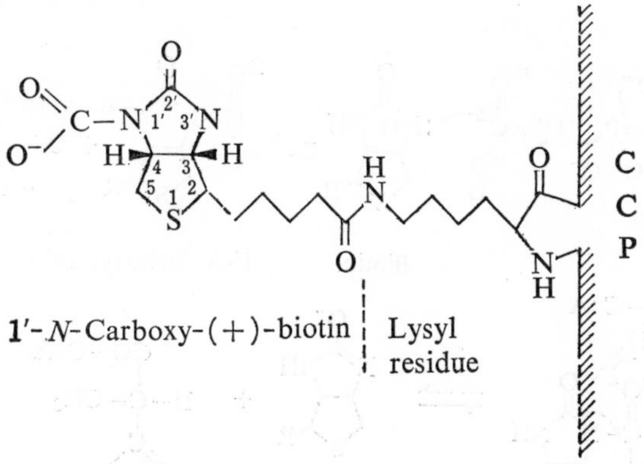

1'-N-Carboxy-(+)-biotin Lysyl residue

Fig. 5.9. Structure of carboxybiotin bound to lysyl residue of carboxyl carrier protein (CCP).

a carboxyltransferase that is specific to each carboxyl acceptor. When acyl CoA (5.37–5.40) pyruvate (5.36) or urea (5.41) is the acceptor, carboxy acyl CoA, oxaloacetate or N-carboxyurea, respectively, is the carboxylated product.

Carboxybiotin is also an intermediate in three other biotin enzymes. One is methylmalonyl CoA: pyruvate transcarboxylase (E C 2.1.3.1). The carboxyl group of methylmalonyl CoA is first transferred to enzyme-bound biotin and propionyl CoA is produced. The resulting carboxybiotin transfers its carboxyl group to pyruvate, forming oxaloacetate. This enzyme plays an important role in propionate formation by propionic acid bacteria. The other two enzymes are methylmalonyl CoA (*Micrococcus lactilyticus*) and oxaloacetate (*Aerobacter aerogenes*) decarboxylases (E C 4.1.1.41, 4.1.1.3). In these reactions, the carboxyl group of the substrate is transferred to biotin, then carbon dioxide is released from the enzyme-bound carboxybiotin.

Biotin binds covalently to a lysyl ϵ-amino group of the carboxyl carrier protein and the site of carboxylation with biotin is the ureido-1'-N position (Fig. 5.9). When ^{18}O-labeled bicarbonate is used in reaction (5.38), one atom of [18]O is found in orthophosphate and two atoms in the free carboxyl of the methylmalonyl CoA. If CO_2 is a reactant no incorporation of the [18]O into phosphate would be expected and this is one piece of evidence for the carbon dioxide substrate with biotin carboxylases. A possible mechanism of the carboxylation with biotin and the subsequent transfer of the carboxyl to the acceptor is shown in Fig. 5.10.[28,40] For a further discussion of the reaction mechanism of biotin enzymes, see the review by Moss and Lane.[28]

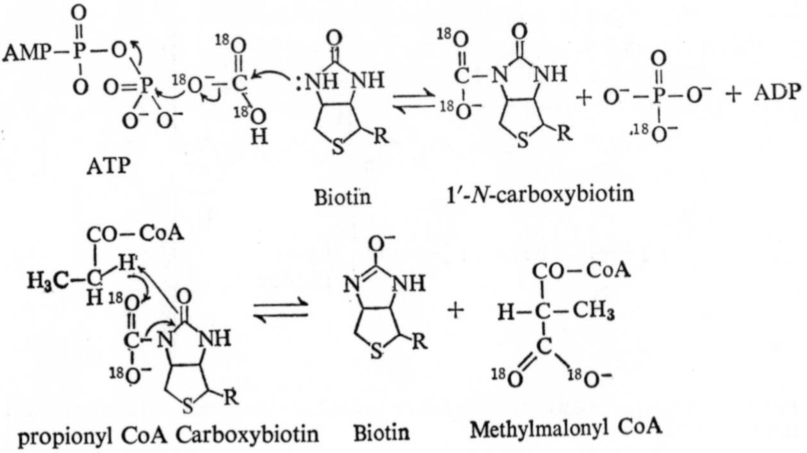

Fig. 5.10.   Distribution of ^{18}O from $HC^{18}O_3^-$ in the reaction products from propionyl CoA carboxylase, and a possible reaction mechanism.[40]

Because biotin is biosynthesized in mammals, symptoms of biotin-deficiency are not observed. However, by administration of avidin, a glycoprotein in raw egg white that has an extraordinary affinity for biotin, a biotin-deficiency symptom such as hair loss can be induced. Biotin is an essential growth factor for yeast and bacteria. These observations indicate the metabolic significance of biotin-dependent carboxylations.

Pyruvate carboxylase has been found in mitochondria and in micro-organisms. It produces oxaloacetate, from which phosphoenolpyruvate is formed by phosphoenolpyruvate carboxykinase through the dicarboxylate shuttle. This shuttle functions in gluconeogenesis from pyruvate in animal tissues. Malonyl CoA produced by acetyl CoA carboxylase is the key intermediate of fatty acid biosynthesis from acetyl CoA.

Propionyl CoA carboxylase (Eq. 5.38) catalyzes the formation of methyl malonyl CoA, which is further converted to succinyl CoA, then to oxaloacetate. It yields hexose through phosphoenolpyruvate, and thus, this carboxylase is a key enzyme for gluconeogenesis from the propionate that arises from the oxidation of odd-numbered fatty acids and the catabolism of branched-chain amino acids. The role of β-methylcrotonyl CoA carboxylase (Eq. 5.39) is the catabolism of leucine, which produces the substrate for the carboxylation. The carboxylated product is converted to β-methylglutaryl CoA, which is cleaved to acetyl CoA and acetoacetate.

In contrast to reactions (5.36)–(5.39), reactions (5.40) and (5.41) are limited to specific micro-organisms. Geranoyl CoA carboxylase is a component of the enzyme system for the bacterial degradation of isoprenoid compounds. Farnesyl CoA is also carboxylated by this enzyme. Carboxylation (5.41) has been found in algae and yeast that can use urea as the sole source of nitrogen but that lack urease. The allophanate formed is hydrolyzed by a hydrolase to CO_2 and NH_3, and the resulting NH_3 is assimilated by the cells.

ii) Non-biotin carboxylases

$$NH_3 + ATP + HCO_3^- \xrightleftharpoons[\substack{\text{carbamate kinase}\\(2.7.2.2)}]{Mg^{2+}} NH_2\text{-}\overset{\overset{\displaystyle O}{\|}}{C}\text{-}O\text{-}\textcircled{P} + ADP \qquad (5.44)$$
$$\text{carbamoylphosphate}$$

$$NH_3 + 2\,ATP + HCO_3^- \xrightarrow[\substack{\text{carbamoylphosphate}\\\text{synthase (ammonia)}\\(6.3.4.16)}]{\substack{Mg^{2+},\ N\text{-acetyl-}\\\text{glutamate}}} NH_2\text{-}\overset{\overset{\displaystyle O}{\|}}{C}\text{-}O\text{-}\textcircled{P} + 2\,ADP + P_i$$
$$\text{carbamoylphosphate} \qquad (5.45)$$

$$\text{Glutamine} + 2\,ATP + \text{``}CO_2\text{''} \xrightarrow[\substack{\text{carbamoylphosphate}\\\text{synthase(glutamine)}\\(6.3.5.5)}]{Mg^{2+}} H_2N\text{-}\overset{\overset{\displaystyle O}{\|}}{C}\text{-}O\text{-}\textcircled{P} + \text{glutamate}$$

$$\text{Carbamoylphosphate}$$
$$+ 2\,ADP + P_i \qquad (5.46)$$

$$CH_3\text{-}B_{12}\text{-}E + ATP + \text{``}CO_2\text{''} \longrightarrow {}^-OOC\text{-}CH_2\text{-}B_{12}\text{-}E + ADP + P_i \qquad (5.47)$$
Methylcorrin enzyme complex   Carboxymethylcorrin enzyme complex

Carbamoylphosphate is a key intermediate in the biosynthesis of pyrimidines, arginine, and urea. Reaction (5.44) occurs in bacteria and plants, and its equilibrium favors ATP synthesis, which is the major physiological function of carbamoyl kinase. The actual substrate for carbamoylphosphate synthesis is carbamic acid (NH_2COO^-), produced non-enzymatically from ammonia and bicarbonate.[30] In contrast to carbamate kinase, the synthasis of carbamoylphosphate catalyzed by carbamoylphosphate synthases (Eqs. 5.45 and 5.46) is essentially irreversible. The enzyme which uses ammonia (Eq. 5.45) is localized in mitochondria and is a key enzyme in the ornithine cycle for nitrogenous excretion. On the other hand, the enzyme that uses glutamate as an amino donor (Eq. 5.46) is found in bacteria, fungi, and mammalian cytosol and is mainly functional for the biosynthesis of pyrimidines.

Clostridium thermoaceticum ferments glucose, producing 3 moles of acetate from 1 mole of substrate. Two moles of the acetate are derived from glucose, but the remaining 1 mole is biosynthesized totally from carbon dioxide. In the biosynthesis of acetate from carbon dioxide, CO_2 is first reduced to formate by formate dehydrogenase. The formate is then transferred to B_{12} enzyme, forming methyl B_{12} (corrin) via an intermediate of formyl- and methyltetrahydrofolates. Reaction (5.47) is the second CO_2 fixation. The carboxylation product, carboxymethyl B_{12} complex, is reduced to form acetate.[41)]

5.4.   Photosynthetic Carboxylations

5.4.1.   Photosynthetic Carbon Cycle and Energy for CO_2 Fixation

The previous section discussed biological carboxylations in autotrophic and heterotrophic organisms and the physiological functions of these reactions. Most of the reactions play important roles in metabolic control rather than as carbon sources and only a few participate in the production of organic compounds or bio-mass on the globe. The production of bio-mass by plants and algae is a source of our food and energy and, in this section, carboxylations in photosynthetic organisms are described.

Although there are several variations in the carboxylations of photosynthetic organisms, depending on the stage of evolution and on ecological adaptation, the fundamental type of C–C bond formation is the binding of "CO_2" to the CO_2 acceptor which is generated in the carbon cycle for CO_2 fixation using a portion of the CO_2 fixation product (Fig. 5.11). The assimilation products are released from the carbon cycle according to the amounts of CO_2 fixed. The simplest CO_2 fixation carbon cycle is the reduction of carbon dioxide to formate in *Clostridium* (Fig. 5.8), although no C–C bond is formed. The driving force for the rotation of the carbon cycle and the generation of the CO_2 acceptor is biochemical energy, such as reduced ferredoxin, $NADPH_2$, or ATP, produced in the chloroplast thylakoids by chlorophyll-photosensitized reactions using water as the electron donor.

Thus, the over-all reaction of photosynthesis in plants and algae is the reduction of carbon dioxide to carbohydrate by water, and the energy gap for the thermodynamically unfavorable reduction is filled by light.

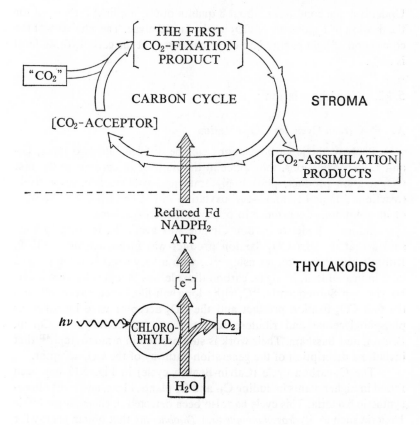

Fig. 5.11. The photosynthetic carbon cycle for CO_2 fixation, and the energy supply for the cycle in plant and algal chloroplasts.

The E_0' of $CO_2/[(CH_2O) + H_2O]$ and $1/2\ O_2/H_2O$ are -0.40 V and 0.82 V, respectively.

$$CO_2 + 2\ H_2O \xrightarrow{h\nu} (CH_2O) + O_2 + H_2O - 115\ kcal \qquad (5.48)$$

Photosynthetic bacteria are not able to use water as the electron donor and do not evolve molecular oxygen. Instead, they use H_2S and organic acids as the electron donors. The general formula for this type of photosynthesis is represented as shown in Eq. (5.49), where H_2A is the electron donor for the reduction of carbon dioxide.

$$CO_2 + 2\ H_2A \xrightarrow{h\nu} (CH_2O) + A_2 + H_2O \qquad (5.49)$$

Under optimal conditions, about 8 quanta of 680 nm light is required for the fixation of 1 molecule of CO_2 in plants and algae. The efficiency of the conversion of light energy into chemical energy at the carbohydrate level is 36%.

5.4.2. C_3-carboxylation

A. C_3-Carbon Cycle for CO_2-Fixation

Photosynthetic carboxylations can be divided into two types, C_3- and C_4-carboxylations, that differ in the number of carbons in the first CO_2 fixation product, phosphoglycerate and oxaloacetate, respectively (reactions 5.17 and 5.18). C_3-carboxylation will be described first because of its ubiquitous occurrence in photosynthetic organisms.

Before radioactive carbon dioxide was available, investigators assumed that the first CO_2 fixation product was formaldehyde. In 1939, Ruben's pioneering studies using ^{11}C, with a very short half-life of 20.5 min, indicated that, in algae, carbon dioxide was incoporated into a carboxyl group. Subsequently, ^{14}C, with a long half-life, became available, and the first CO_2 fixation product and the CO_2 acceptor were identified as phosphoglycerate and ribulose bisphosphate, respectively, by Calvin, Benson, and Bassham. Their work is summarized in a monograph[42] that includes a description of the generation pathway of the CO_2 acceptor.

The C_3-carbon cycle (Calvin-Benson cycle) in Fig. 5.12 has been found in higher plants including C_3- and C_4-plants, algae, and most photosynthetic bacteria. This cycle has also been detected in chemoautotrophic bacteria such as *Hydrogenomonas* and *Thiobacillus* that obtain energy for CO_2 reduction through the oxidation of H_2S or H_2. Thus, the C_3-carbon cycle has been the major route for supplying organic compounds on Earth. This cycle is composed of three parts. The first part is the fixation of carbon dioxide to D-ribulose 1,5-bisphosphate, producing D(−)-3-phosphoglycerate. The second part is the phosphorylation of the CO_2 fixation product by ATP and the subsequent reduction of glycerate-1,3-bisphosphate by $NADPH_2$ to glyceraldehyde-3-phosphate, from which dihydroxyacetone phosphate is produced. Fructose 1,6-bisphosphate is formed by aldolase from glyceraldehyde phosphate and dihydroxyacetone phosphate. The third part is the rearrangement of hexose and triose to pentose, via sedoheptulose and erythrose as the intermediates, to regenerate the CO_2 acceptor. During the rearrangement, hexose is released from the cycle as an assimilation product, corresponding to the amount of fixed CO_2; thus, the size of the pool of sugar phosphates in the cycle is kept constant under the steady state. However, in the transient state, especially

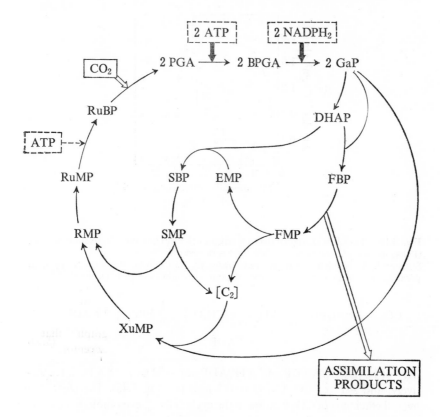

Fig. 5.12. The C_3-carbon cycle for CO_2 fixation (Calvin-Benson cycle). PGA: 3-phosphoglycerate; BPGA: glycerate 1,3-bisphosphate; GaP: glyceraldehyde 3-phosphate; DHAP: dihydroxyacetonephosphate; F: fructose; E: erythrose; S: sedoheptulose; R: ribose; Xu: xylulose; Ru: ribulose; MP: monophosphate; BP: bisphosphate; [C_2]: transketolase-bound thiaminepyrophosphate glycolaldehyde.

when light is turned on or off, the pool size of the acceptor and the fixation product in chloroplasts changes dramatically because of the initiation or termination of the production of ATP and NADPH$_2$ in thylakoids. These results provide evidence for the identities of the CO_2 acceptor (RuBP) and the first fixation product (PGA) (Fig. 5.13).

The cycle is driven by NADPH$_2$ and ATP which are consumed in the reduction of glycerate-bisphosphate and in the phosphorylation of phosphoglycerate and ribulose monophosphate. The over-all reaction of the cycle which reduces CO_2 to fructose bisphosphate (FBP) is formulated as shown in Eq. (5.50).

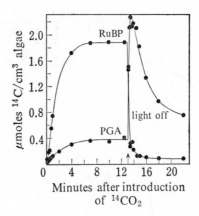

Minutes after introduction
of $^{14}CO_2$

Fig. 5.13.   Transient change of the CO_2 acceptor (RuBP) and the CO_2 fixation product (PGA) in *Chlorella pyrenoidosa* when light is turned off.[43]
Source: J. A. Bassham, *Physiol. Plantarum*, 19, 219 (1966). Reproduced by kind permission of Munksgaard International Bookseller and Publisher, Denmark.

$$CO_2 + 2\ NADPH_2 + 3\ ATP \longrightarrow (CH_2O)\left(= \frac{1}{6}\ FBP\right) + 2\ NADP$$

$$+ 3\ ADP \tag{5.50}$$

$\Delta G^{0\prime}$ of $NADPH_2/NADP$ and ATP/ADP are -52 and -8.6 kcal, respectively, and $\Delta G^{0\prime}$ of CO_2/CH_2O is 115 kcal (cf. Eq. 5.48). Therefore, the overall efficiency of CO_2 fixation in the cycle is 89 %, even though the cycle involves many reaction steps.

Since the C_3-carbon cycle has been found even in primitive, prokaryotic autotrophs, it has been supposed that it developed at an early evolutionary stage in autotrophic organisms. However, heterotrophic prokaryotes appeared on Earth before the autotrophic prokaryotes did, so the prototype of the C_3-cycle has been sought among the heterotrophic prokaryotes. In the C_3 carbon-cycle, the reduction of CO_2 is a unique reaction, but the second part of the cycle is the reverse of glycolysis and the rearrangement portion is a modified pentose phosphate cycle.

In terms of evolution, the assimilation pathway in the microorganisms that use C_1-compounds such as methanol (*Methylococcus capsulatus*) and methane (*Pseudomonas methanica*) as the sole carbon source is interesting. These C_1-assimilating organisms have been called methylotrophs,[44] in contrast to autotrophs that use CO_2 as the sole carbon source. In these organisms, the reduced C_1-compounds are first oxidized to formaldehyde, which is then "fixed" by the serine or ribulose monophosphate pathway.

The first reaction in the serine pathway is the "fixation" of formaldehyde to glycine, producing serine via methylenetetrahydrofolate as the intermediate. In the other pathway, formaldehyde is "fixed" to ribulose monophosphate (RuMP), forming D-*arabo*-3-hexulose-6-phosphate (HuMP) which is subsequently isomerized to fructose monophosphate (FMP) (Eq. 5.51).

$$
\text{HCHO} +
\begin{array}{c}
\text{CH}_2\text{OH} \\
| \\
\text{C}=\text{O} \\
| \\
\text{HCOH} \\
| \\
\text{CH}_2\text{O}\textcircled{P}
\end{array}
\longrightarrow
\begin{array}{c}
\text{CH}_2\text{OH} \\
| \\
\text{HOCH} \\
| \\
\text{C}=\text{O} \\
| \\
\text{HCOH} \\
| \\
\text{HCOH} \\
| \\
\text{CH}_2\text{O}\textcircled{P}
\end{array}
\longrightarrow
\begin{array}{c}
\text{CH}_2\text{OH} \\
| \\
\text{C}=\text{O} \\
| \\
\text{HOCH} \\
| \\
\text{HCOH} \\
| \\
\text{HCOH} \\
| \\
\text{CH}_2\text{O}\textcircled{P}
\end{array}
\qquad (5.51)
$$

$$\text{RuMP} \qquad\qquad \text{HuMP} \qquad\qquad \text{FMP}$$

HuMP synthase has been purified from several methylotrophs but the molecular size of the enzyme and its cellular localization differ from species to species. Particulate enzyme from *Methylococcus capsulatus* is composed of six subunits (molecular weight, 320,000), and the soluble enzyme from *Methylomonas* M15 is composed of two subunits (molecular weight, 43,000).[44] Further characterization of the molecular properties of HuMP synthase and a comparison with the properties of RuBP carboxylase/oxygenase will provide important data for understanding the evolutionary relationship between autotrophs and methylotrophs.

The equilibrium constant for the aldol condensation reaction which forms HuMP is 4.5×10^{-5} M, and that for the subsequent isomerization by HuMP isomerase has been estimated to be 2×10^{-2} M; thus, the equilibrium of reaction (5.51) favors the formation of FMP.[44] In addition to the similarity between the HCHO acceptor in methylotrophs and the CO_2 acceptor in autotrophs, the resemblance in the regeneration pathways (RuMP cycle, Fig. 5.14) of the CO_2 and HCHO acceptors also suggests an evolutionary relationship between the organisms. Fructose monophosphate is cleaved into triose through the glycolysis pathway, and FMP and the triose phosphate are the materials for the regeneration of the HCHO acceptor by a modified pentose phosphate pathway. Triose or hexose is released from the cycle, corresponding to the amount of "fixed" HCHO, as the assimilation product.

The over-all reaction of the RuMP cycle is shown in Eq. (5.52).

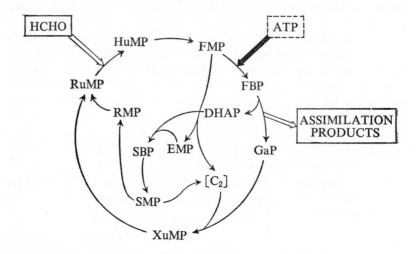

Fig. 5.14.   Ribulose monophosphate cycle in methylotrophs. HuMP: *arabo*-3-hexu-lose-6-phosphate. Other abbreviations are the same as these in Fig. 5.12.

$$HCHO + \frac{2}{3} ATP \longrightarrow (CH_2O) \; (=1/6 \; FBP) + \frac{2}{3} ADP \qquad (5.52)$$

Compared to the C_3-carbon cycle (Eq. 5.50), the driving force required for the RuMP cycle to "fix" 1 molecule of formaldehyde is very low, reflecting the fact that HCHO is a reduced form of C_1. Because of the low energy requirement for the assimilation of C_1, the RuMP cycle is an advantageous pathway for constituting cellular components in bacteria that have an underdeveloped capacity for the production of $NADPH_2$ and ATP. These considerations are another indication that methylotrophs are prototypes of autotrophs.

B.   Ribulose Bisphosphate Carboxylase/Oxygenase

Ribulose bisphosphate carboxylase/oxygenase (RuBP C/O) is a major protein in chloroplast stroma and constitutes 30 to 40% of soluble leaf protein. Wildman and Bonner first discovered this protein in 1947 and characterized it as a fraction I protein[45], but he did not find its enzymatic activity. In 1954, Weissbach *et al.* showed that fraction I protein was the carboxylating enzyme of the C_3-carbon cycle.[46] Because of the universal occurrence of the C_3-carbon cycle we can say that this protein has been the key enzyme for producing C–C bonds from the time of the first appearance of autotrophs to the present, and that all heterotrophs depend on the products of this enzyme. Owing to its widespread distribution in

almost all autotrophs, especially plants, fraction I protein is the most a-
bundant single protein on Earth, and attempts have been made to use
fraction I protein as a supplemental protein in foods. Much work has been
done on the molecular properties and reaction mechanism of RuBP C/O;
recent studies have been summarized by Siegelman and Hind.[47]

a) Molecular properties

RuBP C/O is composed of large (L) and small (S) subunits; their
molecular weights are 56,000 and 12,000, respectively. The molecular
weight of the entire enzyme is about 560,000, and a structure with eight
of each subunit, L_8S_8, has been proposed. This has been confirmed by
X-ray crystallographic analysis and by the electron microscopic image of
the crystalline tobacco enzyme. A model of the enzyme proposed by
Baker *et al.*[48] is shown in Fig. 5.15. The L_8S_8 enzyme clusters in two layers,
perpendicular to a fourfold axis of symmetry.[48] Only the enzyme from
Pseudomonas oxaliticus, a facultative autotroph which assimilates carbon
via the C_3-carbon cycle when grown on formate, has a structure of L_6S_6.[49]

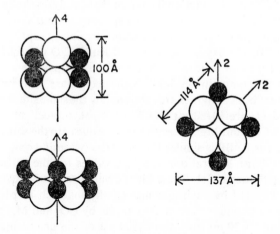

Fig. 5.15.   A model of ribulose bisphosphate carboxylase/oxygenase. ○: large subunit;
●; small subunit.[48]

The large subunit has the catalytic site for both the carboxylase and
the oxygenase activity,[50] as well as the activation site.[51] As described
below, the enzyme is activated by CO_2 and Mg^{2+} through carbamate formation with the enzyme's amino group and subsequent complex formation
with Mg^{2+}. The modification of the enzyme during the formation of the
carbamate-Mg^{2+} complex may induce a conformational change in the
enzyme which increases the catalytic turnover for both activities. All

RuBP C/O enzymes from plants and eukaryotic algae have the L_8S_8 structure, as do almost all the enzymes from photosynthetic bacteria, chemoautotrophic bacteria, and cyanobacteria. However, the enzymes from several prokaryotes lack the small subunit, and enzymes with L_2, L_6, and L_8 structures have been isolated. Even enzymes lacking the small subunit have both catalytic activities which proceed at rates similar to these of L_8S_8 enzymes and which are activated by CO_2 and Mg^{2+},[52] confirming the presence on the large subunit of both the catalytic and activation sites. No enzymatic function has been found in the small subunit; it may play an as-yet unknown role in the regulation of the enzyme in chloroplasts. The large subunit is biosynthesized in chloroplasts under the control of chloroplast DNA, but the precursor of the small subunit is formed in the cytoplasm by nuclear DNA, then transferred to the chloroplasts to form the enzyme after the precursor is cleaved into the small subunit and a protein fragment.

b) Catalytic properties

RuBP C/O is a bifunctional enzyme that catalyzes the carboxylation of RuBP, producing 2 molecules of phosphoglycerate (PGA) (Eq. 5.53 in Fig. 5.17), and the oxygenation of RuBP, forming phosphoglycolate and PGA (Eq. 5.54). The oxygenase activity of RuBP C/O was found by taking advantage of the competitive inhibition of the carboxylase activity by O_2. Inversely, CO_2 competitively inhibits the oxygenase activity; thus, it is very likely that both activities are catalyzed by the same site. The K_m values for CO_2 and O_2 are 10–20 μM and 100–200 μM, respectively, in the form activated by CO_2 and Mg^{2+}. Respective K_i values are the same as those of K_m. The K_m for RuBP is similar for the carboxylase and oxygenase activities, about 100 μM. The ratio of the V_{max} values for the carboxylase and oxygenase activities is 5:1. These kinetic values of RuBP C/O, however, are greatly affected by environmental conditions such as pH and Mg^{2+} concentration. The activation of the enzyme by CO_2 and Mg^{2+} lowers the K_m for CO_2 to 10–20 μM from about 500 μM of the unactivated enzyme.

c) Reaction mechanism

The carbon dioxide substrate for the carboxylation is not HCO_3^-, but CO_2, as can be determined by using the properties of slow hydration or dehydration of CO_2 or HCO_3^-(Section 5.2.2). Carboxylation proceeds more rapidly when CO_2 is added to the reaction mixture than when HCO_3^- is added. The addition of carbonic anhydrase, which accelerates the hydration of CO_2, retards the carboxylation (Fig. 5.16).[20]

As discussed in Section 5.3.1, enzymatic carboxylations involve the binding of an electrophilic carbon atom of carbon dioxide to a nucleophilic carbon atom of the CO_2 acceptor molecule. In the case of RuBP C/O,

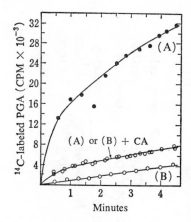

Fig. 5.16. Rate of RuBP carboxylase-catalyzed labeling of phosphoglycerate with $^{14}CO_2 + H^{12} CO_3^-$ (A), with $^{12}CO_2 + H^{14}CO_3^-$ (B), or with (A) or (B) and carbonic anhydrase (CA).[20]
Source: ref. 53. Reproduced by kind permission of the American Society of Biol. Chemists.

^{14}CO is incorporated exclusively into the carboxyl group of PGA, and the ^2H from ^2H_2O and the ^{14}C from $2\text{-}^{14}C$-RuBP are both incorporated into the 2-position of the same PGA molecule (Fig. 5.17).[53] Therefore, CO_2 binds to C-2 of RuBP, then the C-C bond between C-2 and C-3 of RuBP is cleaved to form PGA. However, the C-2 of RuBP is electrophilic. The transformation in the CO_2 acceptor that produces a nucleophilic C-2 before the binding of CO_2 was first proposed by Calvin in 1954.[54] He suggested that RuBP is rearranged to an enediol form (I in Fig. 5.17) or a 3-ketopentose form (II).

The following observations have sustained Calvin's proposal. The inhibition of both carboxylase and oxygenase activities by the carboxylated substrate analogue, 2-carboxyribitol-1,5-bisphosphate (CRBP, IV in Fig. 5.17),[55] and 2-carboxy-D-hexitol 1,6-bisphosphate (CHBP, V in Fig. 5.17),[56] suggests that the enediol or 3-ketopentose form is carboxylated. Furthermore, the proposed carboxylated intermediate, 2-carboxy-3-keto-ribitol-1, 5-bisphosphate (III), was synthesized by Lane *et al.*, and incubation with RuBP C/O produced PGA.[57]

RuBP C/O is a simple protein that contains no metals or cofactor for the catalysis. One binding site for RuBP ($K_d = 10^{-6}\text{--}10^{-7}$ M) or CRBP ($K_d = 10^{-8}$ M) has been found in the large subunit. The ϵ-amino group of lysine and the guanidyl group of arginine have been shown to participate in the binding of substrate and in the catalysis. Affinity labeling of these groups, and inhibition of carboxylase and oxygenase activities by the reagent

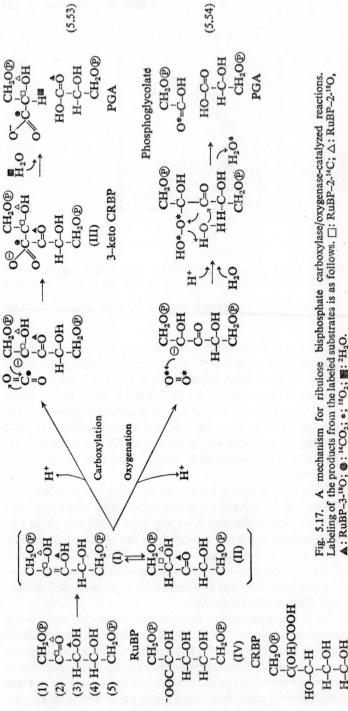

Fig. 5.17. A mechanism for ribulose bisphosphate carboxylase/oxygenase-catalyzed reactions. Labeling of the products from the labeled substrates is as follows. □: RuBP-2-¹⁴C; △: RuBP-2-¹⁸O, ▲: RuBP-3-¹⁸O; ●: ¹⁴CO₂; ∗: ¹⁸O₂; ■: ²H₂O. CRBP: 2-carboxyribitol-1,5-bisphosphate; CHBP: 2-carboxy-D-hexitol 1,6-bisphosphate. Other abbreviations are the same as those in Fig. 5.12.

for affinity labeling, are retarded by RuBP, CRBP, and CO_2. Indirect participation of the sulfhydryl group has also been suggested.[58,59]

The formation of ketimine (Schiff base) between the ϵ-amino group of the lysine residue and the C-2 or C-3 (3-keto form) of RuBP has been proposed as the enzyme substrate complex. On the other hand, if the sulfhydryl group participates in the catalysis, the thiohemiketal form may be the reaction intermediate. However, the observation that ^{18}O of RuBP-2-^{18}O or -3-^{18}O is retained in the reaction product, in C-2 or C-1 of phosphoglycerate, respectively[60,61] (Fig. 5.17), provides additional evidence for the cleavage between C-2 and C-3 of RuBP. Retention of the ^{18}O in PGA indicates that neither ketimine nor the thiohemiketal form is an intermediate, because ^{18}O is lost in the formation of covalent intermediates. Thus, how the ϵ-amino group of the lysine residue and the guanidyl groups of arginine participate in the catalysis is not known yet. These groups may interact ionically with the phosphate groups of RuBP.

Figure 5.17 shows a reaction mechanism for the carboxylation and oxygenation of RuBP. Evidence which indicates that both activities are catalyzed by the same site on the large subunit includes; competitive inhibition of the carboxylase and oxygenase activities by O_2 and CO_2, respectively; the parallel inhibition of both activities by pyridoxal phosphate, which modifies the ϵ-amino group of lysine residues through the formation of a Schiff base; and the parallel activation of both activities by CO_2 and Mg^{2+}. It may therefore be reasonable to assume that both reactions of RuBP C/O have the same intermediate. All oxygenases other than this one contain transition metals or a cofactor such as flavin or pteridines especially for the activation of molecular oxygen. In view of the fact that a reaction between triplet ground-state molecular oxygen and a singlet molecule is spin-forbidden, the mechanism by which oxygen is activated in the oxygenation of RuBP remains an interesting puzzle.

C. Activation of Ribulose Bisphosphate Carboxylase/Oxygenase by Carbon Dioxide and Mg^{2+}

Sugar phsophates of the C_3-carbon cycle, NADPH, and orthophosphate activate slightly RuBP C/O. The most distinct activation is induced by the incubation of the enzyme with CO_2 and Mg^{2+}. Both carboxylase and oxygenase activities are concurrently activated, and the K_m values for the substrates are lowered.[62,63] This activation is reversible; the CO_2-Mg^{2+}-activated enzyme is deactivated by removal of Mg^{2+} and CO_2. The actual species of carbon dioxide for the activation is CO_2.[63] The activation site is different from the catalytic site,[64,65] but both sites are in the large subunit.

The degree of activation is high at a high pH (Fig. 5.18), and affinity

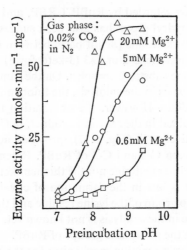

Fig. 5.18. Dependence of the carboxylase activity of RuBP C/O on the preincubation pH at a constant CO_2 concentration (10 μM) and the Mg^{2+} concentration indicated.[63] Source: ref. 63. Reproduced by kind permission of the American Chemical Society.

labeling of the ϵ-amino group of the lysine residue causes a loss of activation.[59] These observations suggest that CO_2 interacts with the amino group of the enzyme to form carbamate, because, with CO_2 carbamate is formed only from unprotonated amino group. The formation of carbamate in the active enzyme has been confirmed by the ^{13}C-NMR spectrum of *Rhodospirillum* RuBP C/O, which contains only the large subunit (L_2). The enzyme incubated in Mg^{2+} and $^{13}CO_2$ showed a signal of ^{13}C-carbamino derivatives.[66] Which residue of the large subunit binds with CO_2 has not

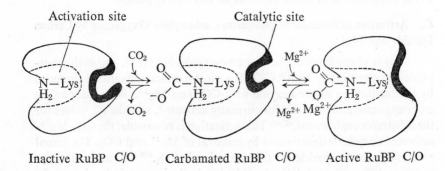

Fig. 5.19. A schematic illustration of the activation of ribulose bisphosphate carboxylase/oxygenase by CO_2 and Mg^{2+}.

yet been determined, but the most probable group is the ϵ-amino group of the lysine residue or the amino group of the NH_2-terminal amino acid residue.

Magnesium ion is required for the activation. Complex formation of the carbamated enzyme with Mg^{2+} may cause a conformational change in the enzyme which lowers the K_m for the substrates and increases the catalytic turnover. The activation of RuBP C/O by CO_2 and Mg^{2+} is schematically shown in Fig. 5.19.

D. Effect of Carbon Dioxide on Enzymes

The modifying influence of CO_2 on RuBP C/O is very similar to its effect on hemoglobin. The affinity of hemoglobin for O_2 decreases when the concentration of CO_2 is high (Bohr effect), a characteristic which fits well with its physiological function as an oxygen carrier in the blood. The effect of carbon dioxide is reversible, and the affinity for O_2 again increases in the lung, where the concentration of CO_2 is low. Mammalian hemoglobin is composed of two kinds of subunits, α and β with a subunit structure of $\alpha_2\beta_2$. Each subunit has an oxygen binding site, heme. The Bohr effect is attributable to the binding of CO_2 to the amino group of the NH_2-terminal residue of β-subunit, forming carbamate ions.[67] This has been confirmed by ^{13}C-NMR[68] and X-ray analysis.[69] The formation of carbamate induces the conformational change in the heme-binding site of hemoglobin which lowers the affinity of heme for O_2.

Thus, it has been established that the biological activities of RuBP C/O and hemoglobin are affected by CO_2 through its capacity to form carbamates by interaction with the amino group of their amino acid residues. Other enzymes or systems are known to be affected by carbon dioxide, although it is not known how carbon dioxide works. An ionic effect of HCO_3^- and the carbamation of enzyme proteins by CO_2 may be involved. The enzymatic activities of cytochrome c oxidase,[70] succinate dehydrogenase[71] malic enzyme,[72] glucose-6-phosphate dehydrogenase,[73] glucose-6-phosphate phosphohydrolase,[74] glycolate oxidase[75], ATP ase[76], and catalase[77] are stimulated or inhibited by carbon dioxide.

In chloroplasts, in addition to its effect on RuBP C/O in the stroma, carbon dioxide also stimulates the electron transport activity in thylakoids. When isolated thylakoids were treated with salts (0.1 M NaCl and 0.1 M Na formate) at pH 5, the Hill reaction activity was lost. These "CO_2"-depleted thylakoids recovered their activity only with carbon dioxide. The reducing side of photosystem II, especially around plastoquinone, is likely to be the site affected by the "CO_2" depletion.[78] Photophosphorylation activity in chromatophores from photosynthetic bacteria and in spinach chloroplasts is also stimulated by carbon dioxide.[79,80].

5.4.3. C_4-carboxylation

The wide distribution of the C_3-carbon cycle in autotrophs ranging from chemoautotrophs to higher plants had led investigators to think that this cycle operated in all plants. In 1965, however, Kortschak *et al.* found that the first $^{14}CO_2$ fixation product in sugar cane was not PGA, but a C_4-dicarboxylic acid such as malate or aspartate.[81] Further study in this plant has made it clear that the reaction is catalyzed by phosphoenolpyruvate carboxylase and the primary product is oxaloacetate (Eq. 5.18); the same labeling has been found in many other plants, especially in plants of tropical origin such as maize and sorghum. These plants have been called C_4-plants because the primary product is a C_4-dicarboxylic acid. In contrast to C_4-plants, C_3-plants are usually of temperate origin and include rice and wheat.

C_3- and C_4-plants are distinguished by their leaf structures. In C_3-plants, chloroplasts are found in mesophyll cells, but in C_4-plants, they occur in both mesophyll and bundle sheath cells (Kranz anatomy). The bundle sheath cells surround the leaf bundle sheath and are surrounded in turn by the mesophyll cells. CO_2 first enters the mesophyll cells through the stomata and is fixed by the C_4-dicarboxylic acid pathway. The assimilation products are subsequently transported to the bundle sheath cells, where they are decarboxylated. The released carbon dioxide is refixed via the C_3-carbon cycle, which is exclusively localized in the bundle sheath cells.

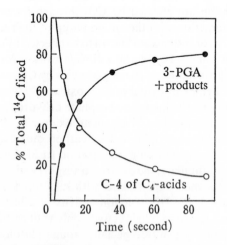

Fig. 5.20.   Time course of labeling the C-4 of C_4-acids and phosphoglycerate and its products during $^{14}CO_2$ assimilation in maize.[82]
Source: ref. 82. Reproduced by kind permission of University Park Press, USA.

The time course of $^{14}CO_2$ incorporation into the C_4-acids and PGA indicates no direct CO_2 fixation via the C_3-carbon cycle in C_4-plants (Fig. 5.20). Thus, in C_4-plants, carbon dioxide is fixed twice with different carboxylases and in different cells.

A. C₄-Dicarboxylic Acid Pathway

The CO_2 acceptor in the C_4-dicarboxylic pathway is phosphoenolpyruvate (PEP). The first CO_2 fixation product, oxaloacetate (OAA), is reduced to malate or converted to aspartate with aminotrasferase; which major reaction occurs depends upon the species. Malate or aspartate is transported to the bundle sheath cells, then decarboxylated. The decarboxylated products in the bundle sheath cells are converted to pyruvate or alanine which is again transported into the mesophyll cells. Alanine is converted to pyruvate with aminotrasnferase. The CO_2 acceptor is regenerated from pyruvate by phosphorylation with pyruvate orthophosphate dikinase (Eq. 5.55).

$$
\begin{array}{ccc}
\underset{\substack{|\\ \text{COOH}}}{\overset{\substack{\text{CH}_3 \\ |}}{\text{C=O}}} + P_i + ATP \rightleftharpoons \underset{\substack{|\\ \text{COOH}}}{\overset{\substack{\text{CH}_2 \\ \parallel}}{\text{C-®}}} + PP_i + AMP & & (5.55)
\end{array}
$$

Pyruvate   PEP   Pyrophosphate

The equilibrium of this reaction favors the formation of ATP or pyruvate, but removal of pyrophosphate and AMP with pyrophosphatase and adenylate kinase shifts the reaction to the right. Both enzymes are contained in the mesophyll cells in high concentrations.

The first carboxylation, the regeneration of the CO_2 acceptor, and the refixation via the C_3-carbon cycle are common to all C_4-plants, and the basic C_4-pathway is shown in Fig. 5.21. However, C_4-plants show considerable diversity with respect to 1) the C_4-acid transported to the bundle sheath cells, malate or aspartate; 2) the decarboxylation in the bundles sheath cells; 3) the pathway of pyruvate formation. Accordingly, C_4-plants can be divided into three types.[82]

I) NADP-malic enzyme type. Malate is transported into the chloroplasts of the bundle sheath cells and is decarboxylated to pyruvate by NADP-malic enzyme (Eq. 5.31). Pyruvate reenters the mesophyll cells and provides the CO_2 acceptor.

II) PEP carboxykinase type. Aspartate is transported to the bundle sheath cells and converted to OAA by aminotransferase. OAA is decarboxylated by PEP carboxykinase (Eq. 5.19), then PEP is converted to alanine via pyruvate. Alanine transported to the mesophyll cells is converted to pyruvate with aminotransferase.

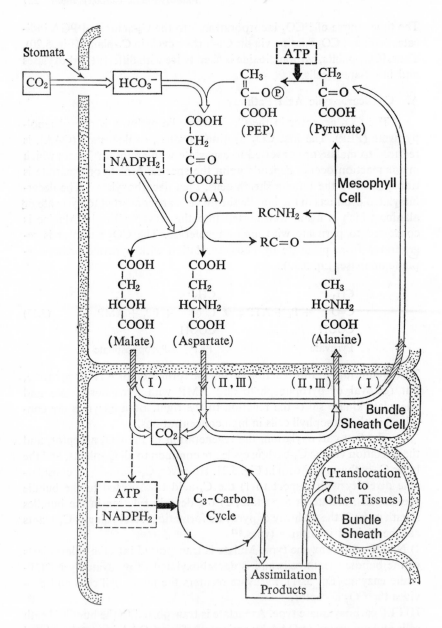

Fig. 5.21. Carbon dioxide fixation pathway in C_4-plants. ⟹ indicates the inter-cellular transport. Roman numerals in brackets correspond to types I, II, and III in the text.

III) NAD-malic enzyme type. The pathway from aspartate to OAA is the same as in type II, but OAA is reduced to malate with NAD-malate dehydrogenase. Malate is then decarboxylated as in type I, but with NAD-malic enzyme (Eq. 5.31) in the mitochondria of the bundle sheath cells. The pyruvate formed is converted into alanine, which then reenters the mesophyll cells. Alanine becomes the precursor of the CO_2 acceptor as in type II.

These pathways, including those in Fig. 5.21, have been confirmed by the intercellular and subcellular distributions of the enzymes and by the incorporation of $^{14}CO_2$ and other labeled substrates into the metabolites. In all types, CO_2 is donated to the bundle sheath cells through these processes. In type I plants, although $NADPH_2$ is consumed for the reduction of OAA in the mesophyll cells, $NADPH_2$ is supplied to the C_3-carbon cycle in the bundle sheath cells by malic enzyme. In types II and III, the amino group is transferred between keto acids and amino acids with aminotransferase, and no net change of the amino group occurs in the over-all process. Therefore, C_4-plants require 2 more molecules of ATP in reaction (5.55) for the reduction of carbon dioxide to carbohydrate, in addition to the ATP and $NADPH_2$ required for the C_3-carbon cycle (Eq. 5.50). Since the product is AMP, 2 molecules of ATP are required for 1 molecule of the CO_2 acceptor in reaction (5.55). The over-all CO_2 reduction in C_4-plants is represented by (Eq. 5.56), which has an energy efficiency of 78%, in contrast to 89% in the C_3-carbon cycle.

$$CO_2 + 2\ NADPH_2 + 5\ ATP \longrightarrow (CH_2O)\ (=1/6\ FBP)$$
$$+\ 2\ NADP + 5\ ADP \qquad (5.56)$$

B. Phosphoenolpyruvate Carboxylase

PEP carboxylase is localized in the mesophyll cells of C_4-plants in a concentration at least ten fold higher than that of C_3-plants on the basis of chlorophyll. No RuBP C/O is detectable in the mesophyll cells of C_4-plants. PEP carboxylases from both C_3- and C_4-plants have similar properties; they are composed of four subunits of equal size (molecular weight, 99,000), with a total molecular weight around 400,000. No prosthetic has been found, but Mg^{2+} is an essential cofactor for the carboxylation.[83,84] In contrast to RuBP C/O, the substrate for the carboxylation is HCO_3^-,[27,85,86] and molecular oxygen does not affect the carboxylation. The K_m for HCO_3^- is 20 μM.[84] The enzyme has been shown to be activated by glycine and glucose-6-phosphate.[84]

The K_m value for HCO_3^- of PEP carboxylase is apparently similar to the K_m for CO_2 of RuBP C/O (10 to 20 μM in the activated form). However, under the conditions of chloroplast stroma pH (7 to 8), the content of HCO_3^- is five- to fifty fold higher than that of CO_2(Table 5.1); therefore, more carbon dioxide is available for PEP carboxylase than for RuBP C/O. These properties of PEP carboxylase, together with the insensitivity of the carboxylation to molecular oxygen, contribute to a higher CO_2 fixation rate in C_4-plants than is found in C_3-plants.

The carboxylation mechanism with PEP carboxylase may be similar to the mechanisms of PEP carboxykinase and carboxyphosphotransferase (Fig. 5.7) which involve the binding of the electrophilic carbon atom of carbon dioxide to the nucleophlic C-3 of PEP. However, PEP carboxylase utilizes HCO_3^- rather than CO_2, and transfers the phosphate from PEP to H_2O.

5.4.4.   Comparison of CO_2 Fixation by C_3- and C_4-plants

Leaf anatomy and intercellular distributions of RuBP C/O and PEP carboxylase suggest that C_4-plants developed from C_3-plants. This hypothesis is supported by the following observations. C_4-plants have been found only among angiosperms, the most developed land plants, and no C_4-plants have been identified among gymnosperms, ferns, and mosses. The $^{13}C/^{12}C$ ratio in fossil fuels is the same as that of C_3-plants, which shows that the coal plants were the C_3-type (section 5.4.5). We are then led to ask why C_4-plants added the C_4-dicarboxylic acid pathway to the C_3-carbon cycle.

Before turning to a discussion of this question, however, another difference in the leaf anatomy of C_3- and C_4-plants should be mentioned. The resistance to the passage of gas through the leaf stomata of C_4-plants is higher than that of C_3-plants. This high resistance retards the transport of carbon dioxide from the atmosphere to the chloroplasts, but the loss of water by transpiration is suppressed. As a consequence, the concentration of carbon dioxide in the mesophyll chloroplasts of C_4-plants is usually lower than that in C_3-plants; the concentration of CO_2 in the chloroplasts of mesophyll cells has been estimated to be about 6 μM in C_3-plants and about 1 μM in C_4-plants when the leaves are equilibrated with air.[82] The CO_2 concentration in an air-equilibrated solution is 11 μM (Table 5.1); the difference from the concentrations of CO_2 found in chloroplasts is due to the resistance of the stomata.

The concentration of CO_2 in the chloroplasts of C_3-plants (about 6 μM) permits carboxylation to proceed at the rate observed in the leaves because it is similar to the K_m for CO_2 of RuBP C/O. However, in C_4-plants, the concentration of CO_2 in the mesophyll cells (about 1 μM) is an order of magnitude lower than the K_m for CO_2 of RuBP C/O. PEP carboxylase in mesophyll cells uses HCO_3^- rather than CO_2 as described in the previous section, and the concentration of HCO_3^- in 1 μM CO_2-equilibrated solution is 4.5 μM at pH 7 and 45 μM at pH 8, which are similar to the K_m for HCO_3^- of PEP carboxylase (20 μM). Thus, the capacity of PEP carboxylase to fix HCO_3^- as the substrate allows the fixation of carbon dioxide at a reasonable rate, even when the concentration of CO_2 is low, as in the mesophyll cell chloroplasts of C_4-plants. Therefore, the over-all reaction in the mesophyll cells of C_4-plants is the fixation of carbon dioxide in low concentrations and the transport of the carbon dioxide in the form of malate or aspartate to the bundle sheath cells. A high concentration of CO_2 (approximately 20 μM)[82] is maintained in the bundle sheath cells by the decarboxylations, and the released CO_2 is fixed by the C_3-carbon cycle. In other words, the physiological function of the C_4-pathway is the concentration of carbon dioxide from the atmosphere in the bundle sheath cells. Because of the high resistance to the passage of gas through the stomata, C_4-plants can grow while conserving more water than C_3-plants can.

These considerations suggest that C_4-plants are a form of C_3-plants that adapted to low concentrations of CO_2 in the atmosphere and to water stress. In fact, the compensation point for CO_2 (that concentration of CO_2 when the fixation of CO_2 is equal to the release of CO_2 by respiration) is nearly zero ppm for C_4-plants, but it is 15 to 150 ppm for C_3-plants. This indicates that PEP carboxylase can fix carbon dioxide even from an atmosphere containing low CO_2. In addition, it shows that C_4-plants do not lose the assimilation products by photorespiration. This is because the enzymes participating in photorespiration are localized in the bundle sheath cells, where the concentration of carbon dioxide is high. A high concentration of CO_2 suppresses the production of glycolate, a substrate of photorespiration, by RuBP C/O. The structural features of C_4-plants prevent the escape of carbon dioxide and make possible the refixation of the released carbon dioxide.

The CO_2 fixation rate by plants is determined by the concentration of carbon dioxide available to the carboxylases, as discussed above, and also by the amounts of the carboxylases and their turnover rates (V_{max}). PEP carboxylase in C_4-plants has a higher potential than has RuBP C/O, and as a consequence, C_4-plants have a high maximum rate of CO_2 fixation, about twice that of C_3-plants. However, since the CO_2 transport to the chloroplasts through the leaf stomata is affected by the water supply to the plants, the CO_2 fixation rate of rice (a C_3-plant) on wet paddy land is similar to that of maize (a C_4-plant) on dry land.

5.4.5. Isotopic Effects on CO_2 Fixation by C_3- and C_4-plants

Atmospheric CO_2 is mainly composed of $^{12}CO_2$ (98.855%) and $^{13}CO_2$ (1.145%); $^{14}CO_2$ is also present in the atmosphere, but as a very minor component (6×10^{-12} Ci/g C). One might expect that CO_2 fixation processes would be subject to isotapic effects, that is, a preferential fixation of $^{12}CO_2$ over $^{13}CO_2$. Little isotopic difference between ^{12}C and ^{13}C has been observed during decarboxylations in plants,[87] so the $^{13}C/^{12}C$ ratio in plant materials reflects the isotopic effect on photosynthetic CO_2 fixation. The isotopic ratio $^{13}C/^{12}C$, $\delta^{13}C$, is expressed by Eq. (5.57).

$$\delta^{13}C \ (‰) = \left[\frac{^{13}C/^{12}C \ \text{(sample)}}{^{13}C/^{12}C \ \text{(standard)}} - 1 \right] \times 1,000 \qquad (5.57)$$

A Cretaceous belemnite from the PeeDee formation in North Carolina, USA, used as a standard ($^{13}C/^{12}C = 0.0112372$). The $\delta^{13}C$ value of atmospheric CO_2 is $-7 \pm 0.3‰$.

The $\delta^{13}C$ values of over 100 species of angiosperms are shown in Fig. 5.22. As expected, plant values are always lower than the atmospheric

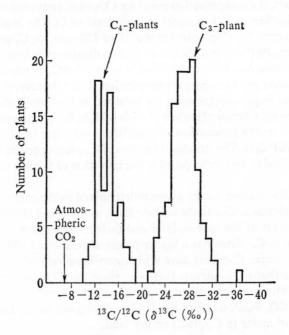

Fig. 5.22. Histogram depicting the bimodal distribution of δ^{13} C values for seed plants.[88]

CO_2 value. Plants can be divided into two groups with respect to $\delta^{13}C$, those with $\delta^{13}C$ of -10 to $-18\permil$ and those with -21 to $-33\permil$. Interestingly, high $\delta^{13}C$ plants are the C_4-type, and low $\delta^{13}C$ plants are the C_3-type.[88]. This means that the carbons of C_3- and C_4-plants are 10 to $18\permil$ (1.0 to 1.8 %) and 21 to $33\permil$ (2.1 to 3.3 %), respectively, enriched in ^{12}C compared to the carbon standard. Thus, the isotopic effect on photosynthetic CO_2 fixation is greater in C_3-plants than in C_4-plants; C_3-plants fix more ^{12}C from the atmosphere than C_4-plants do.

The fractionation between $^{12}CO_2$ and $^{13}CO_2$ in photosynthetic CO_2 fixation may occur in two steps; the first is the transport of CO_2 to chloroplasts and includes hydration, if necessary, and the second is carboxylation. In the steady photosynthetic state, the $\delta^{13}C$ values of dissolved HCO_3^- in leaf cytoplasm has been determined to be -9 to $-14\permil$.[89] Therefore, -2 to $-7\permil$ fractionation ($\Delta\delta^{13}C$) occurs during the transport and hydration of the atmospheric CO_2. However, the same fractionation has been observed when CO_2 is hydrated in solution, so the fractionation during transport is negligible. The $\Delta\delta^{13}C$ value for the carboxylation with isolated PEP carboxylase from HCO_3^- is $-2\permil$.[86,90] The $\delta^{13}C$ value of C_4-plants is equal to the sum of $\Delta\delta^{13}C$ for the carboxylation by PEP carboxylase and $\delta^{13}C$ for the CO_2-hydration; $-2 + (-9 \sim -14) = -11 \sim -16\permil$. C_3-plants fix CO_2 directly by RuBP C/O, so the $\Delta\delta^{13}C$ for the carboxylation by RuBP C/O is predicted to be the same as the $\Delta\delta^{13}C$ for C_3-plants with respect to atmospheric CO_2 if no fractionation occurs during the transport of CO_2; that is, $(-21 \sim -33) - (-7) = -14 \sim -26\permil$. The $\delta^{13}C$ values for the carboxylation by isolated RuBP C/O has been determined to be $-27.1\permil$.[91] The $\Delta^{13}C$ for the carboxylation by RuBP C/O is $(-27) - (-7) = -20\permil$. A large isotopic effect on the carboxylation with RuBP C/O may relate to the reaction mechanism. The difference in the ratios of the molecular weights of isotopic CO_2 and HCO_3^-, $^{13}CO_2/^{12}CO_2 (=1.0227)$ and $H^{13}CO_3^-/H^{12}CO_3^- (=1.0164)$, should influence the isotopic fractionation of carbon dioxide by the carboxylases, because the isotopic effect is caused by the difference in the mass of the reactants. It is, therefore, reasonable that RuBP C/O, using CO_2 as the substrate ($\Delta\delta^{13}C = -20\permil$), promotes more isotopic fractionation of carbon dioxide than does PEP carboxylase, using HCO_3^- as the substrate ($\Delta\delta^{13}C = -2\permil$). In a C_4-plant, the second carboxylation in the bundle sheath cells is catalyzed by RuBP C/O, but the $\delta^{13}C$ value of the whole plant is the value expected from carboxylation by PEP carboxylase alone. This is probably due to little or no exchange between the atmospheric CO_2 and the carbon dioxide released from aspartate or malate in the bundle sheath cells (Fig. 5.21).

The $\delta^{13}C$ values of gymnosperms, ferns, and mosses are similar to those of C_3-plants, reflecting the development of the C_4-dicarboxylic acid pathway during the same evolutionary stage as the angiosperms. Furthermore, the $\delta^{13}C$ values of coals from the Permian to Pliocene periods are between -25 and -28‰, showing that the coal plants were the C_3-type.[88,92] In addition to these applications to the study of evolution, the two distinct $\delta^{13}C$ values have been used to determine the source of food. For example, honey containing cane sugar or its products (a C_4-plant) has a higher $\delta^{13}C$ value than honey made by bees because bees collect nectar mainly from C_3-plants. Whiskies made from barley (a C_3-plant) and corn (a C_4-plant) are easily distinguished each other by their $\delta^{13}C$ values.

5.4.6.   Photorespiration

Respiration in mitochondria produces biological energy together with the oxidation of substrates. In plants, mitochondrial dark respiration provides the energy required for the growth and maintenance of biological activities of non-chlorophyll tissues, and even in leaves, dark respiration is necessary for cellular activities at night. However, a high rate of respiration lowers the productivity of plants, hence the dark respiration level is an important consideration in the breeding of plants for crops.

Plants have an other respiratory process for oxidizing early photosynthetic CO_2 fixation products, photorespiration. Photorespiration does not produce biological energy, and it consumes ATP, $NAD(P)H_2$, and the assimilation products. And it is more directly related to the productivity of plants than dark respiration. In this section we will describe photorespiration pathway and discuss its physiological function in relation to photooxidative damage in plants.

A. Glycolate Pathway

The precursor of the photorespiratory substrate is phosphoglycolate, a product of the oxygenase activity of ribulose bisphosphate carboxylase/oxygenase (RuBP C/O) (Eq. 5.54). As discussed in Section 5.4.2.B, the oxygenase and carboxylase activities of RuBP C/O are competitive with respect to O_2 and CO_2, so the production of phosphoglycolate and photorespiratory activity increase when the environmental or chloroplast CO_2/O_2 ratio is low. Phosphoglycolate is hydrolyzed by a specific phosphatase in the chloroplasts, yielding glycolate which is then transported to the peroxisomes. Glycolate is oxidized by glycolate oxidase, then the oxidized product, glyoxylate, is converted to glycine with an aminotransferase. Glycine is subsequently transported into the mitochondria, where it is decarboxylated by glycine decarboxylase (Eq. 5.58), then serine is formed by serine hydroxymethyltransferase (Eq. 5.59).

$$\text{Glycine} + O_2 \longrightarrow CO_2 + C_1\text{-tetrahydrofolate} + NH_3 \qquad (5.58)$$

$$\text{Glycine} + \text{methylenetetrahydrofolate} \longrightarrow \text{serine} + \text{tetrahydrofolate} \quad (5.59)$$

Serine is transported again into the peroxisomes and is converted to hydroxypyruvate by an aminotransferase, then to glycerate by hydroxypyruvate reductase with $NADH_2$. Glycerate is transported either to the chloroplasts or to the cytoplasm, where it is phosphorylated, yielding phosphoglycerate (PGA).

The photorespiratory oxidation pathway of phosphoglycolate is summarized in Fig. 5.23. The glycolate pathway has been established from observations of the incorporation of labeled substrates into the metabolites on the pathway, the effects of inhibitors, and the distribution of enzymes in the three organelles. For details, see recent reviews.[93,94]

When glycine is oxidized in isolated mitochondria, ATP formation by oxidative phosphorylation has been observed along with oxygen uptake.[95] Thus, in reactions (5.58) and (5.59), $NADH_2$ or ATP is produced (Eq. 5.60).

$$2 \text{ Glycine} \longrightarrow \text{serine} + CO_2 + NH_3 + NADH_2 \text{ (or 3 ATP)} \qquad (5.60)$$

In the glycolate pathway, oxygen uptake occurs at three steps, the oxygenation of RuBP, the oxidation of glycolate, and the oxidation of glycine, if $NADH_2$ is oxidized in the mitochondria. CO_2 is produced during the decarboxylation of glycine in the mitochondria. Reaction (5.60) yields biological energy but the glycolate pathway consumes $NADH_2$ for the reduction of hydroxypyruvate and ATP for the phosphorylation of glycerate. Furthermore, for the reassimilation of the NH_3 released in (Eq. 5.58), ATP and $NAD(P)H_2$ (or reduced ferredoxin) are required in the glutamine synthase and gltamate synthase system. Taking these energy transactions into account, the over-all reaction for the glycolate pathway can be represented as in Eq. (5.61).

$$RuBP + 2 O_2 + ATP + 0.5 \text{ NAD(P)}H_2 \longrightarrow 0.5 \text{ } CO_2 + 1.5 \text{ PGA}$$
$$+ \text{ ADP} + 0.5 \text{ NAD(P)} \qquad (5.61)$$

Thus, this pathway apparently offers no advantage to plants and lowers the net CO_2 fixation rate.

B. Environmental Factors Affecting Photorespiration

Plants have the potential to convert light energy into chemical energy

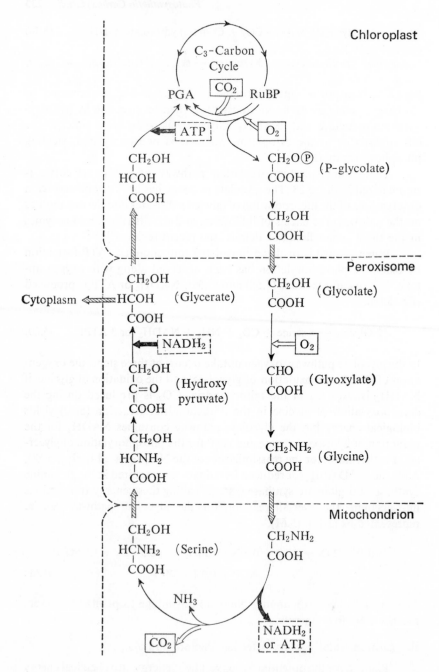

Fig. 5.23. The C$_3$-carbon cycle and glycolate pathway. ⟺ indicates interorganelle transport

in the form of carbohydrates with an efficiency of 36% (Section 5.4.1), but this efficiency has been observed only under conditions of low light intensity and high CO_2 and low O_2 concentrations. Natural conditions on Earth deviate widely from these conditions which give the maximum efficiency, and this is one of the main explanations for the low efficiency in the conversion of sunlight energy into agricultural products (below 1%).

Photorespiration lowers the conversion efficiency through a loss of the assimilation products. The following conditions induce an increase in the photorespiration rate: high light intensity, high temperature, low CO_2 and high O_2 concentrations. These conditions are just the reverse of those needed for maximum conversion efficiency, and thus photorespiration is another cause of low conversion efficiency on farms. Table 5.3 summarizes the effect of photorespiration on net CO_2 fixation rates in C_3- and C_4-plants. Under conditions of low O_2 concentration, the net CO_2-fixation of C_3-plants is 1.5 times that in air.

Table 5.3. Effect of environmental concentration of oxygen on the net CO_2-fixation rates of C_3- and C_4-plants.[†]

Plant	Relative CO_2-fixation rate (the value in air $= 1$)		
	1–2% O_2	21% O_2	100% O_2
C_3-plants			
Soybean	1.5	1.0	0.16
Wheat	1.5	1.0	
Tobacco	1.4	1.0	
C_4-plants			
Maize	1.2	1.0	0.36
Sugar cane	1.0	1.0	

[†] Estimated from the results in references 96–98).

The high rate of photorespiration in low CO_2 and high O_2 concentrations is a result of the increased supply of phosphoglycolate to the glycolate pathway. Table 5.4 summarizes the K_m for CO_2 and O_2 of the carboxylase and oxygenase activities of RuBP C/O, and K_m for CO_2 of PEP carboxylase, together with the concentration of carbon dioxide in leaf cells. V_{max} of the oxygenase is one-fifth that of the carboxylase, so in air-equilibrated solution about one-fifth of the RuBP is oxygenated, forming phosphoglycolate. O_2 is produced in thylakoids under illumination, so the concentration of O_2 inside the cells is higher than that in the atmosphere. However, the concentration gradient of O_2 between the cells and the atmosphere would be more gentle than that of CO_2 because of the high basal concentration of O_2; the O_2 gradient is probably 10% below that of CO_2.

Thus, in mesophyll cells of C_3-plants, the concentration of O_2 would

Table 5.4. The K_m values for carbon dioxide and oxygen of ribulose bisphosphate carboxylase/oxygenase and phosphoenolpyruvate carboxylase, and the concentration of carbon dioxide and oxygen in leaf cells.

K_m and concentrations	CO_2 (μM)	HCO_3^- (μM)	O_2 (μM)
K_m for CO_2 and O_2 of RuBP C/O[1]	10–20 (carboxylase)	—	100–200 (oxygenase)
K_m for HCO_3^- of PEP carboxylase[2]	4.5 (pH 7) 0.45 (pH 8)	20	—
Concentration in air (0.033% CO_2 and 21% O_2)-equilibrated solution at 25°C[3]	11	50 (pH 7) 501 (pH 8)	250
Concentration in air-equilibrated leaf cells[4]			
Mesophyll cells of C_3-plants	~6		
Mesophyll cells of C_4-plants	~1	~4.5 (pH 7) ~45 (pH 8)	
Bundle sheath cells of C_4-plants	~20		

[1] See Section 5.4.2.B.
[2] See Section 5.4.3.B.
[3] See Table 5.1.
[4] See Section 5.4.4.

approach 250 μM and that of CO_2, 6 μM (Table 5.4). Therefore, in C_3-plants in air, more than 30% of the RuBP will leak out of the C_3-carbon cycle to the glycolate pathway through oxygenation. Under a low concentration of environmental O_2, as in Table 5.3, the oxygenase activity is suppressed and the net CO_2 fixation is increased. In C_4-plants, the enzymes and organelles for the glycolate pathway are exclusively localized in the bundle sheath cells, where the concentration of CO_2 is high (Table 5.4). A high concentration of CO_2 in the bundle sheath cells suppresses the oxygenase activity of RuBP C/O, and even when CO_2 released by photorespiration leaks to the mesophyll cells, CO_2 is again fixed by PEP carboxylase, which has a high affinity for carbon dioxide (Section 5.4.3.B). These are the reasons why C_4-plants do not apparently show photorespiration and why the net CO_2 fixation rate is scarcely affected by low concentrations of O_2 (Table 5.3).

The high rate of photorespiration found at high temperatures has been ascribed to the differing solubilities of O_2 and CO_2.[99] O_2 is relatively more soluble than CO_2 with increasing temperature, so that the solubility ratio of O_2/CO_2 increases with rising temperature; the molar ratio in air-equilibrated solution is 22 at 10°C and 29 at 35°C. The inhibition of CO_2 fixation in C_3-plants by 21% oxygen (= photorespiration) is highly cor-

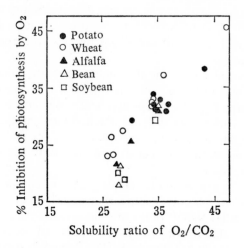

Fig. 5.24. Relationship between molar solubility ratio of O_2/CO_2 and percentage inhibition of net CO_2 fixation by O_2 for five species of C_3-plants.[99]

$$\% \text{ Inhibition} = \frac{\text{Rate of } CO_2 \text{ fixation at } 1.5\% \ O_2 - \text{The rate at } 21\% \ O_2}{\text{The rate at } 1.5\% \ O_2}$$

Source: ref. 99. Reproduced by kind permission of the American Society of Plant Physiologists.

related with the solubility ratio of O_2/CO_2. Figure 5.24 illustrates this relationship when the O_2/CO_2 ratio is changed by environmental temperature and gas composition. Thus, the induction of photorespiration at high temperatures is a result of the increased solubility ratio of O_2/CO_2 and the enhanced production of phosphoglycolate.

The mechanism by which photorespiration is stimulated at high light intensity has remained obscure. However, it may be that, under high light intensity, the concentration of CO_2 in the chloroplasts is reduced by a high rate of CO_2 fixation in the C_3-carbon cycle because of a high rate production of ATP and $NADPH_2$ in the thylakoids. Lowering the CO_2 concentration would result in an increase in the oxygenase activity of RuBP C/O.

C. Physiological Function of Photorespiration

Photorespiration appears to be nonfunctional for plants in Eq. (5.61), which indicates the loss of the CO_2 fixation products, ATP, and $NAD(P)H_2$ in the glycolate pathway. However, it seems likely that photorespiration is an essential process for plants since all known RuBP C/O enzymes, ranging from these in photosynthetic bacteria, even anaerobic ones, to these in higher plants, have the oxygenase activity. Preservation of this oxygenase activity of RuBP C/O throughout a long evolutionary history suggests that phosphoglycolate and its oxidizing pathway are necessary for photosynthetic organisms.

Environmental conditions that induce photorespiration in plants include: a shortage of the light energy (or electron) acceptor (low CO_2), excess light energy (high light intensity), and a high concentration of O_2. These conditions are similar to those which induce damage to plant leaf tissues under light, and this fact provides a clue for understanding the function of photorespiration. Let us consider what happens in plants under these conditions.

When CO_2 is deficient in the chloroplasts, the consumption of ATP and $NADPH_2$ by the photosynthetic carbon cycle is reduced and the ATP/ADP and $NADPH_2/NADP$ ratios are increased. NADP is reduced in thylakoids of the chloroplasts with ferredoxin-NADP reductase in photosystem I through P-430 and ferredoxin (Fig. 5.25). When pyridine nucleotide is in a reduced form, ferredoxin also remains in a reduced form. Under such conditions. P-430, the primary electron acceptor in photosystem I, reduces molecular oxygen, yielding superoxide anion radicals

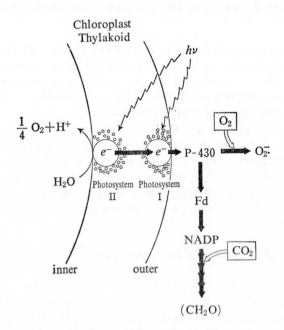

Fig. 5.25. Electron transport system in chloroplast thylakoids, and the competition for a photo-reductant in (P-430) in photosystem I between O_2 and CO_2.

(O_2^-). The production of O_2^- in illuminated chloroplasts has been demonstrated by O_2^--induced reactions, and the inhibition of the radical-induced reaction by superoxide dismutase.[100] Superoxide dismutase catalyzed the disproportionation of O_2^-, producing H_2O_2 and O_2(Eq. 5.62) and thus removing O_2^- from the reaction mixture.

Figure 5.26 shows the oxidation of epinephrine by illuminated chloroplasts and the inhibition of oxidation by superoxide dismutase,[100] confirming the light-dependent production of O_2^-. The primary electron acceptor in photosystem I is the most probable photoreductant of molecular oxygen.[100] The reduction of O_2 in photosystem I is not an artifact of isolated chloroplasts. Competition between CO_2 and O_2 for the primary electron acceptor in photosystem I has been observed in isolated intact chloroplasts and in algal cells. Figure 5.27 shows that, in intact chloro-

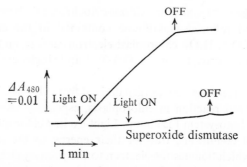

Fig. 5.26. Univalent reduction of molecular oxygen by illuminated spinach chloroplast thylakoids. The oxidation of epinephrine by O_2^- to form adrenochrome (A_{480}), and the inhibition of oxidation by superoxide dismutase.[101]

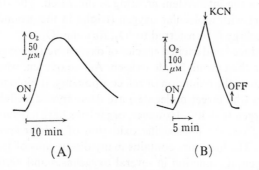

Fig. 5.27. Effect of bicarbonate and cyanide on the O_2 evolution and reduction by intact spinach chloroplasts. A) Contained only 0.14 mM HCO_3^-. After the consumption of HCO_3^-, O_2 was reduced. B) Contained 0.5 mM HCO_3^-, but 1 mM KCN was added at the indicated point. In both cases, the O_2 uptake is due to a peroxidative decomposition of H_2O_2 from O_2^- using an endogeneous electron donor.[102]

plasts, the O_2 uptake (O_2 reduction) starts when CO_2 has been consumed (A) or when RuBP C/O has been inhibited by cyanide (B).[102] The light-dependent $^{18}O_2$ uptake by algal cells is high under conditions where CO_2 is not fixed.[103] Although CO_2 deficiency causes an increased production of O_2^- in chloroplasts, it has been suggested that the reduction of O_2 is essential to the production of ATP by pseudocyclic photophosphorylation, and thus, to adjusting the $ATP/NADPH_2$ ratio for the C_3-carbon cycle.[104] For this reason O_2^- may be produced at a low rate even when the physiological electron acceptor, CO_2, is sufficiently supplied to chloroplasts.

Most O_2^- thus produced in chloroplasts is disproportionated to H_2O_2 and O_2 in a reaction catalyzed by superoxide dismutase (Eq. 5.62).

$$2\,O_2^- + 2\,H^+ \longrightarrow H_2O_2 + O_2 \tag{5.62}$$

A portion of the O_2^- reduces or oxidizes cytochrome f, plastocyanin, Mn^{2+}, ascorbate, and reduced glutathione contained in the chloroplasts.[100] Interaction of O_2^-, H_2O_2, chloroplast electron carriers, and other components produces hydroxyl radicals $(OH\cdot)$[105] and singlet excited molecular oxygen $(^1O_2)$.[106]

In addition to the active oxygen derived from O_2^-, $^1O_2$ is produced by a photosensitized reaction in chloroplasts, especially when the electron acceptor for photosystem I is absent. Light-excited chlorophyll molecules, under normal conditions, transfer their energy to the reaction center chlorophyll, which donates the electron to the primary electron acceptors of photosystems I and II after the molecular charge separation. However, in the absence of an electron acceptor because of deficient CO_2, the probability that singlet excited chlorophyll will make the transition to the triplet state via the intersystem crossing is increased. The triplet excited molecules react with molecular oxygen (triplet in the ground state), and their excited energy is transferred to $^3O_2$, forming $^1O_2$.

These reduced and excited species of oxygen, active oxygen, are more highly reactive than ground-state oxygen. Active oxygen, when generated in the cells, causes the oxidation of cell components. If biologically important molecules (the target molecules) are decomposed, cellular activity is disturbed (oxygen toxicity). However, organisms have a defense system to protect them from the nonspecific oxidation of cellular components by active oxygen. The literature contains many discussions of the formation of active oxygen, its function in several oxygenases and peroxidases, the scavenging of active oxygen, and oxygen toxicity.[107–112] These relationships are summarized in Fig. 5.28. Thus, oxygen toxicity or damage is observed when active oxygen produced by cellular activity, radiation, or light cannot be removed by the scavenging system; that is, under condi-

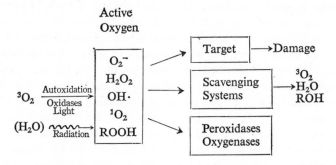

Fig. 5.28. Formation and scavenging of active oxygen and their biological effects.

tions when the production rate of active oxygen is high or the scavenging capacity is low.

In chloroplasts, the scavengers of active oxygen are superoxide dismutase for O_2^-,[113] a peroxidase system for H_2O_2,[102,114] and tocopherol and carotenoids for $^1O_2$. Under normal conditions, these scavengers are able to protect the chloroplasts from photo-oxidative damage by active oxygen. However, under conditions when almost all the photoreductant in photosystem I reduces O_2, the chloroplast scavenging system cannot remove all the active oxygen and the leaf tissues are damaged. When methyl viologen (MV or paraquat, 1,1'-dimethyl-4,4'-bipyridylium dichlorides) is applied to leaves, MV is photoreduced in place of ferredoxin by the primary electron acceptor in photosystem I (Fig. 5.25). Reduced MV cation radicals are auto-oxidized very rapidly ($7.7. \times 10^8 \ M^{-1} \ s^{-1}$)[115] to produce the O_2^- radicals. Because of its ability to induce a high O_2^- production rate in leaf tissues, methyl viologen has been used as a herbicide. Similar photo-oxidative damage has been observed in the absence of CO_2 and the presence of 100% O_2, probably due to a high rate of active oxygen production.[116]

As described previously, photorespiration is induced when the light energy acceptor, CO_2, is not sufficiently supplied to the chloroplasts. Under these conditions, active oxygen is produced, causing irreversible damage to leaf tissues. From these relationships, the present author proposes, in agreement with Björkman,[117] that the function of photorespiration is the protection of leaf tissues from photo-oxidative damage by providing CO_2, ADP, and NAD(P) through the glycolate pathway. In other words, photorespiration supplies the electron donor to chloroplasts and reduces the production of active oxygen. It seems likely that the protection of leaf tissues from irreversible photo-oxidative damage is essential, even when it necessitates wasting a portion of the CO_2 fixation products.

Under natural conditions, the diurnal change of sunlight intensity does not always correspond to that of the CO_2 supply to the chloroplasts. Instead, the CO_2 supply is usually limited when the light intensity is at a maximum in the daytime. The concentration of CO_2 in the atmosphere around the leaves is diminished by absorption into the leaves, especially when the air is not circulating. In addition, water stress closes the stomata to conserve water, which arrests the transport of CO_2 to the chloroplasts. With the stomata closed, the leaf temperature increases because less heat is lost through transpiration and in turn, the solubility ratio of O_2/CO_2 increases in the chloroplasts (Section 5.4.6.B). Since a CO_2 deficiency is unavoidable under conditions of high light intensity, photorespiration is an essential process by which plants prevent irreversible photo-oxidative damage by active oxygen; it is a physiological defense system against oxygen toxicity. It has been shown that the CO_2 fixation rate in leaves decreases after a depression of photorespiration,[118,119] an observation which supports the above hypothesis.

In accordance with this hypothesis, chemical inhibition of the glyco-late pathway leads to photo-oxidative damage of the leaves rather than to an increase in the net CO_2 fixation rate. In fact, several chemicals have been shown to regulate photorespiration, but no one report the increase of crop yields. Depression of photorespiration by an enrichment of CO_2 on farms is a possible approach, but technical problems in the supply of the gas have limited its application only to special cases such as green-houses.

D.   Interaction of Carbon Dioxide with Active Oxygen

The relationship between carbon dioxide and active oxygen in the photorespiration of plants is an indirect one. There also occurs a direct interaction between active oxygen and carbon dioxide, although its biological function is uncertain. When HCO_3^- is added to the O_2^--generating system, chemiluminescence has been observed.[120] This luminescence is not due to $^1O_2$, but has been thought to be a result of peroxycarbonate.[121] H_2O_2 is produced by reaction (5.62). Hydroxyl radicals ($OH\cdot$) are produced from O_2^- and H_2O_2 in the presence of traces of metal ions in the reaction mixture (Haber-Weiss reaction) (Eq. 5.63). The interaction of $OH\cdot$ with bicarbonate produces carbonate radicals (Eq. 5.64), then the radicals are dimerized, yielding peroxycarbonate (Eq. 5.65).

$$O_2^- + H_2O_2 \longrightarrow O_2 + OH\cdot + OH^- \qquad (5.63)$$

$$HCO_3^- + OH\cdot \longrightarrow OH^- + HCO_3\cdot \qquad (5.64)$$

$$2 \text{ HCO}_3 \cdot \longrightarrow \underset{\substack{| \\ \text{OH}}}{\text{O=C}}-\text{O}-\text{O}-\underset{\substack{| \\ \text{OH}}}{\text{C=O}} \tag{5.65}$$

The biological function of these reactions has not been determined, but there is a possibility that carbon dioxide traps the most reactive active oxygen, OH·, and thus protects cells from oxidative damage.

5.4.7. CO_2-fixation in Crassulacean Acid Metabolism Plants

Transpiration in plants has an intimate relation with the CO_2 supply to chloroplasts because both water and CO_2 are transported through the leaf stomata. Under conditions of water stress, plants close the stomata and the transport of CO_2 is retarded. In this respect, C_4-plants are an adaptive form of C_3-plants to mild water stress. High resistance in the stomata conserves water, although CO_2 transport is arrested. This dilemma has been solved, however, by introducing an effective CO_2-fixing enzyme, PEP carboxylase (Section 5.4.4). Photorespiration protects leaf cells from photo-oxidative damage that might result from CO_2 deficiency caused by the stomata closure (Section 5.4.6.C).

Plants such as cactus that grow under the most severe water stress have another adaptive form of CO_2 fixation. These desert plants close the stomata during the day to reduce water loss and open only at night. As a consequence, the CO_2 supply to leaf cells is limited at night. These plants have been referred to as Crassulacean acid metabolism plants (CAM plants) because most of them are a family of Crassulaceae.

The most characteristic feature of the metabolism of CAM plants is the diurnal change in the malic acid content of the leaves; it accumulates at night and disappears during the day. That CO_2 is available to the leaf cells only at night suggests the production of malate by carboxylation, and $^{14}CO_2$ incorporation studies have shown that this is the case. The CO_2 acceptor is phosphoenolpyruvate (PEP), which is carboxylated with PEP carboxylase (Eq. 5.18). The fixation product is reduced to malate by malate dehydrogenase. Malate is stored in the cell vacuoles and is decarboxylated in the daytime to supply CO_2 to the C_3-carbon cycle in the chloroplasts. The daytime fixation product is stored in the form of starch, which provides the CO_2 acceptor for the CO_2 fixation at night.

Figure 5.29 shows an outline of CO_2 fixation in CAM plants. For details, see a recent review.[122] Carboxylations in CAM plants closely resemble those in C_4-plants. The nighttime and daytime carboxylations of CAM plants correspond to the reactions in the mesophyll and bundle sheath cells of C_4-plants, respectively. In C_4-plants, the two carboxylations

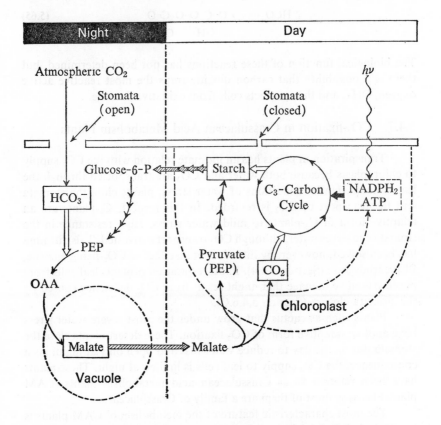

Fig. 5.29. CO_2 fixation pathway in Crassulacean acid metabolism plants.

are carried out in different cells at the same time (Fig. 5.21); CAM plants fix CO_2 in the same cells, but at different times.

In C_4-plants, three types of decarboxylation of C_4-acids occur in the bundle sheath cells (5.4.3.A). CAM plants apparently have one or both of the following pathways, depending on the species.[122] One is the decarboxylation of malate by NADP-malic enzyme (Eq. 5.31). The other is the oxidation of malate to oxaloacetate by malate dehydrogenase, after which oxaloacetate is decarboxylated by PEP carboxykinase (Eq. 5.19). Nishida has shown that malic enzyme is inhibited by carbon dioxide,[123] in accordance with Walker.[72] He has suggested that this inhibition explains why malate is not decarboxylated at night.

The $\delta^{13}C$ values of CAM plants were expected to approximate those of C_4-plants because of the similarities in their CO_2-fixation pathways, but the observed values varied between those of C_3- and C_4-plants.[124, 125] This

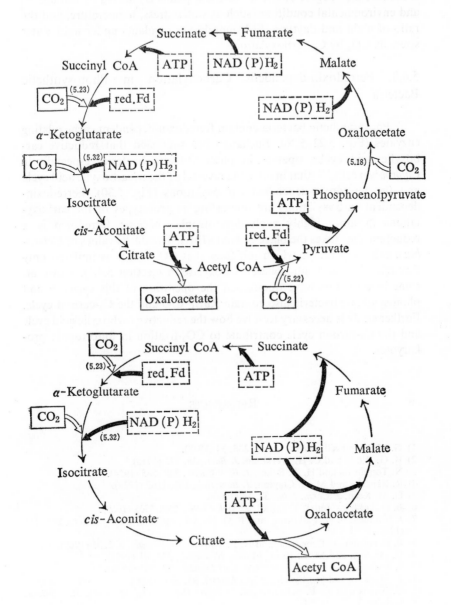

Fig. 5.30. The reductive carboxylic acid cycle in photosynthetic bacteria.[127] red. Fd: reduced ferredoxin. Equation numbers of carboxylations in the text are shown in brackets.

is because the "degree of CAM" of CAM plants is affected by leaf age[126] and environmental conditions such as water stress, temperature, and the ratio of night and daytime.[122] Young leaves or plants under mild water stress fix CO_2 by C_3-carboxylation.

5.4.8. Ferredoxin-dependent Carboxylation in Photosynthetic Bacteria

Photosynthetic bacteria contain ferredoxin-dependent carboxylating enzymes (Eqs. 5.22–5.26). Buchanan has suggested that reductive carboxylic acid cycles operate in photosynthetic bacteria, especially in *Chlorobium* cells,[127] that involve the reversal of the tricarboxylic acid cycle and the ferredoxin-dependent carboxylations (Fig. 5.30). Ferredoxin-dependent carboxylations are interesting as prototypes of the carboxylations in more highly evolved-autotrophs, because ferredoxin is a reductant that takes the place of $NAD(P)H_2$. The $\varDelta\delta^{13}C$ values of *Chlorobium* cells is $-12.5\permil$ which indicates that CO_2 is not assimilated only through ribulose bisphosphate carboxylase[128] (section 5.4.5). However, some investigators have questioned the operation of this cycle,[129] and photosynthetic bacteria also contain RuBP C/O and the C_3-carbon cycle. Further study is necessary to solve how the reductive carboxylic acid cycle and the C_3-carbon cycle contribute to CO_2-fixation in autotrophic prokaryotes.

REFERENCES

1) G. M. Woodwell, *Scient. Amer.*, **238**, 34 (1978).
2) H. G. Wood and C. H. Werkman, *J. Bact.*, **30**, 332 (1935).
3) N. Tomlinson and H. A. Baker, *J. Biol. Chem.*, **209**, 585 (1954).
4) R. Repaske and M. A. Clayton, *J. Bacteriol.*, **135**, 1162 (1978).
5) D. M. Kern, *J. Chem. Edu.*, **37**, 14 (1960).
6) B. H. Gibbons and J. T. Edsall, *J. Biol. Chem.*, **238**, 3502 (1963).
7) J. Brundell, S. D. Falkbring, and P. O. Nyman, *Biochim. Biophys. Acta*, **284**, 311 (1972).
8) E. B. Nelson, A. Cenedella, and N. E. Tolbert, *Phytochem.*, **8**, 2305 (1969).
9) D. Graham and M. L. Reed, *Nature, New Biol.*, **231**, 81 (1971).
10) S. Miyachi and Y. Shiraiwa, *Plant Cell Physiol.*, **20**, 341 (1979).
11) Y. Shiraiwa and S. Miyachi, *FEBS Letters*, **95**, 207 (1978).
12) G. Nahas and K. E. Schaefer (ed.), *Carbon Dioxide and Metabolic Regulations*, Springer-Verlag (1974).
13) R. G. Khalifah, *Proc. Natl. Acad. Sci. USA*, **70**, 1986 (1973).
14) Y. Pocker and S. Sarkanen, *Adv. Enzymol.*, **47**, 149 (1978).
15) Y. Pocker and D. W. Bjorkquist, *Biochemistry*, **16**, 5698 (1977).
16) Y. Pocker and R. R. Miksch, *Biochemistry*, **17**, 1119 (1978).

17) B. S. Jacobson, F. Fong, and R. L. Heath, *Plant Physiol.*, **55**, 468 (1975).
18) J. E. Coleman, *Inorganic Biochemistry* (ed. G. L. Eichhorn), p. 488, Elsevier (1973).
19) B. H. Jonsson, H. Steiner, and S. Lindskog, *FEBS Letters*, **64**, 310 (1976).
20) T. G. Cooper, D. Filmer, M. Wishnick, and M. D. Lane, *J. Biol. Chem.*, **244**, 1081 (1969).
21) T. G. Cooper, T. T. Tchen, H. G. Wood, and C. R. Benedict, *J. Biol. Chem.*, **243**, 3857 (1968).
22) S. Asami, K. Inoue, K. Matsumoto, A. Murachi, and T. Akazawa, *Arch. Biochem. Biophys.*, **194**, 503 (1979).
23) K. Dalziel and J. C. Londesborough, *Biochem. J.*, **110**, 223 (1968).
24) R. H. Villet and K. Dalziel, *Biochem. J.*, **115**, 633 (1969).
25) R. K. Thauer, B. Käufer, and G. Fuchs, *Eur. J. Biochem.*, **55**, 111 (1975).
26) J. P. Jones, E. J. Gradner, T. G. Cooper and R. E. Olsen, *J. Biol. Chem.*, **252**, 7738 (1977).
27) T. G. Cooper and H. G. Wood, *J. Biol. Chem.*, **246**, 5488 (1971).
28) J. Moss and M. D. Lane, *Adv. Enzymol.*, **35**, 321 (1971).
29) P. P. Cohen, *The Enzyme*, vol. 6, p. 477, Academic Press (1962).
30) M. E. Jones and L. Spector, *J. Biol. Chem.*, **235**, 2897 (1960).
31) M. F. Utter and H. M. Kolenbrander; *The Enzymes*, 3rd ed., vol. 6 p. 117, Academic Press (1972).
32) P. A. Scherer and R. K. Thauer, *Eur. J. Biochem.*, **85**, 125 (1978).
33) R. K. Thauer, E. Rupprecht, and K. Jungermann, *FEBS Letters*, **8**, 304 (1970).
34) N. K. Ramaswamy and P. M. Nair; *Eur. J. Biochem.*, **67**, 275 (1976).
35) F. R. Williams and L. P. Hager, *J. Biol. Chem.*, **236**, PC 36 (1961).
36) Y. Motokawa and G. Kikuchi, *Arch. Biochem. Biophys.*, **135**, 402 (1969).
37) S. M. Klein and R. D. Sagers, *J. Biol. Chem.*, **242**, 301 (1967).
38) J. Stenflo, *Adv. Enzymol.*, **46**, 1 (1978).
39) P. A. Friedman, M. A. Shia, P. M. Gallop, and A. E. Griep, *Proc. Natl. Acad. Sci. USA*, **76**, 3126 (1979).
40) Y. Kaziro, L. F. Hass, P. D. Boyer, and S. Ochoa, *J. Biol. Chem.*, **237**, 1460 (1962).
41) A. Y. Sun, L. Ljungdahl, and H. G. Wood, *J. Bacteriol.*, **98**, 842 (1969).
42) M. Calvin and J. A. Bassham, *The Photosynthesis of Carbon Compounds*, W. A. Benjamin Inc., *N.Y.* (1962).
43) T. A. Peterson, M. Kirk and J. A. Bassham, *Physiol. Plantarum*, **19**, 219 (1966).
44) J. R. Quayle and T. Ferenci, *Microbiol. Rev.*, **42**, 251 (1978).
45) S. G. Wildman and J. Bonner, *Arch. Biochem.* **14**, 381 (1947).
46) A. Weissbach, B. L. Horecker and J. Hurwitz, *J. Biol. Chem.*, **218**, 795 (1956).
47) H. W. Siegelman and G. Hind (ed.), *Photosynthetic Carbon Assimilation*, Plenum Press (1978).
48) T. S. Baker, S. W. Suh, and D. Eisenberg, *Proc. Natl. Acad. Sci. USA*, **74**, 1037 (1977).
49) V. B. Lawelis, G.L.R. Gordon, and B. A. McFadden, *J. Bacteriol.*, **139**, 287 (1979).
50) M. Nishimura and T. Akazawa, *Biochem. Biophys. Res. Commun.*, **54**, 842 (1973).
51) H. Kobayashi, T. Takabe, M. Nishimura, and T. Akazawa, *J. Biochem.*, **85**, 932 (1979).
52) B. A. McFadden and K. Puohit, *Photosynthetic Carbon Assimilation* (ed. H. W. Siegelman and G. Hind), p. 179, Plenum Press (1978).
53) G. Müllhofer and I. A. Rose, *J. Biol. Chem.*, **244**, 1081 (1969).
54) M. Calvin, *Federation Proc.*, **13**, 697 (1954).
55) M. I. Siegel and M. D. Lane, *Biochem. Biophys. Res. Commun.*, **48**, 508 (1972).
56) G. L. R. Gordon, V. B. Lawlis and B. A. McFadden, *Arch. Biochem. Biophys.*, **199**, 400 (1980).
57) M. I. Siegel and M. D. Lane, *J. Biol. Chem.*, **248**, 5486 (1973).
58) I. L. Norton, M. H. Welch and F. C. Hartman, *J. Biol. Chem.*, **250**, 8062 (1975).
59) C. Paech and N. E. Tolbert, *J. Biol. Chem.*, **253**, 7864 (1978).
60) J. M. Sue and J. R. Knowles, *Biochemistry*, **17**, 4041 (1978).

61) G. H. Lorimer, *Eur. J. Biochem.*, **89**, 43 (1978).
62) M. R. Badger and G. H. Lorimer, *Arch. Biochem. Biophys.*, **175**, 723 (1976).
63) G. H. Lorimer, M. R. Badger, and T. J. Andrews, *Biochemistry*, **15**, 529 (1976).
64) H. M. Miziroko, *J. Biol. Chem.*, **254**, 270 (1979).
65) G. H. Lorimer, *J. Biol. Chem.*, **254**, 5599 (1979).
66) M. H. O'Leary, R. J. Jaworski, and F. C. Hartman, *Proc. Natl. Acad. Sci. USA*, **76**, 673 (1979).
67) C. Bauer, R. Baumann, U. Engels, and B. Pacyna, *J. Biol. Chem.*, **250**, 2173 (1975).
68) J. S. Morrow, P. Keim, R. B. Visscher, R. C. Marshall, and F.R.N. Gurd, *Proc. Natl. Acad. Sci. USA*, **70**, 1414 (1973).
69) A. Arnone, *Nature*, **247**, 143 (1974).
70) D. S. Bendall, S. L. Ranson, and D. A. Walker, *Biochem. J.*, **76**, 221 (1960).
71) J. Swierczynski and J. K. Davis, *Biochem. Biophys. Res. Commun.*, **85**, 173 (1978).
72) D. A. Walker, *Biochem. J.*, **74**, 216 (1960).
73) W. B. Anderson and R. C. Nordlie, *Biochemistry*, **7**, 1479 (1968).
74) J.E.D. Dyson, W. B. Anderson, and R. C. Nordlie, *J. Biol. Chem.*, **244**, 560 (1969).
75) R. Brändén and S. Styring, *Biochem. Biophys. Res. Commun.*, **89**, 607 (1979).
76) A. Senser, I. Valverde, and W. J. Malaisse, *FEBS Letters*, **105**, 40 (1979).
77) K. Yasumoto, Ph. D. thesis, Kyoto Univ. (1961).
78) Govindjee and J.J.S. Van Rensen, *Biochim. Biophys. Acta*, **505**, 183 (1978).
79) T. Murai and T. Akazawa, *Plant Physiol.*, **50**, 568 (1972).
80) W. S. Cohen and A. T. Jagendorf, *Plant Physiol.*, **53**, 220 (1974).
81) H. P. Kortschak, C. E. Hartt and G. O. Burr, *Plant. Physiol*, **40**, 209 (1965).
82) M. D. Hatch, CO_2 *Metabolism and Plant Productivity* (ed. R. H. Burris and C. C. Balck), p. 59, University Park Press, (1976).
83) I. P. Ting and C. B. Osmond, *Plant Physiol.*, **51**, 439 (1973).
84) K. Uedan and T. Sugiyama, *Plant Physiol.*, **57**, 906 (1976).
85) J. Coombs, S. L. Maw, and C. W. Baldry, *Plant Science Letters*, **4**, 97 (1975).
86) P. H. Reibach and C. R. Benedict, *Plant Physiol.*, **59**, 564 (1977).
87) R. Park and S. Epstein, *Plant Physiol.*, **30**, 133 (1961).
88) J. H. Troughton, *Photosynthesis and Photorespiration* (ed. M. D. Hatch, C. B. Osmond, and R. O. Slatyer), p. 124, Wiley-Interscience (1971).
89) R. Park and S. Epstein, *Geochim. Cosmochim. Acta*, **21**, 110 (1960).
90) T. Whelan, W. M. Sackett, and C. R. Benedict, *Plant Physiol.*, **51**, 1051 (1973).
91) W. W. Wong, C. R. Benedict, and R. J. Kohel, *Plant Physiol.*, **63**, 852 (1979).
92) B. N. Smith and S. Epstein, *Plant Physiol.*, **47**, 380 (1971).
93) N. E. Tolbert and F. J. Ryan, CO_2 *Metabolism and Plant Productivity* (ed. R. H. Burris and C. C. Black), p. 141, University Park Press (1976).
94) B. Halliwell, *Prog. Biophys. Molec. Biol.*, **33**, 1 (1978).
95) A. L. Moore, C. Jackson, B. Halliwell, J. E. Dench, and D. O. Hall, *Biochem. Biophys. Res. Commun.*, **78**, 483 (1977).
96) M. L. Forrester, G. Krotokov, and C. D. Nelson, *Plant Physiol.*, **41**, 428 (1966).
97) J. Hesketh, *Planta*, **76**, 371 (1967).
98) T. A. Bull, *Crop Sci.*, **9**, 276 (1969).
99) S. B. Ku and G. E. Edwards, *Plant Physiol.*, **59**, 986 (1977).
100) K. Asada, M. Takahashi, K. Tanaka, and Y. Nakano, *Biochemical and Medical Aspects of Active Oxygen* (ed. O. Hayaishi and K. Asada), p. 45, Japan Scientific Societies Press (1977).
101) K. Asada, K. Kiso, and K. Yoshikawa, *J. Biol. Chem.*, **249**, 2175 (1974).
102) Y. Nakano and K. Asada, *Plant Cell Physiol.*, **21**, (in press) (1980).
103) R. J. Radmer and B. Kok, *Plant Physiol.*, **58**, 336 (1976).
104) H. Egneus, U. Heber, U. Mathiesen, and M. Kirk, *Biochim. Biophys. Acta*, **408**, 502 (1975).
105) E. F. Elstner and J. R. Konze, *FEBS Letters*, **45**, 18 (1974).
106) U. Takahama and M. Nishimura, *Plant Cell Physiol.*, **17**, 111 (1976).
107) I. Fridovich, *Ann. Rev. Biochem.*, **44**, 147 (1975).

108) K. Asada, *Jap. J. Biochem.*, **48**, 226 (1976).
109) O. Hayaishi and K. Asada (ed.), *Biochemical and Medical Aspects of Active Oxygen*, Japan Scientific Societies Press (1977)
110) A. M. Michelson, J. M. McCord, and I. Fridovich (ed.), *Superoxide and Superoxide Dismutases*, Academic Press (1977).
111) A. Singh and A. Petkau (ed.), *Singlet Oxygen and Related Species in Chemistry and Biology, Photochem. Photobiol.*, **28**, Nos. 4 and 5 (1978).
112) I. Fridovich (ed.), *Oxygen Free Radicals and Tissue Damage*, Excerpta Medica (1979).
113) K. Asada, M. Urano, and M. Takahashi, *Eur. J. Biochem.*, **36**, 257 (1973).
114) K. Asada and Y. Nakano, *Photochem. Photobiol.*, **28**, 917 (1978).
115) J. A. Farrington, M. Ebert, E. J. Land, and K. Fletcher, *Biochim. Biophys. Acta*, **314**, 372 (1973).
116) A. Abeliovich, G. Kellenberg, and M. Shilo, *Photochem. Photobiol.*, **19**, 379 (1974).
117) O. Björkman, *Photophysiology* (ed. A. C. Gieses), vol. 8, p. 1, Academic Press (1973).
118) M. G. Cornic, *C. R. Acad. Sci. Paris*, **282D**, 1955 (1976).
119) S. B. Powles, C. B. Osmond and S. W. Thorne, *Plant Physiol.*, **64**, 982 (1979).
120) E. K. Hodgson and I. Fridovich, *Arch. Biochem. Biophys.*, **172**, 202 (1976).
121) J. P. Henry and A. M. Michelson, *Biochemical and Medical Aspects of Active Oxygen* (ed. O. Hayaishi and K. Asada), p. 135, Japan Scientific Societies Press (1977).
122) C. B. Osmond, *Ann. Rev. Plant Physiol.*, **29**, 379 (1978).
123) K. Nishida, *Plant Cell Physiol.*, **18**, 927 (1977).
124) J. C. Lerman, E. Deleens, A. Nato, and A. Moyse, *Plant Physiol.*, **53**, 581 (1974).
125) E. Deleens and J. Garnier-Dardart, *Planta,* **135**, 241 (1977).
126) K. Nishida, *Plant Cell Physiol.*, **19**, 935 (1978).
127) B. B. Buchanan, *Iron-Sulfur Proteins* (ed. W. Lovenberg), vol. 1, p. 129, Academic Press (1973).
128) R. Sirevag, *Arch. Microbiol.*, **112**, 35 (1977).
129) N. Benscher and G. Gottschalk, *Z. Naturforschg.*, **27b**, 967 (1972).

6

Model Reactions of Biochemical Carbon Dioxide Fixations

Shohei INOUE

Faculty of Engineering, the University of Tokyo
Bunkyo-ku, Tokyo 113, Japan

As described in previous chapters, there are several types of carbon dioxide fixation reactions that are important in biochemical processes including photosynthesis. The enzyme carbonic anhydrease plays an important part in the exchange of carbon dioxide between the biosphere and the geosphere by catalyzing the reversible hydration of carbon dioxide. This reaction is essential for the excretion of carbon dioxide during respiration, and probably also participates in the uptake of carbon dioxide by plants prior to photosynthesis. Another type of biochemical carboxylation provides key intermediates for metabolic pathways such as the tricarboxylic acid (TCA) cycle and fatty acid synthesis, where the active species for the carboxylation is a coenzyme biotin. Carbon dioxide fixation reactions are also utilized by some organisms for the excretion of unnecessary products; urea formation in mammals, for example, removes ammonia from the system.

In this chapter, the emphasis is placed on "models" and reactions related to these biochemical carboxyations. The previous chapter discussed photosynthesis in details, so here the discussion is limited to some related photochemical reactions. We will also touch on the possible role of carbon dioxide reactions in chemical evolution.

6.1.   CARBONIC ANHYDRASE

Carbonic anhydrase catalyzes the reversible hydration of carbon dioxide (Eq. 6.1.).[1,2]

$$CO_2 + H_2O \rightleftharpoons H_2CO_3 \rightleftharpoons HCO_3^- + H^+ \qquad (6.1)$$

The enzyme is extremely efficient, with the highest turnover numbers known for any enzyme. It is widely distributed in nature, and one of the best-known examples is the human erythrocyte enzyme, which plays an important role in respiration, where the rapid interconversion between carbon dioxide and bicarbonate is essential. Human erythrocyte carbonic anhydrase contains one zinc atom per molecule. According to X-ray crystallo-graphic analysis, the secondary structure of the enzyme is that illustrated in Fig. 6.1.[3] As can be seen in the figure, the zinc atom is thought to be coordinated with three imidazole groups of histidine residues in the enzyme protein.

A hydration mechanism has been postualted which involves the nucleophilic attack on carbon dioxide by a hydroxide ion bound to zinc (Eq. 6.2).[4,5]

$$Enz\text{–}Zn\text{–}OH + CO_2 \rightleftharpoons Enz\text{–}Zn\text{–}O\text{–}\overset{\displaystyle \|}{\underset{\displaystyle O}{C}}\text{–}O\text{–}H \qquad (6.2)$$

Although infrared spectra of the carbonic anhydrase–CO_2 complex indicate the binding of CO_2 in the hydrophobic cavity of the enzyme in an unstrained, linear configuration,[5] the coordination of CO_2 to the zinc in the transition state cannot be excluded, since such coordination is thought to greatly enhance the electrophilicity of the carbonyl carbon.

Several metal ions were examined for their activity when coupled with the cooresponding apoenzyme, and cobalt(II) ion exhibited an activity of a magnitude similar to that of the native zinc enzyme. Related to this is the finding that hydroxopentaamminecobalt(III) ion rapidly converts CO_2 to cobalt-bound bicarbonate (Eq. 6.3)[6]

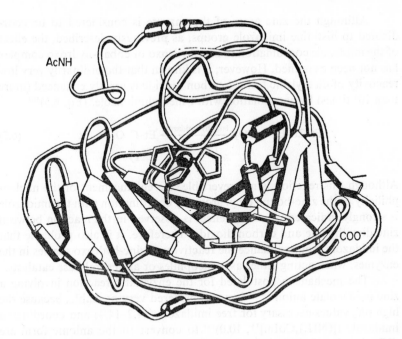

AcNH

COO⁻

Fig. 6.1. The secondary structure of human carbonic anhydrase C. Cylindrical shapes represent helical parts, and arrows the β-strand structures pointing from the amino end toward the carboxyl end.[3] (Source: ref. 3. Reproduced by kind permission of Academic Press, Inc., USA).

$$[Co(NH_3)_5OH]^{2+} + CO_2 \rightleftharpoons [Co(NH_3)_5OCO_2H]^{2+} \qquad (6.3)$$

The solvent isotope effects for the same reaction carried out in H_2O and D_2O are 1.0 and 1.1 for forward and backward processes, respectively.[7] Thus, the reaction does not involve proton transfer in the rate-determining step. On the other hand, the hydration of CO_2 catalyzed by bovine carbonic anhydrase exhibits a large solvent isotope effect (around 3),[8] reflecting the action of the enzyme as a general base rather than the action of a zinc-hydroxo species as a nucleophile.

The formation of thermally stable crystalline bicarbonate complexes of iridium(I) and rhodium(I) from the corresponding hydroxo complexes are represented in Eq.(6.4).[9]

$$[(Ph_3P)_2(CO)M(OH)] + CO_2 \longrightarrow [(Ph_3P)_2(CO)M(OCO_2H)] \qquad (6.4)$$

in
C_2H_5OH

$$M = Rh, Ir$$

Although the zinc atom of the enzyme is considered to be coordinated to histidine imidazole groups, as previously described, the effect of the imidazole coordination to zinc-hydroxo or cobalt-hydroxo complex has not been examined. However, it is known that the ordinarily very low reactivity of diethylzinc toward carbon dioxide is greatly enhanced (more than 10^4 times) by coordination with l-methylimidazole. (Eq. 6.5)[10]

$$\text{Et–Zn–Et} + CO_2 \xrightarrow{\text{l-methylimidazole}} \underset{\overset{\|}{O}}{\text{Et–C–O–Zn–Et}} \tag{6.5}$$

Although this reaction is not reversible, the enhancement of the nucleophilicity of the zinc-bound ethyl group by coordination with the imidazole is strongly indicated. A similar effect is observed in the reaction between zinc dimethoxide and carbon dioxide. Therefore, it is also probable that the imidazole groups enhance the reactivity of zinc-hydroxo species in the enzymes, when acting either as nucleophiles or as general base catalysts.

The mechanism postulated for the enzymatic reaction involving a zinc imidazolate anion[11–13] is not considered very probable, because the high pK_a values necessary for free imidazole (14.2–14.4) and coordinated imidazole ($[(NH_3)_4CoIm]^{3+}$, 10.0)[14] to convert to the anionic form are difficult to justify with the optimal pH (7–8) of the enzyme.

6.2.   BIOTIN AND MODEL REACTIONS

The coenzyme biotin (Formula 6.1) plays an essential role in certain biological carbon dioxide fixation reactions.[15,16] There are two types of biotin-containing enzymes for carboxylation, acyl coenzyme A carboxylase and pyruvate carboxylase. Biotin is bound to the enzyme by an amide linkage formed from its carboxyl group and the amino group of a lysine residue of the enzyme protein.

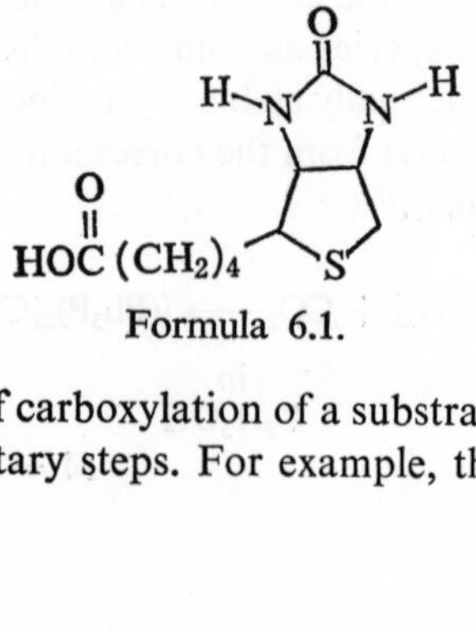

Formula 6.1.

The mechanism of carboxylation of a substrate by the biotin enzyme consists of two elementary steps. For example, the conversion of acetyl coenzyme A to malonyl coenzyme A is shown in Eqs. (6.6) and (6.7).

$$\text{E–biotin} + \text{HCO}_3^- + \text{ATP} \rightleftharpoons \text{E–biotin–CO}_2^- + \text{ADP} + \text{Pi} \qquad (6.6)$$

$$\text{E–biotin–CO}_2^- + \text{CH}_3\text{–CO–CoA}$$
$$\rightleftharpoons \text{E–biotin} + {}^-\text{OOC–CH}_2\text{–CO–CoA} \qquad (6.7)$$

E-biotin represents enzyme-bound biotin; E-biotin-CO$_2$, its carboxylated form; ATP, adenosine triphosphate; ADP, adenosine disphosphate; and Pi, inorganic phosphate.

Acetyl coenzyme A carboxylase from *E. coli* consists of three kinds of subunits, carboxyl carrier protein (CCP), biotin carboxylase (BC), and carboxyl transferase (CT), biotin being contained in the CCP subunit.

Figure 6.2 illustrates the "ping-pong" mechanisum suggested to explain the action of acetyl coenzyme A carboxylase.[17] First, the biotin group is carboxylated in the BC subunit according to reaction (6.6), then the carboxylated biotin unit is transferred to the CT subunit, where the carboxyl group is transferred to acetyl coenzyme A, regenerating a free

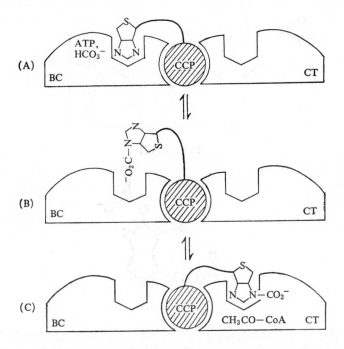

Fig. 6.2. "Ping-pong" mechanism for the action of acetyl coenzyme A carboxylase. (Source: ref. 17. Reproduced by kind permission of National Academy of Science.)

biotin group. Pyruvate carboxylase catalyzes the carboxylation of pyruvic acid to form oxalacetic acid by a similar "ping-pong" mechanism that requires magesium and maganese(II) ions as cofactors.

As can be seen in these two carboxylation reactions, the role of biotin is to be a catalyst in the formation of key intermediates in metabolic pathways, rather than to fix a bulk of carbonaceous material for metabolism. Thus, the carboxylation of acetyl coenzyme A is a key step in the synthesis of fatty acid, but the product, malonyl coenzyme A, loses carbon dioxide in a subsequent reaction. Carboxylation of pyruvic acid to oxaloacetic acid supplies this key compound in the TCA cycle.

6.2.1.   Acyl Coenzyme A Carboxylase

Although carboxylation has been suggested to occur at the nitrogen of the imidazolidone group in biotin (Formula 6.1), as depicted in Formula 6.2, there is still some controversy about whether the carboxylation is a N-carboxylation (Formula 6.2) or an O-carboxylation (Formula 6.3).

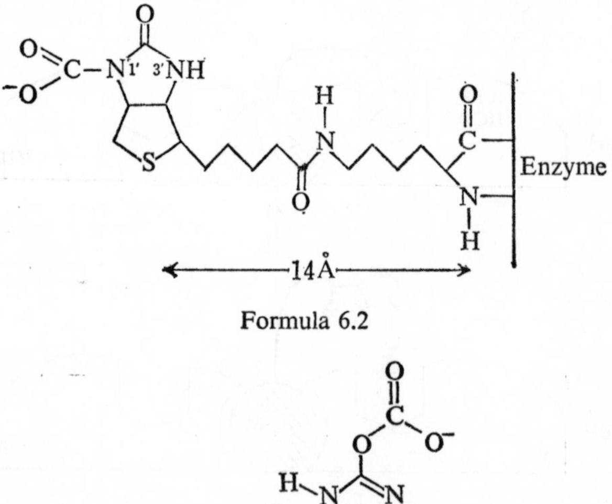

Formula 6.2

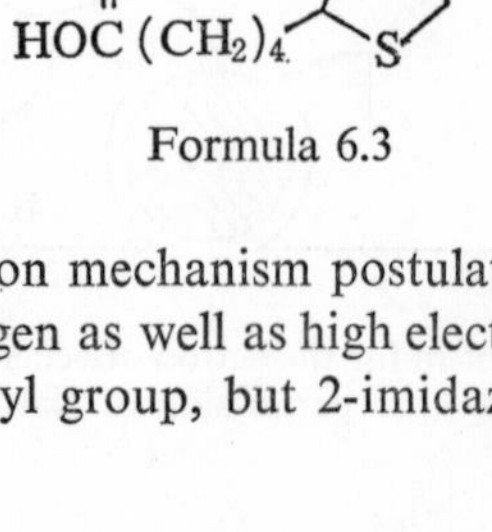

Formula 6.3

The N-carboxylation mechanism postulates high nucleophilicity of the imidazolidone nitrogen as well as high electrophilicity of the carbonyl carbon in the N-carboxyl group, but 2-imidazolidone (Formula 6.4), a

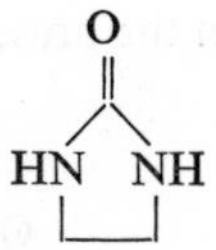

$$\text{O}$$
$$\|$$
$$\text{HN} \quad \text{NH}$$
$$\boxed{}$$

Formula 6.4

model compound, is not acylated by active acylating agents such as *p*-nitrophenyl acetate, 1-acetylimidazole, and 1-acetyl-3-methylimidazolium chloride.[18] Furthermore, alkaline hydrolysis of imidazolidone-1-carboxylic acid ester is even slower than that of methyl acetate.[18] Thus, low nucleophilicity of the imidazolidone nitrogen and low electrophilicity of the *N*-carbonyl group in this model compound, at least, do not conform to the hypothesized *N*-carboxylation mechanism.

Consequently, the *O*-carboxylation mechanism was proposed,[19] although the instability of *O*-carboxybiotin and its methyl ester makes direct evidence impracticable. However, studies of model reaction (6.8),

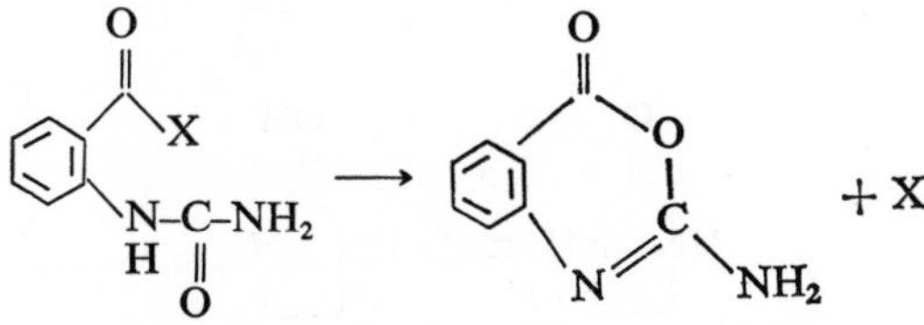

$$\text{(6.8)}$$

Formula 6.5

confirmed the possibility of *O*-acylation of a urea.[20] Furhtermore, an *O*-acylated compound (Formula 6.5) is more susceptible to alkaline hydrolysis than the corresponding *N*-acylated derivative (Formula 6.6),[21] favoring the *O*-carboxylation mechanism.

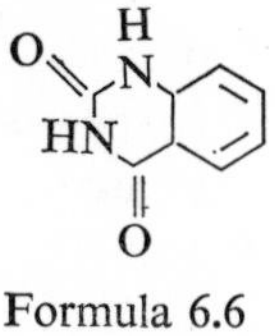

Formula 6.6

Examples of facile O → N,[22] S → N,[23] and C → N[24] acyl transfers in related systems help to account for the isolation of *N*-carboxylated derivatives of biotin.[25,26]

On the other hand, the carboxylation of acetyl coenzyme A by 1-*N*-

carboxybiotinol (Formula 6.7) in the presence of the CT subunit (Fig. 6.2) of the carboxylase[27] (Eq.6.9)

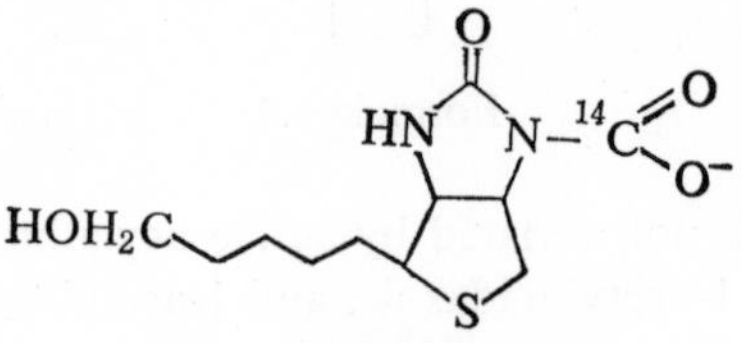

Formula 6.7

Formula 6.7 + CH_3CO–CoA

$$\longrightarrow \text{biotinol} + \underset{\underset{^{14}COO^-}{|}}{CH_2} -CO-CoA \qquad (6.9)$$

appears to confirm the N-carboxylation mechanism.

Although the nucleophilicity of imidazolidone nitrogen is low, as described above, it is much enhanced in the enol form (Formula 6.8, Eq. 6.10)

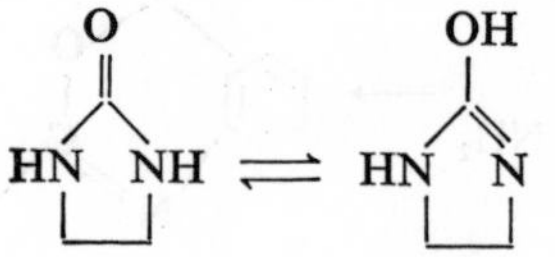

$$(6.10)$$

Formula 6.8

since an ether corresponding to Formula 6.8 is readily acylated by p-nitrophenyl acetate.[28] In connection with this, a mechanism for the first step (Eq. 6.6) of the biotin reaction has been proposed that involves the nucleophilic attack by an O-phosphorylated enolated imidazolidone on bicarbonate (Scheme 6.1)

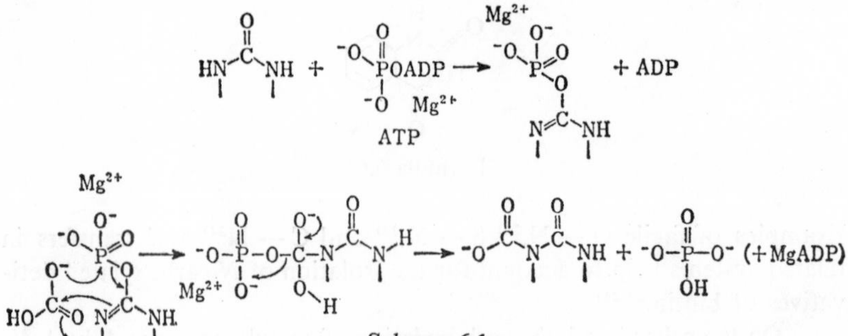

Scheme 6.1

on the basis of the rapid hydrolysis of a phosphonate ester (Formula 6.9, Eq. 6.11).[29]

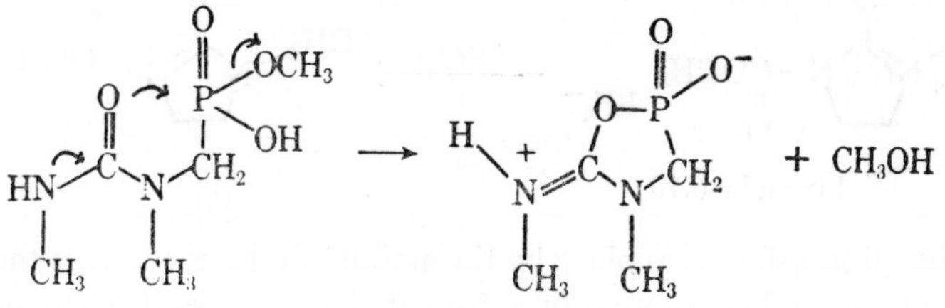

Formula 6.9

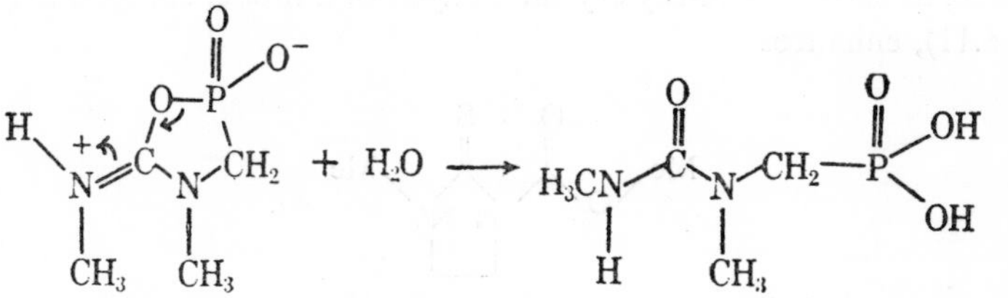

(6.11)

As the second step, Scheme 6.2 has been proposed,[30]

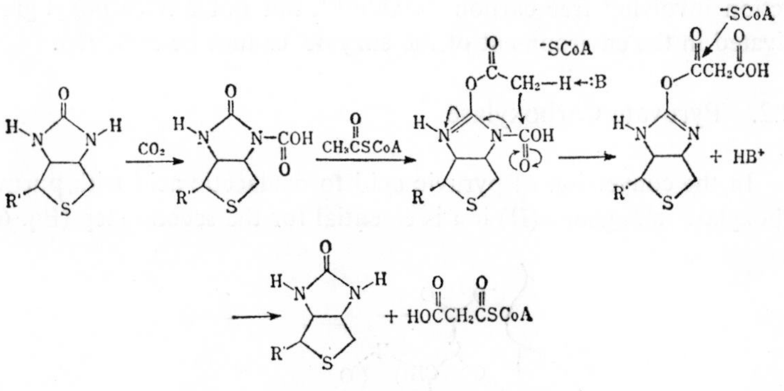

Scheme 6.2

in which *O*-acylation of the carboxylated biotin by acetylcoenzyme A takes place, followed by the intramolecular nucleophilic attack by active methylene on the *N*-carboxyl group. Here, the protonation of imidazolidone nitrogen is considered important, since in a model compound (Formula 6.10, Eq. 6.12),

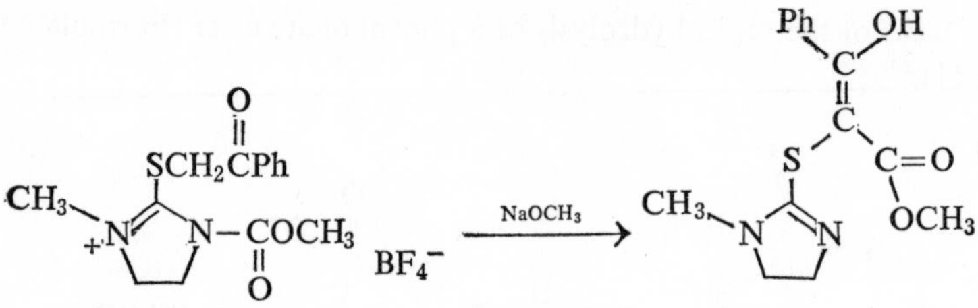

Formula 6.10

the carboxyl transfer takes place by the action of a base, whereas the corresponding neutral compound (not N-methylated) undergoes cleavage at the methoxyl-carbonyl linkage. Nonplanarity of a carboxylated biotin molecule, as shown by X-ray crystal analysis of a model compound (Formula 6.11), enhances

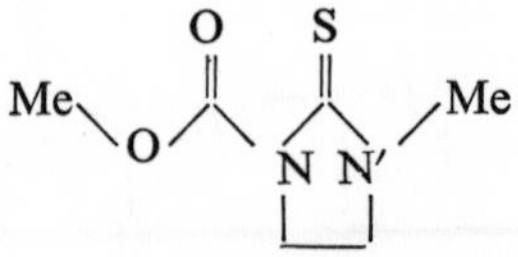

Formula 6.11

the electrophilicity of the carboxy-carbonyl carbon.[31] The possibility of a reaction involving free carbon dioxide [32], but not a N-carboxyl group, activated in the environment of the enzyme cannot be excluded.

6.2.2.   Pyruvate Carboxylase

In the conversion of pyruvic acid to oxalacetic acid with puruvate carboxylase manganese(II) ion is essential for the second step (Eq. 6.7)

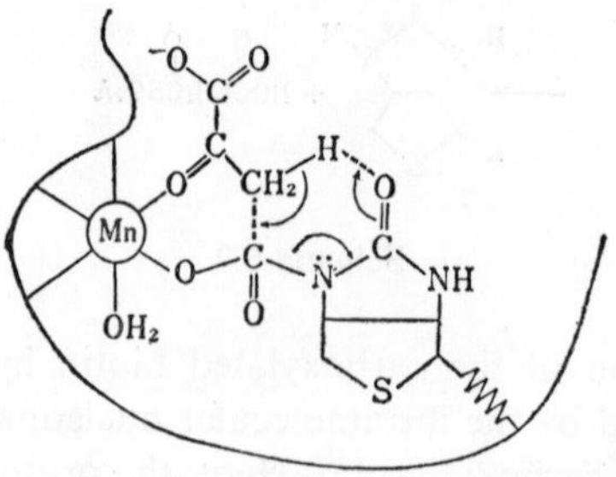

Fig. 6.3.   A mechanism for the action of pyruvate carboxylase.[33] (Source: ref. 33. Reproduced by kind permission of American Society of Biological Chemists, Inc.)

and the mechansim illustrated in Fig. 6.3 has been suggested.[33] The electron-withdrawing Mn(II) ion facilitates proton abstraction from the methyl group of pyruvate and the nucleophilic reaction at the carboxyl-carbonyl carbon. On the other hand NMR studies of the enzyme indicate the occurrence of pyruvate in the second coordination sphere of Mn(II)[34] differing with the scheme in Fig. 6.3. However the coordination behavior of the metallic ion in the presence of carboxy-biotin is as yet unkonwn.

6.2.3. Model Reactions

Metalation at the nitrogen of imidazolidone and analogous compounds such as pyrrolidone greatly enhances the nucleophilicity of the nitrogen, and the reaction with carbon dioxide preceeds to form metal carbamate. Upon further reaction with an active hydrogen compound, the carboxyl group may be transferred by fixed to form a carbon–carboxyl bond (Eqs. 6.13 6.14).[35,36]

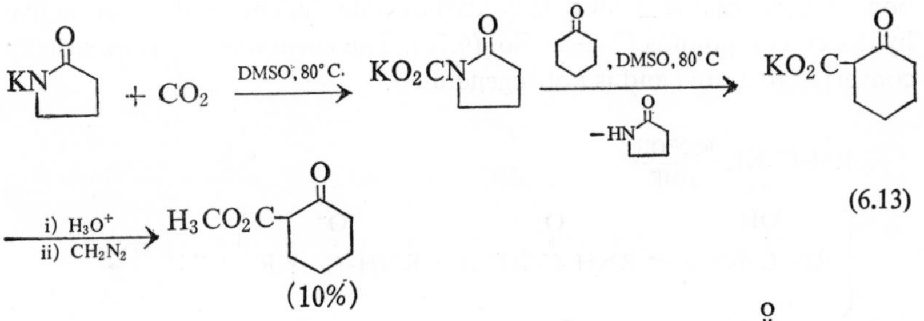

$$(6.13)$$

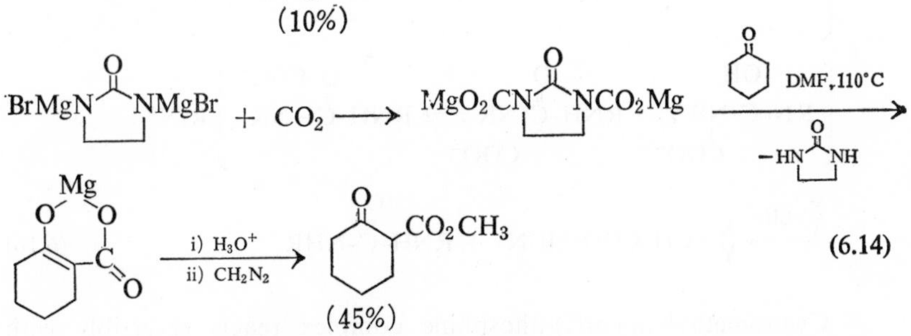

$$(6.14)$$

A similar reaction sequence has been investigated using mono-lithiated imidazolidone.[37] These reactions may be regarded as a model for the action of biotin although the mechansims of the carboxylation of the metalated cyclic amide or cyclic ures (*N*- or *O*-carboxylation) and of the carboxyl transfer to cyclohexanone have not been elucidated yet.

When we extend the definition of "model," there are more examples of a reaction sequence involving carboxylation followed by carboxyl transfer. Methyl methoxymagensium carbonate, obtained from magnesium methoxide and carbon dioxide, when reacted with active hydrogen compounds such as nitromethane, cyclohexanone and acetophenone, gives the corresponding carboxylated products,[38–40] as shown in Eq. (6.15).

$$
\begin{array}{ccc}
\underset{\underset{Mg}{O\diagdown \diagup O}}{\overset{\overset{O}{\parallel}}{\underset{|}{CH_3\ \ C-OCH_3}}}
& \xrightarrow[-CH_3OH]{+C_6H_5COCH_3}
& \underset{\underset{Mg}{O\diagdown \diagup O}}{\overset{\overset{CH_2\ \ O}{\diagup\!\!\diagup \ \ \parallel}}{\underset{|}{C_6H_5-C^{\diagdown}\!\!-\!\!\diagup C-OCH_3}}}
& \xrightarrow{-CH_3OH}
& C_6H_5-C\overset{\overset{H}{\overset{|}{C}}}{\diagup}\!\!\diagdown C=O
\end{array}
$$

$$ (6.15) $$

Potassium phenolate-carbon dioxide complex, the intermediate of the Kolbe-Schmitt reaction, can also carboxylate active hydrogen compounds such as cyclohexanone, nitromethane, etc.[41] Dicyclohexylcarbodiimide in the presence of a tetraalkylammonium hydroxide is considered to be a carbon dioxide carrier which may carboxylate indene and other active hydrogen compounds,[37] as in Eq. (6.16). The carbodiimide is eventually converted to a urea and is not regenerated.

$$
RN=C=NR \xrightarrow[DMF]{R'_4N^+OH^-}
$$

$$
\left[\underset{}{RN=\overset{OH}{\overset{|}{C}}-\bar{N}R} \ \rightleftarrows \ RNH-\overset{O}{\overset{\parallel}{C}}-\bar{N}R \ \rightleftarrows \ RNH-\overset{O^-}{\overset{|}{C}}-NR \right] R'_4N^+ \xrightarrow{CO_2}
$$

$$
\left[\underset{COO^-}{RN=\overset{OH}{\overset{|}{\underset{|}{C}}}-NR} \ \rightleftarrows \ \underset{COO^-}{RNH-\overset{O}{\overset{\parallel}{\underset{|}{C}}}-NR} \ \rightleftarrows \ RNH-\overset{O-COO^-}{\overset{|}{C}}-NR \right] R'_4N^+
$$

$$
\xrightarrow{\overset{X}{\underset{Y}{>}}CH_2} \ \overset{X}{\underset{Y}{>}}CH-COO^- \ R'_4N^+ + RNH-\overset{O}{\overset{\parallel}{C}}-NHR
$$

$$ (6.16) $$

Cyanomethylcopper(I)-phosphine complex reacts reversibly with carbon dioxide, and the resulting copper (I) cyanoacetate in the presence of a phosphine carboxylates cyclohexanone,[42] but the carboxylation actually takes place with free carbon dioxide. (Eq. 6.17)

$$
\underset{}{\overset{\overset{O}{\parallel}}{\bigcirc}} \xrightarrow[\text{ii)} \diagup\!\!\diagdown\!\!\diagup Br]{\text{i) } NCCH_2CO_2Cu\cdot PBu^n{}_3,\ DMF,\ 50°C} \underset{CO_2 \diagup\!\!\diagdown\!\!\diagup}{\overset{\overset{O}{\parallel}}{\bigcirc}}\!\!\diagup\!\!\diagdown\!\!\diagup
$$

$$ (6.17) $$

6.3. UREA FORMATION

Mammals make use of a carbon dioxide fixation reaction for the excretion of the final product of the metabolism of protein and other nitrogenous compounds; the product is ammonia, which is excreted in the form of urea (Eq. 6.18).

$$2 NH_3 + CO_2 \longrightarrow H_2N{-}CO{-}NH_2 + H_2O \qquad (6.18)$$

Although urea is produced industrially from carbon dioxide and ammonia, the reaction requires rather rigorous conditions such as high pres-

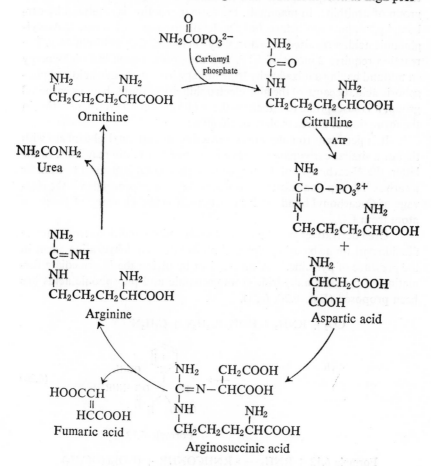

Fig. 6.4.   The urea cycle.

sure and high temperature. In contrast, biochemical urea formation takes place under mild, physiological conditions, and the reaction proceeds via a sequence of reaction steps called the urea cycle (Fig. 6.4) which starts with the formation of carbamyl phosphate (Eq. 6.19).

$$2\,NH_3 + CO_2 \rightleftharpoons [H_2N\text{--}COO^-NH_4{}^+] \overset{ATP}{\rightleftharpoons} H_2N\text{--}\underset{\underset{O}{\|}}{C}\text{--}O\text{--}PO_3H_2 \qquad (6.19)$$

Ammonium carbamate   Carbamyl phosphate

The key steps with respect to carbon dioxide fixation are the formation of carbamyl phosphate and its reaction with the side-chain amino group of ornithine. In mammals, the former reaction is catalyzed by carbamyl phosphate synthetase, but the details are not yet known. N-Acetylglutamic acid, a reguired cofactor, may "activate" CO_2 as biotin does. The reaction requires 2 moles of ATP, and the product itself is a high-energy compound having a mixed anhydride linkage from carbamic acid and phosphoric acid. Because of the high electrophilicity of the carbamate carbonyl group, it is susceptible to attack by the ornithine amino group, resulting in the formation of a urea moiety of citrulline.

It is pertinent to compare the reaction of carbamyl phosphate with that of a similar carbamate species encountered in biotin reactions. In the latter, the N-carbonyl bond of carbamate (N-CO-O) is cleaved to transfer a carboxyl group to a substrate, whereas the former proceeds via the cleavage of O-carbonyl bond, and the net result is the cleavage of 1 oxygen atom from CO_2.

Urea derivatives may be formed under mild conditions from carbon dioxide and amine by using diphenyl phosphite as a dehydrating agent in the presence of pyridine.[43] A mechanism involving the intermediate formation of a mixed anhydride from carbamic and phosphonic acids has been proposed (Eqs. 6.20, 6.21).

$$CO_2 + RNH_2 + HOP(OC_6H_5)_2 + C_5H_5N$$

$$(6.20)$$

Formula 6.12

$$Formula\ 6.12 + RNH_2 \longrightarrow RNHCONHR + (HO)_2P(OC_6H_5)$$
$$+ C_6H_5OH \qquad (6.21)$$

Thus, diphenyl phosphite may be regarded as a model for ATP in the enzymatic formation of carbamyl phosphate. The application of this reaction to the preparation of polymers from carbon dioxide and diamine is described in detail in Chapter 4.

Related to this reation is the formation of oxazalidone from carbon dioxide, 2-aminoethanol, and tetraphenyl pyrophosphate (Eq. 6.22)[44]

$$H_2NCH_2CH_2OH + CO_2 + (PhO)_2P(O)\text{--}O\text{--}P(O)\,(OPh)_2$$

$$\longrightarrow \begin{matrix} CH_2\text{--}O \\ | \\ CH_2\text{--}NH \end{matrix} \Big\rangle C\text{=}O \qquad (6.22)$$

In the absence of CO_2, phosphorylated compounds such as Formulas 6.13 and 6.14 may be isolated

$$\begin{matrix} PhO \\ \\ PhO \end{matrix} \Big\rangle \overset{\overset{\textstyle O}{\|}}{P}\text{--}O\text{--}CH_2CH_2\text{--}NH_2 \qquad\qquad \begin{matrix} PhO \\ \\ PhO \end{matrix} \Big\rangle \overset{\overset{\textstyle O}{\|}}{P}\text{--}NH\text{--}CH_2\text{--}CH_2\text{--}OH$$

Formula 6.13   Formula 6.14

which react with CO_2 to give oxazolidone, indicating the involvement of a carbamyl phosphate as an intermediate of reaction (6.22).

6.4. REACTIONS RELATED TO PHOTOSYNTHESIS

In photosynthesis, as described in detail in Chapter 5, the reaction step in which CO_2 is incorporated is not photochemical, but a dark process. In the Calvin cycle, ribulose-1,5-diphosphate in the form of enolate is thought to attack carbon dioxide (Eq. 6.23).

$$\begin{matrix}
H_2CO\,\circledP & H_2CO\,\circledP & H_2CO\,\circledP & H_2CO\,\circledP \\
C\text{=}O & C\text{--}OH & C\text{--}OH & {}^-O_2C\text{--}COH \\
HCOH & \rightleftharpoons\ C\text{--}OH & \xrightarrow{} C\text{--}O^- & \xrightarrow{CO_2}\ C\text{=}O \\
HCOH & HCOH & HCOH & HCOH \\
H_2CO\,\circledP & H_2CO\,\circledP & H_2CO\,\circledP & H_2CO\,\circledP
\end{matrix} \qquad (6.23)$$

The formation of oxalacetate from phosphoenolpyruvate, important in C_4-plants, is considered to involve a similar path (Eq. 6.24).

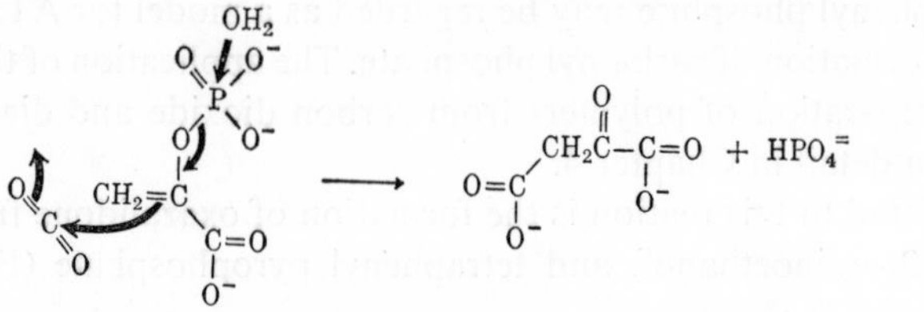

$$\text{(structure)} \longrightarrow O=C\overset{CH_2\overset{O}{\overset{\|}{C}}-C=O}{\underset{O^-}{}} + HPO_4^= \qquad (6.24)$$

Although the reaction between carbon dioxide and alkoxide is widely known, as described in the previous chapters, no particular attention has been paid to the reaction involving enolate. It is relevant to mention that a zinc enolate is more reactive than a simple alkoxide to carbon dioxide.[45,46]

The photochemical portion of photosynthesis functions to excite cholorphyll and to generate electrons from water which eventually reduce ferredoxin, and also convert NADP (nicotinamide adenine dinuclotide phosphate) to the hydrogenated form, NADPH; the high energy compound ATP (adenosine triphosphate) is produced in this electron transfer process. NADPH is used for the reduction of the carboxylated product in Eqs. (6.23) and (6.24), and ATP for the phosphorylation that provides the starting material of these reactions.

Apart from the rather complicated, incompletely understood reaction paths of photosynthesis, a few other, rather simple reactions are thought to be related to some aspects of photosynthesis. For example, tetraphenylporphinatoaluminum ethyl (TPPAlEt, Formula 6.15) coor-

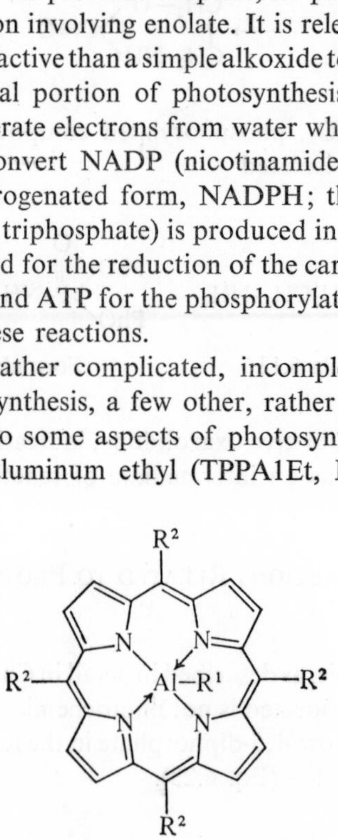

$$R^1 = CH_3CH_2, R^2 = C_6H_5$$

Formula 6.15

dinated to 1-methylimidazole reacts with carbon dioxide only under irradiation by visible light (Eq. 6.25).[47]

$$TPPAlEt + CO_2 \xrightarrow[\text{1-methylimidazole}]{h\nu} TPPAlO_2 CEt \qquad (6.25)$$

Since the aluminum-ethyl bond itself is not excited by visible light, excitation of the porphyrin ring is considered to indirectly activate the aluminum-ethyl bond. This is of much interest in relation to electron transfer in the initial stage of photosynthesis.

A reaction that actually involves electron transfer is the photochemical carboxylation of phenanthrene in the presence of dimethylailine.[48] These two compounds form a donor (D)-acceptor (A) exciplex which is converted to radical-ions (Eq. 6.26), and the phenanthrene radicalanion traps carbon dioxide (Eq. 6.27).

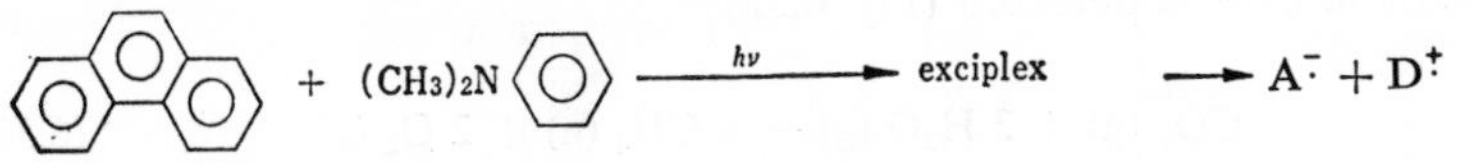

$$\text{(A)} \qquad \text{(D)} \qquad\qquad\qquad\qquad\qquad\qquad (6.26)$$

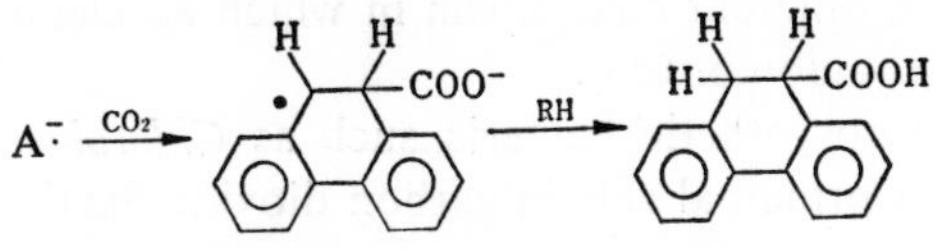

$$(6.27)$$

A similar reaction using perylene and 9,10-dicyanoanthracene as donor and acceptor, respectively, when carried out in aqueous acetonitrile yields formic acid (Eq. 6.28).[49]

$$CO_2 \xrightarrow[\text{CH}_3\text{CN/H}_2\text{O}]{h\nu/\text{D}-\text{A}} HCO_2H \qquad\qquad (6.28)$$

Here, the anion-radical of the acceptor may not combine with CO_2, but instead transfers an electron to the latter, which forms an anionradical that binds a proton to give formic acid. The proton is believed to come from water. This reaction is closely related to the electrolytic reduction of carbon dioxide in aqueous media that produces formic acid (Eqs. 6.29, 6.30), oxalic acid and other products.[50]

$$CO_2 + e^- \longrightarrow CO_2^- \underset{}{\overset{H^+}{\rightleftarrows}} \cdot HCO_2 \qquad\qquad (6.29)$$

$$\cdot HCO_2 + e^- \longrightarrow HCO_2^- \underset{}{\overset{H^+}{\rightleftarrows}} HCO_2H \qquad\qquad (6.30)$$

Of great interest is the photo-assisted electrolytic reduction of aqueous carbon dioxide using *p*-type semiconductor gallium phosphide as a cathode.[51] In this reaction formic acid is further reduced to give formaldehyde and methanol as well. (Eq. 6.31).

$$CO_2 + H_2O \xrightarrow[GaP]{h\nu} HCO_2H + HCHO + CH_3OH \tag{6.31}$$

A similar photo-catalytic reduction of carbon dioxide in aqueous suspensions of semiconductor powders such as SiC and TiO_2 gives formic acid, formaldehyde, methanol and a trace of methane.[52]

Furthermore, when a single crystal of $SrTiO_3$ with a thin layer of platinum is irradiated in mixed vapors of carbon dioxide and water, the formation of methane can be detected (Eq. 6.32)[53]

$$CO_2 \text{ (g)} + 2 H_2O \text{ (g)} \longrightarrow CH_4 \text{ (g)} + 2 O_2 \text{ (g)} \tag{6.32}$$

It should be pointed out that, in photosynthesis, a nucleophile attacks CO_2 (Eqs. 6.23 and 6.24), and therefore these photo-reductions of CO_2 proceed by a different mechanism in which an electrophilic proton attacks the radical-anion of CO_2.

In some photosynthetic bacteria such as *Chlorobium sp.*, reduced ferredoxin palys a principal role in carbon dioxide fixation, as shown in Eq. (6.33),

$$CH_3CO\text{–}SCoA + \text{red. Fd} + CO_2$$
$$\longrightarrow CH_3COCOOH + CoASH + Fd \tag{6.33}$$

where CoA represents coenzyme A, and red-Fe reduced ferredoxin. Light energy is used for the production of reduced ferredoxin, as well as NADPH and ATP for later reactions.

A model reaction between thiol ester and CO_2, using Schrauzer's complex (Formula 6.16) instead of ferredoxin and sodium dithionite as the reducing agent, gives 2-keto acid, though in a low yield (Eq. 6.34).[54]

Formula 6.16

$$CH_3-\underset{\underset{O}{\|}}{C}-S-C_8H_{17} + CO_2 \xrightarrow[\text{Na}_2\text{S}_2\text{O}_4]{\text{Formula 6. 16}} CH_3-\underset{\underset{O}{\|}}{C}-CO_2H \qquad (6.34)$$

Reactions (6.33) and (6.34) probably involve the oxidative addition of thiol ester to a low-valent iron complex, followed by the insertion of CO_2 into the acyl-iron linkage, but the details are not known.

It is pertinent here to mention another biochemical carbon dioxide fixation reaction with a transition metal compound, though it is not related to photochemistry. In the synthesis of acetic acid from carbon dioxide by *Clostridium thermoaceticum,* the reaction of CO_2 with methyl-cobalt bond of a Vitamin B_{12} dervative is involved. A model reaction between methyl (pyridine) cobaloxime (Formula 6. 17) and carbon dioxide in the

Formula 6.17

presence of 1, 4-butanedithiol afforded acetic acid, though in a low yield.[55,56] Although the proposed mechanism (Eq. 6. 35) includes the formation of methyl carbanion,

$$\underset{\underset{Py}{\uparrow}}{\overset{CH_3}{\underset{|}{(Co)}}} \xrightarrow{\text{HS(CH}_2)_4\text{SH}} \underset{\underset{(CH_2)_4}{\underset{|}{\underset{SH}{|}}}}{\overset{CH_3}{\underset{|}{\underset{S}{\underset{|}{(Co)}}}}} \longrightarrow [CH_3^-] \xrightarrow{CO_2} CH_3COO^- \qquad (6.35)$$

the methyl group combines with the metal in an activated state, as seen in recently discovered reactions between CO_2 and transition metal compounds (Chapter 3).

6.5.   CARBON DIOXIDE IN CHEMICAL EVOLUTION

An important subject of bio-organic interest is the origin of organic compounds on Earth and the development that eventually led to the creation of life—chemical evolution. Whether the nature of the atmosphere on the primitive Earth was reductive or oxidative is still under debat, but in any case, carbon dioxide would have been an important raw material in the abiotic formation of simple organic compounds. Through the use of high energy from solar radiation, volcanic action, and so on, carbon dioxide would have reacted with water, ammonia, and other inorganic substances.

Attempts have been made to simulate the processes of chemical evolution in the laboratory. Irradiation of a mixture of carbon dioxide and water vapor with UV light of short wavelength leads to the formation of aldehydes.[57] Similarly, formic acid and formaldehyde are produced when aqueous carbon dioxide containing ferrous (Fe(II)) ion is irradiated with accelerated α-particles[58] or UV light (Eq. 6. 36).[59]

$$CO_2 + H_2O \xrightarrow[Fe^{2+}]{UV} HCO_2H + HCHO \tag{6.36}$$

These reactions are, of course, closely related to the photo-reduction of carbon dioxide described in Section 6. 4.

By the glow discharge in a mixture of carbon dioxide, methane, ammonia, nitrogen, hydrogen, water, and oxygen, the formation of amino acids such as glycine, alanine, β-alanine, sarcosine, and α-aminobutyric acid can be observed (Eq. 6.37),[60] but amino acids are not produced only from carbon dioxide, nitrogen, and water.

$$CO_2 + CH_4 + NH_3 + N_2 + H_2 + H_2O + O_2$$
$$\xrightarrow{\text{glow discharge}} \text{amino acids} \tag{6.37}$$

Amino acid formation can also be observed during the irradiation of a similar mixture by high-energy electrons (X-ray).[61] X-ray irradiation of solid ammonium carbonate leads to the formation of glycine(Eq. 6.38).[62]

$$(NH_4)_2CO_3 \text{ (s)} \xrightarrow{X\text{-ray}} H_2NCH_2COOH \tag{6.38}$$

Carbon dioxide in aqueous ammonia, when heated in the presence of kaolinite, gives cytosine, uracil, and urea.[63]

These reactions could also be of importance in modern times for the production of useful organic raw materials from simple substances.

REFERENCES

1) S. Lindskog, L. E. Henderson, K. K. Kannan, A. Liljas, P. O. Nyman, and B. Stranberg, *Enzymes*, **5**, 587 (1971).
2) J. E. Coleman, *Prog. Bioorg. Chem.*, **1**, 296 (1971).
3) S. Lindskog, L. E. Henderson, K. K. Kannan, A. Liljas, P. O. Nyman, and B. Stranberg, *Enzymes*, **5**, 610 (1971).
4) M. E. Riepe and J. H. Wang, *J. Am. Chem. Soc.*, **89**, 4229 (1967).
5) M. E. Riepe and J. H. Wang, *J. Biol. Chem.*, **243**, 2779 (1968).
6) E. Chaffee, T. P. Dasgupta, and G. M. Harris, *J. Am. Chem. Soc.*, **95**, 4169 (1973).
7) Y. Pocker and D. W. Bjorkquist, *J. Am. Chem. Soc.*, **99**, 6537 (1977).
8) Cited in Y. Pocker and D. W. Bjorkquist, *J. Am. Chem. Boc.*, **99**, 6537 (1977).
9) B. R. Flynn and L. Vaska, *J. Am. Chem. Soc.*, **95**, 5081 (1973).
10) S. Inoue and Y. Yokoo, *J. Organometal Chem.*, **39**, 11 (1972).
11) D. W. Appleton and B. Sarkar, *Bioinorg. Chem.*, **4**, 309 (1975).
12) J. M. Pesando, *Biochemistry*, **14**, 675 (1975).
13) R. K. Gupta and J. M. Pesando, *J. Biol. Chem.*, **250**, 2630 (1975).
14) D. A. Buckingham, *Biological Aspects of Inorganic Chemistry* (ed. A. W. Addison, W. R. Cullen, D. Dolphin, and B. R. James, chap. 5, John Wiley & Sons (1977).
15) T. C. Bruce and S. J. Benkovic, *Bioorganic Mechanisms*, chap. 11, Benjamin (1966).
16) E. Imoto, *Kagaku no Ryōiki Zōkan* (Japanese), **110**, 71 (1976).
17) R. B. Guchhait, J. Moss, W. Sokoloski, and M. D. Lane, *Proc. Natl. Acad. Sci. U.S.A*, **68**, 657 (1971).
18) M. Caplow, *J. Am. Chem. Soc.*, **87**, 5774 (1965).
19) A. F. Hegarty and T. C. Bruce, *J. Am. Chem. Soc.*, **92**, 6568 (1970).
20) A. F. Hegarty and T. C. Bruce, *J. Am. Chem. Soc.*, **92**, 6575 (1970).
21) A. F. Hegarty and T. C. Bruce, *J. Am. Chem. Soc.*, **92**, 6561 (1970).
22) Y. Curtin and L. L. Miller, *Tetrahedron Lett.*, 1965, 1869; *J. Am. Chem. Soc.*, **89**, 637 (1967).
23) R. F. Pratt and T. C. Bruce, *Biochemistry*, **10**, 3178 (1971).
24) Y. Akasaki and A. Ono, *J. Am. Chem. Soc.*, **96**, 1957 (1974).
25) J. Knappe, E. Ringermann, and F. Lynen, *Biochem. Z.,* **335**, 168 (1961).
26) F. Lynen, J. Knappe, E. Lorch, G. Jütting, and E. Ringermann, *Angew. Chem.*, **71**, 481 (1959); *Biochem. Z.,* **335**, 123 (1961).
27) M. D. Lane, J. Moss, and S. E. Polakis, *Current Topics in Cellular Regulation* (ed. B. L. Holecker and E. R. Stadman), vol. 8, p. 153, Academic Press (1974).
28) A. F. Hegarty, T. C. Bruce, and S. J. Benkovic, *Chem. Commun.*, **1969**, 1173.
29) R. Kluger and P. D. Adawadkar, *J. Am. Chem. Soc.*, **98**, 3741 (1976).
30) H. Kohn, *J. Am. Chem. Soc.*, **98**, 3960 (1976).
31) S. F. Watkins, H. Kohn, and I. Bernal, *J. Chem. Soc., Perkin II*, **1978**, 26.
32) C. K. Sauers, W. P. Jencks, and S. Groh, *J. Am. Chem. Soc.*, **97**, 5546 (1975).
33) A. S. Mildvan, M. C. Scrutton, and M. F. Utter, *J. Biol. Chem.*, **241**, 3497 (1966).
34) C. H. Fung, A. S. Mildvan, A. Allerhand, R. Komoroski, and M. C. Scrutton, *Biochemistry*, **12**, 620 (1973).
35) H. Mori, Y. Makino, H. Yamamoto, and T. Kwan, *Chem. Pharm. Bull.*, **21**, 915 (1973).
36) H. Sakurai, A. Shirahata, and A. Hosomi, 3rd Meeting for Chemical Utilization of Carbon Dioxide, Tokyo, 1977 Abstracts, p. 23.

37) Y. Otsuji, M. Arakawa, N. Matsumura, and E. Haruki, *Chem. Lett.*, **1973**, 1193.
38) M. Stiles and H. L. Finkbeiner, *J. Am. Chem. Soc.*, **81**, 505 (1959).
39) M. Stiles, *J. Am. Chem. Soc.*, **81**, 2598 (1959).
40) H. Finkbeiner, *J. Org. Chem.*, **30**, 3414 (1965); *J. Am. Chem. Soc.*, **87**, 4588 (1965).
41) G. Bottaccio and G. P. Chiusoli, *Chem. Commun.*, **1966**, 618
42) T. Tsuda, Y. Chujo, and T. Saegusa, *J. Am. Chem. Soc.*, **100**, 630 (1978).
43) N. Yamazaki, F. Higashi, and T. Iguchi, *Tetrahedron Lett.*, **1974**, 1191.
44) Y. Akasaki, M. Hatano, and M. Fukuyama, 3rd Meeting for Chemical Utilization of Carbon Dioxide, Tokyo, 1977 Abstracts, p. 31.
45) K. Kataoka and T. Tsuruta, *Polymer J.*, **9**, 595 (1977).
46) S. Inoue, M. Kobayashi, H. Koinuma, and T. Tsuruta, *Makromol. Chem.*, **155**, 61 (1972).
47) S. Inoue and N. Takeda, *Bull. Chem. Soc., Japan,* **50**, 984 (1977).
48) S. Tazuke and H. Ozawa, *Chem. Commun.*, **1975**, 237.
49) S. Tazuke and N. Kitamura, *Nature,* **275**, 301 (1978).
50) P. G. Russel, N. Kovac, S. Srinivasan, and M. Steinberg, *J. Electrochem. Soc.*, **124**, 1329 (1977).
51) M. Halmann, *Nature,* **275**, 115 (1978).
52) T. Inoue, A. Fujishima, S. Konishi, and K. Honda, *Nature,* **277**, 637 (1979).
53) J. C. Hemminger, R. Carr, and G. A. Somorajai, *Chem. Phys. Lett.*, **57**, 100 (1978).
54) T. Nakajima, Y. Yabushita, and I. Tabushi, *Nature,* **256**, 61 (1975).
55) G. N. Schrauzer, J. A. Seck, R. J. Holland, T. M. Beckham, E. M. Robin, and I. W. Sibert, *Bioinorg. Chem.*, **2**, 93 (1972).
56) G. N. Schrauzer and I. W. Sibert, *J. Am. Chem. Soc.*, **92**, 3509 (1970).
57) W. Groth and H. Suess, *Naturwiss.*, **26**, 77 (1938).
58) W. M. Garrison, D. C. Morrison, J. G. Hamilton, A. Benson, and M. Calvin, *Science*, **114**, 416 (1951).
59) N. Getoff, *Z. Naturforsch*, **17b**, 87 (1962).
60) K. Heyns, W. Walters, and E. Meger, *Naturwiss.*, **44**, 385 (1957).
61) K. Dose and B. Rajewsky, *Biochim. Biophys. Acta,* **25**, 225 (1957).
62) R. Paschke, R. W. H. Chang, and D. Young, *Science*, **125**, 881 (1957).
63) G. R. Harvey, E. T. Degens, and K. Mopper, *Naturwiss.*, **58**, 624 (1971).

Index

275